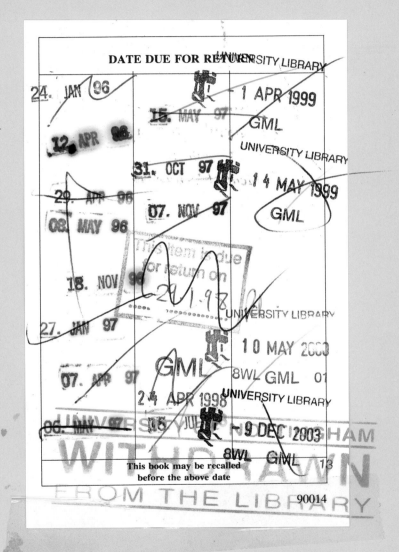

NON-REPRODUCTIVE ACTIONS OF SEX STEROIDS

Ciba Foundation Symposium 191

NON-REPRODUCTIVE ACTIONS OF SEX STEROIDS

1995

JOHN WILEY & SONS

Chichester · New York · Brisbane · Toronto · Singapore

© Ciba Foundation 1995

Published in 1995 by John Wiley & Sons Ltd
Baffins Lane, Chichester
West Sussex PO19 1UD, England
Telephone National Chichester (01243) 779777
International (+44) (1243) 779777

Other Wiley Editorial Offices

John Wiley & Sons, Inc., 605 Third Avenue,
New York, NY 10158-0012, USA

Jacaranda Wiley Ltd, G.P.O. Box 859, Brisbane,
Queensland 4001, Australia

John Wiley & Sons (Canada) Ltd, 22 Worcester Road,
Rexdale, Ontario M9W 1L1, Canada

John Wiley & Sons (SEA) Pte Ltd, 37 Jalan Pemimpin *05-04,
Block B, Union Industrial Building, Singapore 2057

Suggested series entry for library catalogues:
Ciba Foundation Symposia

Ciba Foundation Symposium 191
ix + 307 pages, 35 figures, 6 tables

Library of Congress Cataloging-in-Publication Data

Non-reproductive actions of sex steroids.
        p.   cm.—(Ciba foundation symposium ; 191)
    "Symposium on Non-Reproductive Actions of Sex Steroids, held at
the Ciba Foundation, London, 30 August–1 September 1994"—Contents
p.
    Editors: Gregory R. Bock and Jamie A. Goode.
    Includes bibliographical references and indexes.
    ISBN 0 471 95513 2
    1. Hormones, Sex—Physiological effect—Congresses.   I. Bock,
Gregory.   II. Goode, Jamie.   III. Symposium on Non-Reproductive
Actions of Sex Steroids (1994 : Ciba Foundation)   IV. Series.
    [DNLM: 1. Sex Hormones—physiology—congresses.   2. Steroids—
physiology—congresses.   W3 C161F v. 191 1995 (P)]
QP572.S4N66   1995
612.6—dc20
DNLM/DLC
for Library of Congress   ιοοΟΜΤ6SSΘ   95-6957
                                                                         CIP
British Library Cataloguing in Publication Data

A catalogue record for this book is
available from the British Library

ISBN 0 471 95513 2

Phototypeset by Dobbie Typesetting Limited, Tavistock, Devon.
Printed and bound in Great Britain by Biddles Ltd, Guildford.

Contents

Participants

T. Bäckström   Department of Obstetrics and Gynecology, University Hospital, S-901 58 Umeå, Sweden

E. E. Baulieu (*Chairman*) INSERM U33, 80 rue de Général Leclerc, 92476 Bicêtre Cedex, France

F. Bayard   INSERM U397, Institut Louis Bugnard, CHU Rangeuil, F-31054 Toulouse Cedex, France

M. Beato   Institut für Molekularbiologie und Tumorforschung, Philipps Universität, Emil-Mannkopff-Strasse 2, D-35037 Marburg, Germany

L. F. Bonewald   The University of Texas Health Science Center at San Antonio, Department of Medicine, Division of Endocrinology and Metabolism, 7703 Floyd Curl Drive, San Antonio, TX 78284-7877, USA

L. A. M. Castagnetta   Hormone Biochemistry Laboratories, University of Palermo, Palermo Branch of the National Institute for Cancer Research of Genoa, Cancer Hospital Centre 'M. Ascoli', PO Box 636, Palermo, Italy

M. L. Foegh   Department of Surgery, Georgetown University Medical Center, 4000 Reservoir Road NW, Washington DC 20007, USA

H. S. Fox   Department of Neuropharmacology, CVN-8, The Scripps Research Institute, 10666 N. Torrey Pines Road, La Jolla, CA 92037, USA

K. B. Horwitz   University of Colorado Health Sciences Center, Campus Box B151, 4200 East Ninth Avenue, Denver, CO 80262, USA

S. Inoue   (*Ciba Foundation Bursar*) Department of Geriatrics, Faculty of Medicine, University of Tokyo, 7-3-1 Hongo, Bunkyo, Tokyo 113, Japan

V. H. T. James   Unit of Metabolic Medicine, Department of Clinical Pathology, St Mary's Hospital Medical School, Norfolk Place, London W2 1PG, UK

E. Jensen   Institute for Hormone Fertility Research, University of Hamburg, Grandweg 64, D-22529 Hamburg, Germany

B. S. McEwen   Laboratory of Neuroendocrinology, The Rockefeller University, Box 165, 1230 York Avenue, New York, NY 10021-6399, USA

S. C. Manolagas   Department of Medicine, Division of Endocrinology and Metabolism and the UAMS Center for Osteoporosis and Metabolic Bone Diseases, University of Arkansas for Medical Sciences, 4301 West Markham, Mail Slot 587, Little Rock, AR 72205-7199, USA

M. Muramatsu   Department of Biochemistry, Saitama Medical School, 38 Morohongo, Moroyama, Iruma-gun, Saitama 350-04, Japan

W. K. H. Oelkers   Freie Universität Berlin, Klinikum Benjamin Franklin (Steglitz), Division of Endocrinology, Hindenburgdamm 30, D-12200 Berlin, Germany

M. G. Parker   Molecular Endocrinology Laboratory, Imperial Cancer Research Fund, Lincoln's Inn Fields, London WC2A 3PX, UK

D. W. Pfaff   Neurobiology and Behavior Laboratory, The Rockefeller University, 1230 York Avenue, New York, NY 10021-6399, USA

P. Robel   INSERM U33, 80 rue de Général Leclerc, 92476 Bicêtre Cedex, France

H. Rochefort   INSERM U148, Unité Hormones et Cancer, Université de Montpellier, Faculté de Medecine, 60 rue de Navacelles, 34090 Montpellier, France

Z. Sarnyai   (*Ciba Foundation Bursar*) Alcohol and Drug Abuse Research Center, McLean Hospital, Harvard Medical School, 115 Mill Street, Belmont, MA 02178, USA

M. Schumacher   Lab. Hormones, INSERM U33, 80 rue de Général Leclerc, 92476 Bicêtre Cedex, France

R. L. Sutherland   Cancer Biology Division, Garvan Institute of Medical Research, St Vincent's Hospital, Darlinghurst, Sydney, NSW 2010, Australia

J. H. H. Thijssen   Department of Endocrinology GO2.625, Academisch Ziekenhuis, PO Box 85500, NL-3508 GA Utrecht, The Netherlands

K. Uvnäs-Moberg   Department of Pharmacology, Karolinska Institute, S-17177 Stockholm 60, Sweden

M. R. Vickers   MRC Epidemiology and Medical Care Unit, The Wolfson Institute of Preventive Medicine, St Bartholomew's Hospital Medical College, Charterhouse Square, London EC1M 6BQ, UK

Chairman's introduction

Etienne-Emile Baulieu

INSERM U33, 94276 Le Kremlin-Bicêtre, France

Owing to their involvement in gender identity, sexual behaviour and reproductive processes, sex steroid hormones are among the most famous molecules of contemporary medicine. Synthetic analogues and antagonists have been devised and applied to hormonal contraception and pregnancy interruption. The dissociation of sex and procreation is bringing about profound changes in the psychosocial determinants of contemporary life and social organization. Despite growing acceptance, the use of steroids for family planning still generates worldwide controversies, and these are certainly a cause of the insufficient use of contraception and the resulting demographic problems. Many of our contemporaries fear what we may loosely call the side effects of sex steroids, essentially related to their non-reproductive activities.

Hormonal contraception constitutes a very peculiar example of drug administration to healthy individuals, never previously undertaken on such large scale. In addition to real side effects, opponents of this method have alleged many risks that have not been demonstrated. The same comment applies to another example of steroid administration to healthy individuals, hormone replacement during menopause. Hormones may promote the development of certain cancers. The risk, although limited, is documented in the case of endometrial and prostatic cancer, and discussed in the case of breast cancer. Such risk has to be balanced with the health benefits provided by the prevention of unwanted pregnancies or compensation for the decline of sex steroids after menopause and during ageing. These benefits are related to non-reproductive actions of sex steroids on the cardiovascular and immune systems, and brain, bone, etc.

One step further on, the preventive use of steroid hormone antagonists is now being contemplated, and trials of tamoxifen administration to women at risk of breast cancer are now underway.

The discovery of steroid hormone receptors and the subsequent elucidation of their structure and molecular features as ligand-operated transcription factors, pertain to the most important biomedical advances achieved during the past 25 years. Mutations producing diverse forms of resistance to steroid hormones have been described. However, present knowledge is still confined to

1

the recognition of general principles, whereas the diversity of regulatory interactions applicable to specific cells of, for instance, the genital tract, brain or bone has not been studied enough. The diverse effects of steroids are still poorly understood—therefore selective approaches to the investigation and treatment of diseases in which sex steroids participate are precluded.

Elimination of harmful side effects of sex steroid hormones, in accordance with the general principle of *primum non nocere*, is a mandatory goal whenever replacement or preventive therapy is considered. This is an important objective of this Ciba Foundation meeting.

Sex steroid activities

The activities of sex steroid hormones in their corresponding target organs were the first to be studied, before those of most other steroids and more generally, most other hormones. Oestradiol, progesterone and testosterone display oestrogenic, progestomimetic or androgenic activities, respectively. They share the Δ4-3-oxo-steroid structure (oestradiol and, more generally, phenol steroids are the only apparent exceptions since they are dienones Δ1-2, Δ4-5-19-norsteroids) (Fig. 1). These hormones are secreted by well-defined steroidogenic glands, and transported in the bloodstream towards their target cells where they interact with intracellular receptors and provoke hormone responses, essentially regulating the expression of target genes at the transcriptional level. The regulation of transcription by sex steroid hormones

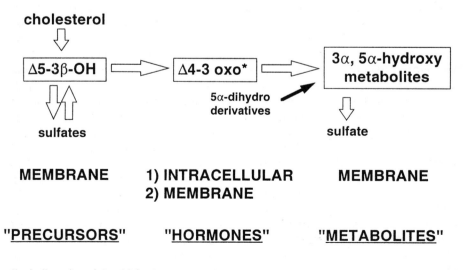

*including phenolsteroidal estrogens

FIG. 1   Steroid activities: precursors, hormones and metabolites.

was described before the discovery of receptors. In the pioneering work of Gerald Mueller on oestradiol action in the mammalian uterus (Mueller et al 1958) this was referred to as 'genomic', and this was soon supported by the description of the corresponding intracellular oestrogen receptor (Jensen & Jacobson 1962). In order to antagonize certain sex steroid actions, Lerner et al (1958) devised the first antihormonal derivatives. Sex steroid hormones have also provided a completely different model of hormone action, exemplified by the progesterone-induced reinitiation of meiosis in amphibian oocytes, which led to the discovery of a new site of steroid action, the plasma membrane (Baulieu et al 1978).

The investigation of sex steroid action was extended to non-reproductive tissues, where both genomic and membrane sites of action have been documented. Many findings made with sex steroids have their counterparts in the field of gluco- and mineralocorticosteroids; indeed, all intracellular steroid receptors are members of the same superfamily, which also includes receptors for thyroid hormones and for the active metabolites of vitamins D and A (Evans 1988).

Precursors and metabolites of sex steroid hormones can also be active

Variations of the (already) classical scheme described above for sex steroid hormones include the effects of precursors or metabolites of sex hormones. Sex steroid hormone precursors are steroids such as oestrone which can be converted to oestradiol; dehydroepiandrosterone (DHEA), androstenediol (a reduced derivative of DHEA) and androstenedione, all potential precursors of testosterone; pregnenolone, the immediate precursor of progesterone by the action of an oxidative enzyme ($\Delta 5$-3, β-hydroxysteroid dehydrogenase, $\Delta 5 \rightarrow 4$ isomerase). These precursors can be released in the bloodstream and transported to target cells, where they can produce active products. This process of 'metabolic activation' is well established for the formation of testosterone from androstenedione (in the liver, skin and prostate), and for the formation of oestradiol from oestrone (in the uterus, mammary gland and adipose tissue) or from DHEA (in human placenta). Testosterone and oestradiol are 17β-hydroxysteroids which, contrary to their 17-oxo counterparts, bind with high affinity to their cognate receptors. Androstenediol is oestrogenic through its binding to the oestrogen receptor. Some hormone precursors, such as pregnenolone and DHEA (in sulphate form) modulate the activity of synaptosomal membrane receptors, namely the γ-aminobutyric acid type A (GABA$_A$) receptors. N-methyl-D-aspartate, σ, acetylcholine and glycine receptors are other targets currently being investigated.

Other active steroids are metabolites of hormones (Baulieu 1970). 5α-dihydrotestosterone is a reduced metabolite of testosterone in many target cells; it has greater affinity for the androgen receptor, and thus is more potent than testosterone itself. Oestradiol is also a metabolite of testosterone in cells of the hypothalamus, granulosa cells, placenta, adipocytes and bone. Metabolites of sex steroid hormones previously considered to be inactive detoxication products have distinct properties, such as the $3\alpha,5\alpha$-tetrahydro derivative of progesterone (TH-progesterone), which is responsible for the sedative effect of progesterone, related to the positive allosteric modulation of the $GABA_A$ receptor by the metabolite. Steroid metabolism may be extremely important in determining the mechanism of action of sex steroids: for instance TH-progesterone, which is active on the membrane $GABA_A$ receptor, can be oxidized to 5α-dihydroprogesterone (DH-progesterone), which binds to the intracellular progesterone receptor.

Paracrine and autocrine modes of distribution and function of steroids

In the testes, testosterone is secreted by Leydig cells, and in part diffuses directly through the blood–testis barrier to Sertoli cells in the seminiferous tubules. The very high testosterone concentration thus established plays an essential role in the maintenance of spermatogenesis. Testosterone produced from DHEA in the liver may be active *in situ*, then converted to inactive metabolites in such a manner that no testosterone enters the bloodstream. Oestrogens formed from androgens in the hypothalamus, or progesterone synthesized *de novo* by myelinating glial cells, are examples of autocrine modes of action, whereas neuroactive metabolites of pregnenolone or progesterone are examples of paracrine distributions.

Clinically, it is practically impossible to evaluate the local synthesis and movement of active steroids, since plasma levels do not faithfully reflect the paracrine/autocrine situations, unless final products of steroid metabolism are in turn released and eliminated via the bloodstream.

The uniqueness of sex steroid hormone receptors among reproductive and non-reproductive organs

In contrast to other members of the superfamily of intracellular receptors, each sex steroid has a receptor encoded by a single gene. However, receptor isoforms, a consequence of different sites of translation initiation in the case of the progesterone receptor or due to alternative splicing in the case of oestrogen and androgen receptors, have been described. The A and B forms of the progesterone receptor (Sherman et al 1970) are widely distributed, and differential activities have been reported. Isoforms of oestradiol and androgen receptors have been described in hormone-responsive tumours, and may even produce constitutively active transcription factors. Since systematic cloning has not been performed in all cells responding to sex steroid hormones, the presence of 'silent' genes cannot be excluded. It is possible that other processes, e.g. RNA editing, may lead to receptor isoforms of yet poorly understood function: the progesterone receptors in glial cells have been suggested as an example. If confirmed, the occurrence of such steroid hormone isoforms would be of the utmost importance, and lead in particular to the synthesis of selective steroid analogues, thus leading the way for novel therapeutic approaches.

New avenues for steroid research

(1) Since intracellular and membrane receptors for a given natural sex steroid generally display distinct ligand binding specificities, and since there are different types of membrane receptors for the same steroid in different tissues, the use of selective synthetic ligands would 'dissociate' the effects of the corresponding hormone. This possibility could be extended to active metabolites formed in some target cells.

(2) The cellular concentrations of receptors vary among cell types and, in a given cell type, according to physiological conditions. This is the case for the progesterone receptor, the concentration of which is increased by oestradiol in several reproductive organs and (curiously) in cultured glial cells, but not in certain cancer cell lines or in cerebrocortical neurons. The molecular mechanisms involved have been explored with cell lines in culture, but extrapolation of these results to target organs might be hazardous. Here again, paracrine mechanisms might complicate the picture. The possibility of modulating receptor levels in certain cell types and not in others might also lead to novel pharmacological approaches.

(3) The response to a given sex steroid may also depend on the mode of administration (peroral, intramuscular or intravenous, local, percutaneous, etc.). The distribution of hormone in the body and the nature of metabolites formed would differ according to the method of administration, thus resulting in different concentrations of active molecules at the level of the target cell.

In summary, current studies are at variance with the classical concept of endocrine control. They offer new possibilities to manipulate the hormonal system and thus find new active drugs.

References

Baulieu EE 1970 The action of hormone metabolites: a new concept in endocrinology. Rev Eur Etud Clin Biol 15:723–726

Baulieu EE, Godeau F, Schorderet M, Schorderet-Slatkine S 1978 Steroid induced meiotic division in *Xenopus laevis* oocytes: surface and calcium. Nature 275:593–598

Evans RM 1988 The steroid and thyroid hormone receptor family. Science 240:889–895

Jensen EV, Jacobson HI 1962 Basic guides to the mechanism of estrogen action. Recent Prog Horm Res 18:387–414

Lerner LJ, Holthaus JF, Thompson 1958 A nonsteroidal estrogen antagonist 1-(p-2-diethylaminoethoxypenyl)-1-phenyl-2-p-methoxyphenyl-ethanol. Endocrinology 63:295–318

Mueller GG, Herranen AM, Jervell KF 1958 Studies on the mechanism of action of estrogens. Recent Prog Horm Res 14:95–139

Sherman MR, Corvol PL, O'Malley BW 1970 Progesterone binding components of chick oviduct. I. Preliminary characterization of cytoplasmic components. J Biol Chem 245:6085–6096

Transcriptional control by steroid hormones: the role of chromatin

Mathias Truss, Reyes Candau*, Sebastián Chávez and Miguel Beato

Institut für Molekularbiologie und Tumorforschung, Philipps Universität, Emil-Mannkopff-Strasse 2, D-35037 Marburg, Germany

Abstract. The mouse mammary tumour virus (MMTV) promoter contains a complex hormone-responsive unit composed of four hormone-responsive elements, a nuclear factor I (NFI) binding site and two octamer motifs. All these sites are required for optimal hormonal induction. Although synergism has been found between hormone receptors and octamer transcription factor 1 (Oct-1/OTF-1), we were unable to detect a positive interaction between receptors and NFI *in vitro*. In chromatin, the MMTV hormone-responsive unit is contained in a phased nucleosome. The precise positioning of the DNA double helix on the surface of the histone octamer precludes binding of NFI and Oct-1/OTF-1 to their cognate sequences, while still allowing recognition of two hormone-responsive elements by the hormone receptors. Hormone treatment leads to a characteristic change in chromatin structure that makes the centre of the nucleosome more accessible to digestion by DNase I and facilitates binding of receptors, NFI and Oct-1/OTF-1 to the nucleosomally organized promoter. The MMTV promoter functions in yeast in a hormone receptor-dependent and NFI-dependent fashion. Depletion of nucleosomes activates hormone-independent transcription from the MMTV promoter. These results imply that nucleosome positioning not only represses hormone-independent transcription, but also enables binding of a full complement of transcription factors to the hormone-responsive unit after hormone induction.

1995 Non-reproductive actions of sex steroids. Wiley, Chichester (Ciba Foundation Symposium 191) p 7–23

Steroid hormones modulate the expression of a variety of genes by influencing the activity of their intracellular receptors which, in turn, either bind to hormone-responsive elements on the DNA or interact directly with other transcription factors to modulate various signal transduction pathways (Beato 1989). The outcome can be induction or repression of specific genes depending

---

*Present address: The Wistar Institute, 3601 Spruce Street, Philadelphia, PA 19104–4268, USA.

on the nature of the interaction with sequence-specific or general transcription factors.

Transcription of the mouse mammary tumour virus (MMTV) promoter is tightly regulated (Beato 1991). Expression of the provirus is mainly observed in mammary epithelial cells of lactating females and is believed to reflect a direct response to elevation of the circulating progestin levels. Prior to and following lactation, transcription of the virus in the mammary gland is not detectable (Bolander & Blackstone 1990). Although sequences important for tissue-specific expression are located throughout the long terminal repeat (LTR) region, all sequences required for efficient hormone-induced transcription reside within the 200 bp immediately flanking the transcriptional start site of the MMTV promoter (Majors & Varmus 1983). In this region, four binding sites for steroid hormone receptors (hormone-responsive elements) have been identified (Scheidereit et al 1983) which are recognized by all members of the glucocorticoid subclass of steroid hormone receptors (Beato 1991). In addition to these hormone-responsive elements, the MMTV promoter contains binding sites for at least two additional sequence-specific transcription factors, nuclear factor I (NFI) and octamer transcription factor 1 (Oct-1/OTF-1). Mutation of any of these binding sites results in a significant decrease in induced transcription, suggesting that these factors play an important role in the process of hormone induction of the MMTV promoter (Brüggemeier et al 1990, 1991, Toohey et al 1990). We therefore define the region from −190 to −37, comprising the four hormone-responsive elements and the binding sites for NFI and Oct-1/OTF-1, as the MMTV hormone-responsive unit.

In this paper, we summarize what is known about the molecular mechanisms controlling basal transcription from the MMTV promoter (i.e. transcription in the non-induced stage) as well as our current ideas on induced transcription following stimulation by steroid hormones. Our results suggest that a detailed description of the regulation of transcription cannot be accomplished without considering the influence of nucleosome positioning on the availability and disposition of binding sites for transcription factors. We propose that low basal transcription is due to the particular structure of the MMTV hormone-responsive unit in nucleosomes, and that a correctly positioned nucleosome is also essential for optimal induction of MMTV transcription.

The free DNA story

Binding of transcription factors to the MMTV hormone-responsive unit

Binding experiments with purified hormone receptors and cloned fragments of the MMTV hormone-responsive unit have demonstrated a stimultaneous binding of the receptors to all four hormone-responsive elements (Scheidereit et al 1983, Scheidereit & Beato 1984, Ahe et al 1985). Although the exact stoichiometry of receptor binding is still unclear (Truss et al 1992), these and other data suggest a low degree of synergism or cooperativity in terms of binding of the hormone receptors to their cognate sites on the hormone-responsive unit. Mutation of the individual receptor binding sites has little effect on the global affinity of the receptors for the hormone-responsive unit, but influences dramatically the response to hormone treatment in transfection assays (Chalepakis et al 1988, Truss & Beato 1993).

Quantitative DNase I footprinting experiments have shown that binding of Oct-1/OTF-1 to the octamer motifs in the MMTV promoter is weak in the absence of steroid hormone receptors, but is strongly enhanced by preincubation of DNA with either glucocorticoid receptor or progesterone receptor. As these results were obtained with purified proteins, they suggest a DNA binding cooperativity which is very likely mediated by direct protein–protein interactions (Brüggemeier et al 1991). This could explain the results of mutations of the octamer motifs on the hormone induction (Brüggemeier et al 1991) and the outcome of cell-free transcription experiments (see below).

The binding site for NFI resides between the octamer motif immediately distal to the promoter and the hormone-responsive element most proximal to the promoter. Mutations of this site reduce MMTV transcription by nine- to 10-fold (Kühnel et al 1986, Miksicek et al 1987, Buetti et al 1989, Brüggemeier et al 1990). In contrast to Oct-1/OTF-1, NFI recognizes its binding site on free DNA with high affinity, even in the absence of steroid hormone receptors with which it competes for binding to the MMTV promoter (Brüggemeier et al 1990, Perlmann et al 1990). Results from exonuclease III protection experiments indicate that binding of NFI to the MMTV promoter *in vivo* only occurs after hormone stimulation (Cordingley et al 1987, Archer et al 1992). However, the DNA-binding ability of NFI is not influenced by hormone treatment, as nuclear extracts from control or hormone-treated cells display equivalent DNA-binding activity (Cordingley & Hager 1988). Thus, the dramatic effect of mutation of the NFI binding site on induced MMTV transcription cannot be accounted for simply in terms of the binding behaviour of hormone receptors and NFI on free DNA. Moreover, binding of NFI interferes with binding of Oct-1/OTF-1 to the adjacent promoter-distal octamer motif, making a cooperation between these two sequence-specific factors on free DNA unlikely (Möws et al 1994).

Cell-free transcription results

To understand the functional relationships between the hormone receptors and other transcription factors on the MMTV promoter, we have studied their interactions in a cell-free transcription system. Using supercoiled DNA templates and a crude nuclear extract from HeLa cells, we showed that transcription is very efficient in the absence of receptors, but can be enhanced by adding purified progesterone receptor or glucocorticoid receptor (Kalff et al 1990, Brüggemeier et al 1991, Möws et al 1994). The effect of the added receptor is dependent on the intact hormone-responsive elements and is completely abolished by mutating the octamer motifs or adding an oligo-nucleotide with the consensus octamer sequence (Brüggemeier et al 1991). On the contrary, mutation of the NFI binding site reduces the basal transcription activity, but has no influence on the receptor-dependent induction of transcrip-tion (Kalff et al 1990, Möws et al 1994). These findings are consistent with a role for NFI in basal transcription of free DNA templates, and confirm the lack of synergism between hormone receptors and NFI in binding assays with free DNA (Brüggemeier et al 1990), but contradict the observed requirement of an intact NFI binding site for optimal induction *in vivo* (see above).

The chromatin story

Nucleosome positioning in vivo and in vitro

In vivo structural analysis has revealed that the chromatin structure over the MMTV-LTR is highly ordered, with individual nucleosomes being translationally positioned (Richard-Foy & Hager 1987). Both in multicopy episomal vectors and when integrated as single copy into the host chromosome, one nucleosome covers the hormone-responsive unit almost completely from position −190 to position −43 (Truss et al 1995). On the surface of this nucleosome, the DNA double helix is rotationally phased in such a way that only the two external hormone-responsive elements have their major grooves exposed for receptor binding, while the two central hormone-responsive elements, as well as the NFI binding site and the octamer motifs, are positioned with the major groove pointing towards the interior of the nucleosome, inaccessible for protein binding (Truss et al 1994). This positioning of the nucleosome is exactly the same as was found in reconstitution experiments with histone octamers and linear or circular MMTV promoter fragments of various lengths (Piña et al 1990a,b,c). These results support the concept that MMTV promoter sequences determine nucleosome positioning (Piña et al 1990c) and that rotational positioning of DNA in this nucleosome constitutively represses the MMTV promoter (Piña et al 1991). In agreement with this hypothesis, inhibition of chromatin assembly in *Xenopus* oocytes seems to correlate with de-repression of microinjected MMTV promoter (Perlmann & Wrange 1991).

Transient transfection with NFI-VP16 construct

It has been argued that transient transfection is not a useful method for studying the involvement of chromatin organization in gene regulation because the transfected DNA is not properly organized in chromatin (Archer et al

1992). However, this concept was developed using calcium phosphate precipitation techniques, whereas in diethylaminoethyl dextran transfection experiments we have obtained transcription results compatible with organized chromatin (Piña et al 1990d). To test the suitability of transient transfection approaches in defining the contribution of nucleosome positioning to MMTV expression, we co-transfected various MMTV reporter plasmids with a chimeric transactivator composed of the NFI protein linked to the activation domain of VP16, the potent transactivator from herpes simplex virus (Sadowski et al 1988). We found that MMTV reporters containing the hormone-responsive elements were not transactivated by NFI-VP16, whereas those truncated at -80 and having only the NFI site and the octamer motifs were activated to a great extent. These results were only observed with diethylaminoethyl dextran transfections and suggest that the hormone-responsive element region precludes access of NFI and possibly Oct-1/OTF-1 to the MMTV promoter (R. Candau, M. Truss & M. Beato, unpublished results). Moreover, under these conditions, co-transfection of an expression vector for NFI does not transactivate the truncated reporters, suggesting that binding of NFI to the promoter is not sufficient to initiate transcription of MMTV.

Binding of factors to MMTV chromatin

In reconstituted MMTV nucleosomes, glucocorticoid receptor and progesterone receptor can bind to the two external hormone-responsive elements, whereas the two central hormone-responsive elements are much less accessible and no binding to the NFI site is detected, even at very high concentrations of protein (Perlmann 1992, Perlmann & Wrange 1988, Piña et al 1990b). So far, we have not been able to displace the nucleosome from MMTV promoter sequences *in vitro* by adding hormone receptors, with or without NFI (Piña et al 1990b).

   In cells carrying a single chromosomally integrated copy of the MMTV promoter, no binding of sequence-specific factors can be detected prior to hormonal stimulation, thus eliminating the possibility that the low level of transcription of MMTV is due to the action of a specific repressor (Truss et al 1995). After hormone induction, a full complement of transcription factors are found bound to the MMTV hormone-responsive unit, as judged by genomic footprints using ligation-mediated polymerase chain reaction (Pfeifer et al 1989). All four hormone-responsive elements, the NFI binding site and the octamer motifs are occupied in the majority of MMTV promoters (Truss et al 1995). These findings are in apparent contradiction to the results obtained on free DNA (steric hindrance between receptor and NFI, and between NFI and Oct1/OTF-1) and with the results obtained on reconstituted nucleosomes (receptor binding only to the two external hormone-responsive elements, no binding of NFI). Most unexpectedly, the nucleosome covering the MMTV hormone-responsive unit is not displaced or removed after induction but appears to remain in place, as determined by low- and high-resolution micrococcal nuclease digestion data (Truss et al 1995). We therefore postulate that the assembly of a full complement of transcription factors is facilitated by their binding to the surface of a positioned nucleosome. This statement presupposes that the organization of the DNA on the surface of the nucleosome must be altered after induction to permit factor binding to the major groove of sites originally masked.

Effect of the ligand

The precise role of the ligand in hormonal gene regulation is not well understood. Recently, evidence has accumulated that hormone receptors can be activated in the absence of their physiological ligands by second messenger signal transduction pathways which probably influence receptor phosphorylation (Power et al 1991). Interestingly, both ligand-induced and ligand-independent receptor activation can be inhibited by hormone antagonists (Gronemeyer et al 1991).

Antiprogestins, like RU486, compete with natural ligands for binding to the progesterone receptor, but why the complex of receptor and antiprogestins is functionally inactive is not clear. It is known that antiprogestins influence the conformation of the receptor, and may impair the dissociation of progesterone receptor from heat shock proteins. Prevalent models assume the existence of two types of antiprogestins (Gronemeyer et al 1991). Members of one class, represented by ZK98299, are considered to be pure antagonists and to prevent binding of the receptor to DNA. Other antiprogestins, such as RU486, can be partial agonists and are supposed to enable DNA binding of the progesterone receptor but to generate receptor complexes unable to activate hormone-responsive genes. This model is based on indirect evidence, such as transfection competition assays, and on *in vitro* DNA binding studies, but conflicting results have also been reported (Delabre et al 1993). Moreover, there is no direct evidence for an effect of antiprogestins on *in vivo* progesterone receptor binding to hormone-responsive elements. Therefore, we decided to study the effect of antiprogestins on *in vivo* DNA binding of progesterone receptor using the T-47D cell line that contains an integrated single copy of the MMTV LTR coupled to a luciferase reporter gene and *in vivo* DMS footprinting (Truss et al 1994). Our results provide unambiguous evidence that agonistic progestins alone, and not antiprogestins, such as RU486 and ZK98299, are able to induce binding of progesterone receptor to the MMTV promoter *in vivo*. Furthermore, antiprogestins cause a rapid disappearance of the agonist-induced footprints of progesterone receptor as well as of footprints of other transcription factors recruited by the receptor to the MMTV promoter. On the other hand, it is known that the partially purified progesterone receptor can bind to hormone-responsive elements in the presence of antiprogestins *in vitro* and, after transient co-transfection of a wild-type progesterone receptor, RU486 treatment prevented transcriptional activation by a constitutive progesterone receptor lacking the steroid binding domains. This was interpreted as a competition for DNA binding (Guiochon-Mantel et al 1988). The apparent contradiction between these findings and the lack of antiprogestin-induced progesterone receptor binding according to *in vivo* footprinting could have a number of explanations related to the changes in receptor conformation. The results of *in vitro* DNA binding studies could have been affected by the artificial induction of the active conformation of the progesterone receptor by *in vitro* manipulations.

Another possible explanation might involve differential effects of agonists and antagonists on the kinetics of receptor binding to DNA. It has been reported that agonists, but not antagonists, accelerate the on- and off-rates of progesterone receptor binding to non-specific DNA (Schauer et al 1989). In *in vitro* binding experiments or in transient transfections, the concentrations of receptors and hormone-responsive elements are high and may compensate for slow binding kinetics. However, under physiological conditions, with single hormone-responsive elements embedded in a large excess of random DNA, the critical parameter determining hormone-responsive element occupancy by the progesterone receptor may be the rate at which the receptor searches the genomic DNA. The efficiency of this process, which involves transfer between DNA sites requiring repeated rounds of association and dissociation, depends on the on- and off-rates of receptor binding to DNA. In such a scenario, only a progesterone receptor in its active conformation, as induced by an agonist, would be capable of efficiently searching chromosomal DNA and binding to hormone-responsive elements.

Remodelling of chromatin following induction: MMTV in yeast

It has been known for several years that hormone induction is accompanied by structural changes of chromatin, as indicated by the appearance of a DNase I-hypersensitive region over the hormone-responsive elements (Cordingley et al 1987, Richard-Foy & Hager 1987, Zaret & Yamamoto 1984). These findings have been interpreted to be a consequence of the removal or disruption of the positioned nucleosome covering the hormone-responsive unit—an interpretation that contradicts our genomic footprinting results. In cells carrying a single integrated copy of the MMTV promoter, even after full loading with transcription factors following hormone treatment, the patterns of DNase I and micrococcal nuclease cleavage sites characteristic of the nucleosome positioning are detectable in the regions not covered by transcription factors (Truss et al 1995). However, hormone induction generates a very narrow zone of hypersensitivity coinciding with the pseudo-dyad axis of the positioned nucleosome over the hormone-responsive unit, indicating a change in conformation of the nucleosomal DNA (Truss et al 1995). The nature of this conformational change remains obscure, but experiments with inhibitors of histone deacetylases and genetic results in *Saccharomyces cerevisiae* suggest a role for modification of the histone tails.

Among the changes in nucleosomes that have been associated with transcriptionally active chromatin is hyperacetylation of lysine residues in the N-terminal tails of all four core histones (Hebbes et al 1988). However, in the case of MMTV, inhibition of histone deacetylase with 5–10 mM sodium butyrate seems to inhibit hormone induction and nucleosome remodelling (Bresnick et al 1990). We have confirmed these results, but find that lower concentrations of butyrate actually activate hormone-independent transcription from single-copy integrated MMTV reporters. Moreover, inducing concentrations of butyrate generate the same type of DNase I hypersensitivity over the pseudo-dyad axis of the regulatory nucleosome that we observed following hormone induction (M. Truss, J. Bartsch, J. Bode & M. Beato, unpublished results). These data suggest that the nucleosome remodelling induced by receptor binding could involve changes in the behaviour of the core histone tails. When introduced into strains of *S. cerevisiae* expressing hormone receptors and NFI, the MMTV promoter is submitted to the same type of transcriptional control as that seen in animal cells. One observes a repression of basal transcription and induction that is dependent on agonist-activated glucocorticoid receptor or progesterone receptor and NFI. These findings suggest that histone H1 does not play an essential role in the response of MMTV to steroid hormones, since yeasts do not contain this class of linker histone. Reduction of the nucleosome density by the use of a mutant histone H4 gene subjected to galactose regulation (Han et al 1988, Kim et al 1988) leads to an activation of basal transcription, suggesting that repression is at least in part due to nucleosomal organization (S. Chávez & M. Beato, unpublished results). In yeast strains carrying a mutated histone H4 gene lacking the N-terminal 23 amino acids, which include all the potentially acetylated lysine residues (Durrin et al 1991), the basal activity of the MMTV promoter is enhanced, again suggesting a participation of the histone tails in transcriptional repression.

Conclusions

The behaviour of the MMTV promoter in transient transfections and in stable transfections (as a multicopy episomal vector, or as a single-copy chromosomally integrated gene) suggests that its primary nucleotide sequence determines the precise organization around a histone octamer, and in this way precludes binding of transcription factors prior to hormone induction. More significantly, the precise organization of the double helix in the nucleosome surface enables full loading with three classes of transcription factors, and is probably essential for their functional synergism. This interplay of transcription factors presupposes a remodelling of the nucleosome structure, which may involve the N-terminal tails of the core histones. As the MMTV promoter works correctly in yeast, the powerful techniques available for yeast genetics should help us to elucidate the molecular basis of the hormone-induced chromatin transition.

Acknowledgements

The experimental work described in this report was supported by grants from the Deutsche Forschungsgemeinschaft and der Fond der Chemischen Industrie.

References

Ahe D, Janich S, Scheidereit C, Renkawitz R, Schütz G, Beato M 1985 Glucocorticoid and progesterone receptors bind to the same sites in two hormonally regulated promoters. Nature 313:706–709

Archer TK, Lefebvre P, Wolford RG, Hager GL 1992 Transcription factor loading on the MMTV promoter: a bimodal mechanism for promoter activation. Science 255:1573–1576

Beato M 1989 Gene regulation by steroid hormones. Cell 56:335–344

Beato M 1991 Transcriptional regulation of mouse mammary tumor virus by steroid hormones. Crit Rev Oncog 2:195–210

Bolander FF, Blackstone ME 1990 Developmental and hormonal regulation of a mouse mammary tumour virus glycoprotein in normal mouse mammary epithelium. J Mol Endocrinol 4:101–106

Bresnick EH, John S, Berard DS, LeFebvre P, Hager GL 1990 Glucocorticoid receptor-dependent disruption of a specific nucleosome on the mouse mammary tumor virus promoter is prevented by sodium butyrate. Proc Natl Acad Sci USA 87:3977–3981

Brüggemeier U, Rogge L, Winnacker EL, Beato M 1990 Nuclear factor I acts as a transcription factor on the MMTV promoter but competes with steroid hormone receptors for DNA binding. EMBO J 9:2233–2239

Brüggemeier U, Kalff M, Franke S, Scheidereit C, Beato M 1991 Ubiquitous transcription factor OTF-1 mediates induction of the mouse mammary tumour virus promoter through synergistic interaction with hormone receptors. Cell 64:565–572

Buetti E, Kühnel B, Diggelmann H 1989 Dual function of a nuclear factor I binding site in MMTV transcription regulation. Nucleic Acids Res 17:3065–3078

Chalepakis G, Arnemann J, Slater EP, Brüller H, Gross B, Beato M 1988 Differential gene activation by glucocorticoids and progestins through the hormone regulatory element of mouse mammary tumor virus. Cell 53:371–382

Cordingley MG, Hager GL 1988 Binding of multiple factors to the MMTV promoter in crude and fractionated nuclear extracts. Nucleic Acids Res 16:609–630

Cordingley MG, Riegel AT, Hager GL 1987 Steroid-dependent interaction of transcription factors with the inducible promoter of mouse mammary tumor virus *in vivo*. Cell 48:261–270

Delabre K, Guichon-Mantel A, Milgrom E 1993 *In vivo* evidence against the existence of antiprogestins disrupting receptor binding to DNA. Proc Natl Acad Sci USA 90:4421–4425

Durrin LK, Mann RK, Kayne PS, Grunstein M 1991 Yeast histone H4 N-terminal sequence is required for promoter activation *in vivo*. Cell 65:1023–1031

Gronemeyer H, Meyer ME, Bocquel MT, Kastner P, Turcotte B, Chambon P 1991 Progestin receptors: isoforms and antihormone action. J Steroid Biochem Mol Biol 40:271–278

Guiochon-Mantel A, Loosfelt H, Ragot T et al 1988 Receptors bound to antiprogestin form abortive complexes with hormone responsive elements. Nature 336:695–698

Han K, Kim UJ, Kayne P, Grunstein M 1988 Depletion of histone H4 and nucleosomes activates the PHO5 gene in *Saccharomyces cerevisiae*. EMBO J 7:2221–2228

Hebbes TR, Thorne AW, Crane-Robinson C 1988 A direct link between core histone acetylation and transcriptionally active chromatin. EMBO J 7:1395–1402

Kalff M, Gross B, Beato M 1990 Progesterone receptor stimulates transcription of mouse mammary tumour virus in a cell-free system. Nature 344:360–362

Kim UJ, Han M, Kayne P, Grunstein M 1988 Effects of histone H4 depletion on the cell cycle and transcription of *Saccharomyces cerevisiae*. EMBO J 7:2211–2219

Kühnel B, Buetti E, Diggelmann H 1986 Functional analysis of the glucocorticoid regulatory elements present in the mouse mammary tumor virus long terminal repeat. A synthetic distal binding site can replace the proximal binding domain. J Mol Biol 190:367–378

Majors J, Varmus HE 1983 A small region of mouse mammary tumor virus long terminal repeat confers glucocorticoid hormone regulation on a linked heterologous gene. Proc Natl Acad Sci USA 80:5866–5870

Miksicek R, Borgmeyer U, Nowock J 1987 Interaction of the TGGCA-binding protein with upstream sequences is required for efficient transcription of mouse mammary tumor virus. EMBO J 6:1355–1360

Möws C, Preiss T, Slater EP et al 1994 Two independent pathways for transcription from the MMTV promoter. J Steroid Biochem Mol Biol 51:21–32

Perlmann T 1992 Glucocorticoid receptor DNA-binding specificity is increased by the organization of DNA in nucleosomes. Proc Natl Acad Sci USA 89:3884–3888

Perlmann T, Wrange Ö 1988 Specific glucocorticoid receptor binding to DNA reconstituted in a nucleosome. EMBO J 7:3073–3079

Perlmann T, Wrange Ö 1991 Inhibition of chromatin assembly in *Xenopus* oocytes correlates with derepression of the mouse mammary tumor virus promoter. Mol Cell Biol 11:5259–5265

Perlmann T, Erikson P, Wrange Ö 1990 Quantitative analysis of the glucocorticoid receptor–DNA interaction at the mouse mammary tumor virus glucocorticoid response element. J Biol Chem 265:17222–17229

Pfeifer GP, Steigerwald SD, Mueller PR, Wold B, Riggs AD 1989 Genomic sequencing and methylation analysis by ligation mediated PCR. Science 246:810–813

Piña B, Barettino D, Truss M, Beato M 1990a Structural features of a regulatory nucleosome. J Mol Biol 216:975–990

Piña B, Brüggemeier U, Beato M 1990b Nucleosome positioning modulates accessibility of regulatory proteins to the mouse mammary tumor virus promoter. Cell 60:719–731

Piña B, Truss M, Ohlenbusch H, Postma J, Beato M 1990c DNA rotational positioning in a regulatory nucleosome is determined by base sequence. An algorithm to model the preferred superhelix. Nucleic Acids Res 18:6981–6987

Piña B, Haché RJG, Arnemann J, Chalepakis G, Slater EP, Beato M 1990d Hormonal induction of transfected genes depends on DNA topology. Mol Cell Biol 10:625–633

Piña B, Barettino D, Beato M 1991 Nucleosome positioning and regulated gene expression. In: Maclean N (ed) Oxford surveys on eukaryotic genes. Oxford University Press, Oxford, vol 7:83–117

Power RF, Mani SK, Codina J, Conneely OM, O'Malley BW 1991 Dopaminergic and ligand-independent activation of steroid hormone receptors. Science 254:1636–1639

Richard-Foy H, Hager GL 1987 Sequence-specific positioning of nucleosomes over the steroid-inducible MMTV promoter. EMBO J 6:2321–2328

Sadowski I, Ma J, Triezenberg S, Ptashne M 1988 GAL4-VP16 is an unusually potent transcriptional activator. Nature 335:563–564

Schauer M, Chalepakis G, Willmann T, Beato M 1989 Binding of hormone accelerates the kinetics of glucocorticoid and progesterone receptor binding to DNA. Proc Natl Acad Sci USA 86:1123–1127

Scheidereit C, Beato M 1984 Contacts between receptor and DNA double helix within a glucocorticoid regulatory element of mouse mammary tumor. Proc Natl Acad Sci USA 81:3029–3033

Scheidereit C, Geisse S, Westphal HM, Beato M 1983 The glucocorticoid receptor binds to defined nucleotide sequences near the promoter of mouse mammary tumour. Nature 304:749–752

Toohey MG, Lee JW, Huang M, Peterson DO 1990 Functional elements of the steroid hormone-responsive promoter of mouse mammary tumor virus. J Virol 64:4477–4488

Truss M, Beato M 1993 Steroid hormone receptors: interaction with DNA and transcription factors. Endocr Rev 14:459–479

Truss M, Chalepakis G, Beato M 1992 Interplay of steroid hormone receptors and transcription factors on the mouse mammary tumor virus promoter. J Steroid Biochem Mol Biol 43:365–378

Truss M, Bartsch J, Beato M 1994 Antiprogestins prevent progesterone receptor binding to hormone responsive elements in vivo. Proc Natl Acad Sci USA 91:11333–11337

Truss M, Bartsch J, Schelbert A, Haché RJG, Beato M 1995 Hormone induces binding of receptors and transcription factors to a rearranged nucleosome on the MMTV promoter in vivo. EMBO J 14, in press

Zaret KS, Yamamoto KR 1984 Reversible and persistent changes in chromatin structure accompanying activation of a glucocorticoid-dependent enhancer element. Cell 38:29–38

DISCUSSION

McEwen: I remember hearing that the nuclear matrix is a site upon which both DNA replication is taking place and also bending of DNA to aid activation of genes. Is anything new known about this?

Beato: I can't say much about the nuclear matrix because I have never worked with it. I have been focusing on the very first level of organization of chromatin, the nucleosome. I haven't even mentioned the linker histones. Nor have I mentioned the many other levels of organization of chromatin, such as the loops and domains, and euchromatin and heterochromatin: I am sure that all these levels of organization will have a dramatic influence on the way proteins can access or modify the behaviour of promoters. The nuclear matrix is a preferred subject of many people because it seems to generate a compartment that will concentrate numerous transcription factors and proteins required for the remodelling of chromatin. Consequently, it may facilitate interactions that would not take place easily by diffusion in the whole nucleus.

Muramatsu: Your finding that the nucleosome structure is important for the MMTV promoter to work properly is very interesting. Viral promoters are especially crowded because of the small size of the viral genome. This may result in steric hindrance caused by the close positioning of the different elements. This may inhibit the proper binding of the various transcription factors. Do you think this requirement for nucleosome structure is unique to viral promoters, or can it be found in any normal mammalian genome?

Beato: That is a very good point: we are worried about it, because the viral system has been forced by evolution to concentrate all the information in a tiny piece of DNA. But if you look at other promoters and enhancers of natural genes, they are also very crowded. In many cases there are overlapping binding sites. It has been postulated that over a single DNA sequence one factor could bind to one side of double helix and another to the opposite side. I do not see how this can happen in chromatin, because only one side of the double helix is exposed. I cannot give you an answer because to my knowledge there is no other system that has been worked out to this level of resolution. We are starting to study the enhancer of the tyrosine aminotransferase in hepatoma cell lines in which we have integrated a single copy of the MMTV promoter. The tyrosine aminotransferase gene is also induced by glucocorticoid and we are going to compare carefully these two inducible promoters.

Muramatsu: You found that the nucleosome structure opens up the major groove of DNA through which some proteins are bound to specific sequences of DNA. But some factors are known to bind through the minor groove.

Beato: In those regions of the DNA double helix pointing to the exterior of the nucleosome, both grooves are wide open. But most of the sequence-specific transcription factors, in particular the three I am dealing with here, interact with the major groove. Now, of course, the TATA box-binding protein interacts with the minor groove, and its widening may be very important. Roger Kingston and Michael Green have shown that the binding of the TATA box binding protein to a nucleosomally organized piece of DNA is very much dependent on the rotational orientation of the DNA, and that this is influenced by the so-called SWI/SNF complex of proteins that is required for remodelling chromatin (Imbalzano et al 1994). You only see binding when you add the purified complex and ATP to the nucleosomes. This is a line we are also pursuing.

Baulieu: Why hasn't more work been done on the positioning of the binding of transcription factors to the hormone-responsive elements?

Beato: Because it's much harder to do than transfections!

Over the past 10 years many *cis*-acting elements and factors have been defined—there was a rush of papers concerning this because it is relatively easy to clone factors and characterize them in transfections. But we don't know how they really interact in the cell because we have been creating very artificial situations *in vitro* and *in vivo*. Now it's time to go back to the cell and look at normal genes, normal receptor concentrations and see what is really going on.

Pfaff: Is the genomic footprinting technique applicable to mixed cell populations?

Beato: One of the main limitations of this technique is that we need a population of promoters behaving in a similar way. For instance, if we have a cell line with 100 copies of the MMTV promoter it is almost impossible to do genomic footprinting with it.

Pfaff: How unusual is it to have terrific transcriptional effects depending on distribution of glucocorticoid-responsive element half sites? You got tremendous transcriptional effects through the MMTV hormone-responsive region and I think I saw only half sites.

Beato: There is no perfect palindromic site in there. Two of the receptor binding sites are imperfect palindromes and two of them are really just half sites. Their fuction may depend on how the promoter is organized in chromatin, because when you look at the nucleosome, one of the half sites is very close to another receptor binding site on the next superhelical turn. This may facilitate receptor–receptor interactions which will not take place on free DNA. But I don't think you can generalize; there are all kinds of combinations of receptor binding sites. We tend to assume that 'anything goes' but I think that this just reflects our ignorance about the real structure of chromatin. If you were to study a particular promoter in detail you would probably find that the organization of the sequences has some kind of functional significance.

Parker: You tended to imply that the role of hormone receptors is to reorganize chromatin to allow other transcription factors to bind. This is unlikely to be the whole story—clearly, receptors can interact with the basic transcription machinery and have roles to play independently of these other transcription factors. Would you agree with this?

Beato: Yes.

Parker: It may be that the transfection experiments that most of us are doing are valid to study those interactions.

Beato: That is true. There is good evidence for this sort of interaction with TFIIB. There is also evidence emerging for direct interaction between progesterone receptor and $TAF_{II}110$, which is a component of TFIID. Also, you can just take a minimum promoter with a TATA box and hormone-responsive elements, and you will get some transactivation. But in terms of the MMTV promoter, this is less than 4% of what you get with a complete set of elements, even in situations where a lot of receptor is present. With normal receptor concentration, the effects on the TATA box will be minimal. You need these synergies with other factors to get significant effects.

Baulieu: You mentioned the interesting activation of steroid receptors by phosphorylation induced by membrane-borne events. What is your opinion about the generality of this phenomenon? It has been described for progesterone, but you mentioned corticosteroids in your slide. Do you believe that the activation of receptors in the absence of hormone for launching transcription occurs frequently?

Beato: I am no authority on this field. My feeling is that it will be used widely, because all the signal transduction pathways are designed to interact and influence each other. It is like a tightly connected network that is always resonating. If you push it in one place, the whole network will respond in some way. I'm sure that receptor phosphorylation or other interactions (e.g. with heat shock proteins) will play a role.

Baulieu: This is a very important point for this meeting, in the sense that we're going to see many non-conventional effects of sex steroids. We should take account of accessory signals other than the classical cognate hormones. You mentioned heat shock proteins. Do you believe that there is a role for the kinetics of the receptor-associated proteins interacting with the receptor to explain the difference of activity between agonists and antagonists? Do you believe that the complex of receptor and heat shock protein (Hsp) found in the chromatin system could play a role in receptor action?

Beato: We have never directly studied interactions with heat shock proteins, but what I can say is that when you add antihormone the receptor changes its behaviour in some way: it is more tightly associated with the cell nucleus. Perhaps it really does dissociate from Hsp90. Our interpretation, which is based on a lot of assumptions, is that the receptor complexed with antihormone moves to the nucleus and binds to DNA. Whether or not there is some kind of kinetic change (as you have postulated) in the efficiency with which agonists and antagonists promote the dissociation from Hsp90, we cannot really judge from our experiments.

Parker: Do hormone antagonists always prevent binding of receptors to DNA? Clearly, some antagonists can also work as agonists. It has been proposed that the antagonist does promote some DNA binding to allow some transcriptional activity. Your model would suggest that, in general, antagonists never allow DNA binding. Do you think this might depend on the DNA binding site, for example?

Beato: It may do. We haven't finished this set of experiments with antagonists. It is not an easy task, because you need to compare reporter genes that are single copy and are integrated in a chromosomal location with similar accessibility. We are now constructing reporters that instead of MMTV promoters have the canonical hormone-responsive element and the tk promoter. We are repeating these experiments with T-47D clones (from Kate Horwitz) without endogenous receptor, and we are trying to introduce recombinant receptor to see if the level of receptor expression is important. All the evidence pointing to binding of the receptors to hormone-responsive elements when occupied by antihormones is indirect. Apart from our own data there is only one other report in which genomic footprinting has been used (Becker et al 1986); in neither case was there any binding of the receptor–antihormone complex to the hormone responsive elements.

If the agonist activity of some antagonists is mediated by hormone-responsive elements, it may be a problem of the receptor conformation. It is very clear that ligand binding induces a change in the structure of the receptor. From protease cleavage and antibody binding experiments, we know that agonists and antagonists appear to induce different changes in the receptor. It may be that in the cell, other parameters influence the type of change in the structure of the receptor that is induced by the ligand—perhaps kinases or phosphatases that are activated by other signal transduction pathways. Under these conditions antagonist could induce the kind of structural transition that normally follows binding of agonist.

Baulieu: Something different should happen because you have a different ligand. The ligand specifically changes the receptor conformation. What is the next step? Does the receptor have a different interaction with other chromatin constituents depending on whether or not there is binding of an agonist?

Beato: In our system, the most frequent type of conversion of antagonists into agonists is by the activation of cAMP. Many studies have shown that when the cAMP levels are high, antagonists such as RU486 behave like partial agonists in mammary cells. This might be explained by changes in the phosphorylation of the receptor or other factors interacting with the receptors. Antagonists could behave as agonists if something other than the ligand is helping convert the receptor to the structure that is normally achieved by the agonist.

Parker: Would you assume, in that case, that the receptor is then bound to DNA?

Beato: I can't say this is what we have found, because if we raise cAMP levels in our cells, RU486 only reaches 10–15% of the normal agonistic activity. With this very small effect we cannot see a footprint.

Baulieu: Can you generalize to oestrogen? There is some evidence, from Katzellenbogen and my colleagues Redeuilh and Sabba, that even in the absence of hormone there is some oestrogen receptor bound to DNA. Do you accept this?

Beato: We cannot generalize: we haven't done work on oestrogen. I don't know whether anybody has looked carefully for genomic footprinting with oestrogen receptor and anti-oestrogens.

Parker: To date, only transfection experiments have been used to assess occupancy of a DNA binding site. This could be criticized on the basis of the copy number of DNA and the amount of receptor.

Jensen: Many years ago, when we knew that the hormone-activated receptor binds in the nucleus to DNA, but before anyone knew about hormone-responsive elements, both Clark and Rochefort showed that nafoxidine and other type I anti-oestrogens differed from oestradiol in that they caused a longer retention of the receptor in the nucleus and prevented or delayed replenishment of the cytosolic receptor (Clark et al 1973, Capony & Rochefort 1975). It was proposed that turnover of nuclear-bound receptor, accompanied by replenishment of the receptor pool, is important for the full oestrogenic effect. But it is clear that type I anti-oestrogens do promote binding and retention of the receptor in the nucleus.

Baulieu: The assumption is that this is due to peripheral metabolism, which is much slower for tamoxifen or nafoxidine than for oestradiol.

Beato: These findings are compatible with the kinetic model I proposed: the agonist-loaded receptor will go into the nucleus and will bind unspecifically to DNA. Since the off-rate is slow, it will stay in the nucleus forever. The agonist-loaded receptor scans the DNA rapidly, finds the hormone-responsive elements, does what it has to do and leaves.

Baulieu: This is just a matter of different metabolism, external to the target cell, of tamoxifen and oestradiol.

Jensen: A few years ago, Pierre Martin, Yolande Berthois and I carried out a collaborative study on the ability of anti-oestrogens, such as tamoxifen and 4-hydroxytamoxifen, to expose a domain in the oestrogen receptor of human breast cancers that reacts specifically with just one of our monoclonal antibodies to the receptor. Because this effect can take place in the presence of excess oestradiol, which should saturate oestrogen binding, we proposed that there is an additional binding site in the receptor that is recognized by the anti-oestrogens but not by oestradiol (Martin et al 1988). At the time, this concept did not receive much acceptance, but now in Hamburg we have found that type II anti-oestrogens cause similar enhancement of receptor immunoreactivity and that the oestrogen receptors from human MCF-7 breast cancer cells binds nearly twice as much of either type I or type II anti-oestrogens as it does oestradiol (A. Hedden, V. Müller, E. V. Jensen, unpublished results). So it is clear that there is additional binding of anti-oestrogens by the receptor, and it would seem likely that this is somehow involved in antagonist action.

Muramatsu: Modifications of histones (e.g. phosphorylation, acetylation) are known to cause structural changes in the nucleosome. Does this affect the accessibility of transcription factors?

Beato: In cells with the MMTV reporter, if you enhance the acetylation of histones by using an inhibitor of the histone deacetylase (the enzyme that removes the acetyl groups from the ε-amino groups of lysines), you can induce transcription to the same extent as with hormone in the absence of an inhibitor of histone deacetylase. If you add it together with a hormone, you get dramatic enhancement. Acetylation of histones activates the promoter. However if you increase the concentration of butyrate to 5–10 mM (the concentration used by most scientists) you get complete acetylation, which results in repression of transcription and inhibition of receptor binding. So moderate acetylation of the histones can mimic the effect of the agonist-activated receptor.

Baulieu: Is this of any physiological significance?

Beato: We really don't know: it is just a potential mechanism. Of greater interest are some results that we have obtained with the MMTV system placed in *S. cerevisiae.* We have produced a nucleosome-depleted state by using a strain of yeast created by Michael Grunstein, which contains a galactose-inducible histone H4 gene. Upon nucleosome depletion, the activity of the MMTV promoter is enhanced in the absence of hormone. When we transfect a chimeric transactivator that has NFI linked to VP16, we see very strong transactivation in the nucleosome-depleted strain, showing that the nucleosome is repressing transcription of the MMTV promoter in yeast. In a histone H4 mutant lacking the amino acids that are acetylated at the N-terminus, the accessibility to NFI is also enhanced. This suggests that it is the acetylation of the histone H4 tail that is responsible for this type of behaviour.

Horwitz: In the sodium butyrate experiment, was the activation receptor dependent in the absence of hormone?

Beato: We have not tested cells without receptor.

References

Becker P, Gloss B, Schmid W, Strähle U, Schütz G 1986 *In vivo* protein–DNA interactions in a glucocorticoid response element require the presence of the hormone. Nature 324:686–688

Capony F, Rochefort H 1975 *In vivo* effects of anti-estrogens on the localization and replenishment of estrogen receptor. Mol Cell Endocrinol 3:233–251

Clark JH, Anderson JN, Peck EJ Jr 1973 Estrogen receptor–anti-estrogen complex: atypical binding by uterine nuclei and effects on uterine growth. Steroids 22:707–718

Imbalzano AN, Kwon H, Green MR, Kingston RE 1994 Facilitated binding of TATA-binding protein to nucleosomal DNA. Nature 370:481–485

Martin PM, Berthois Y, Jensen EV 1988 Binding of antiestrogens exposes an occult antigenic determinant in the human estrogen receptor. Proc Natl Acad Sci USA 85:2533–2537

Non-genomic mechanisms of action of steroid hormones

Etienne-Emile Baulieu and Paul Robel

INSERM U33, 80 rue de Général Leclerc, 92476 Bicêtre Cedex, France

Abstract. Sex steroid hormones are known to act through intracellular receptors and their cognate hormone response elements, located in the promoters of hormone-regulated genes. However, this classical mechanism of action cannot account for a variety of rapid effects of steroids (within seconds or minutes). In this review, non-genomic modes of target cell responses to sex steroids are described. The prototypical example is the resumption of meiosis in amphibian oocytes, triggered by progesterone at the plasma membrane level. Membrane effects of progesterone may also account for sperm maturation. Other membrane-mediated effects of steroids are reviewed. Whether a steroid hormone might elicit responses from a single cell through both genomic and membrane mechanisms remains an open question.

1995 Non-reproductive actions of sex steroids. Wiley, Chichester (Ciba Foundation Symposium 191) p 24–42

General background

The first scientists to be interested in steroid hormone action were largely the chemists who discovered the structures and conformations of the steroid hormones and their precursors and metabolites, and who soon synthesized active analogues. They were struck by the extraordinary range of biological activities of steroids, contrasting with the limited chemical diversity of these small, relatively rigid, lipophilic molecules. When biochemists and molecular biologists also began to study the mechanisms of steroid action, they attempted to resolve this apparent paradox.

Three types of cellular component have been considered as 'primary' molecular targets involved in the cellular responses to steroid hormones—lipids, proteins and nucleic acids. The term 'primary' designates the first molecules that are met by the steroid in a biologically significant manner: this encounter triggers the cascade of events involved in the response of the cell to the hormone.

Intracellular mechanisms

Biochemists first applied to the mode of hormone action a general concept of organic metabolism, derived from the catalytic activity of specific powerful enzymes: steroid hormones were considered as substrates of enzymes involved in hydrogen transfer between pyridine nucleotides (the 'transhydrogenase' hypothesis; Talalay & Williams-Ashman 1960). However, the effective concentrations of steroid hormones in target cells are about three orders of magnitude smaller than those of the substrates of steroid-metabolizing enzymes, which might operate in a transhydrogenation reaction. Moreover, many active synthetic steroid analogues cannot, for chemical reasons, participate in the postulated hydrogen transfer (they lack the appropriate hydroxyl group[s]). The hypothesis was rapidly abandoned. Nevertheless, as yet unproven membrane or intracellular enzymes may well be the primary targets of steroids, but in a completely different context involving their binding to a specific allosteric site and modulating the catalytic function.

The discovery of soluble intracellular receptors for steroid hormones—either cytoplasmic or nuclear—has been the most popular (and important) discovery for the understanding of steroid action. The current developments result from the merging of this description of receptors with the techniques and concepts of molecular biology. Mueller was the first investigator to show that sex steroid hormones control gene transcription, in his work on oestrogenic stimulation of the uterus (Mueller et al 1958). Originally, intracellular steroid receptors were described in target tissues as proteins binding hormone with high affinity and with ligand binding specificity closely correlated with effects (Jensen & Jacobson 1962, Toft & Gorski 1966, Baulieu et al 1967). Receptor proteins perform the two complex functions of binding natural or synthetic hormones and antihormones, and interacting with the nucleic acids and proteins involved in the transcriptional machinery. In early experiments, nuclei of target cells were exposed to crude hormone–receptor complexes that triggered an increase in mRNA synthesis (Raynaud-Jammet & Baulieu 1969, DeSombre et al 1975). Nowadays, the structures of the nuclear receptor superfamily members have been unravelled, and their functions as ligand-operated transcription factors have been firmly established (see Truss et al 1995, this volume, and the reviews by Evans 1988, Beato 1989, Green & Chambon 1988, O'Malley et al 1991, Yamamoto 1985). Steroid action at the gene level depends on the binding of hormone–receptor complexes to the hormone-responsive elements located on the DNA (most often in the promoter region) of target genes.

Another mechanism of transcriptional regulation by steroid hormones has been suggested. In this mechanism, nucleic acids are the primary and direct molecular targets of steroid hormones. The double helical structure of DNA is such that intercalation of the flat steroid molecules between two base pair planes is conceivable and has been documented (Ts'o & Lu 1964). Up to now, no steroid–DNA interaction of biological significance has been demonstrated. However, our present knowledge of chromatin structural diversity, related to DNA base sequences, bending and associated proteins, leaves the way open for specific functional steroid–DNA interactions. They would occur only under defined conditions when steroids have direct access to DNA.

In spite of some early work (Liao et al 1989), there is no definitive demonstration of primary post-transcriptional function of steroid receptors. Nevertheless, steroid hormones often modify post-transcriptional events, such as mRNA stability and efficiency of translation, or provoke post-translational modifications of proteins. However, the synthesis of proteins involved in mRNA function is probably transcriptionally regulated. The interaction of steroids with RNA-binding proteins, which has been proposed as a mechanism of translational control, has not been formally demonstrated, although, for example, the formation of glucocorticosteroid receptor–c-Jun heterodimers may relieve the repression exerted by c-Jun on the translation of its own mRNA (Vig et al 1994).

Plasma membrane mechanisms

The plasma membrane has long been studied as a possible site of steroid action. Several hypotheses have been proposed (Baulieu 1978), and conclusive evidence has been obtained. Indeed, the plasma membrane is a lipid bilayer. Steroids may be integrated at interfaces between lipids and proteins, lying flat or on their edges according to their structural features (Munck 1957). The cholesterol content of cell membranes varies to some extent, and modulates their fluidity. It is conceivable that steroids could play a similar role (Willmer 1961). Pregnenolone, 17-hydroxypregnenolone and dehydroepiandrosterone (DHEA), which are precursors of hormonal $\Delta 4$-3-oxo-steroid hormones, are particularly suited for mimicking cholesterol, since they retain its 3β-hydroxy-$\Delta 5$ structure. Moreover, as with cholesterol itself, they can be conjugated at the 3 position by either acyltransferases or sulphokinases, to form the respective fatty acid or sulphate esters. In this respect, the concentration of DHEA sulphate in primates, particularly the human, may be large enough to compete with cholesterol and its esters in plasma membranes. Anyhow, steroid insertion into the membrane may modify its permeability, flexibility, or the movement of integral proteins, with consequences on the activity of plasma membrane proteins, such as receptors, voltage- or ligand-gated ion channels, or enzymes, responsible for regulatory processes occurring at the cell surface. Some of these proteins are direct targets for steroids, and we will provide examples of such mechanisms. A matter which has not been resolved is the possibility of both membrane and genomic mechanisms of steroid action in the same cells, and of their possible cooperation or coordination.

The unconjugated steroids, in contrast to their hydrophilic and acidic sulphate and glucuronide derivatives, may diffuse 'freely' through the lipidic plasma membranes. Nevertheless, a protein-mediated, not energy-dependent transport system has been described in uterine cells (Milgrom et al 1973), hepatocytes and pituitary cells (reviewed in Duval et al 1983). None of these membrane components has been fully characterized. Non-receptor, steroid-binding plasma proteins such as transcortin (CBG) and sex steroid-binding plasma protein (SBP; also known as sex hormone-binding globulin, SHBG), have been found in the cytosol of several target organs. Their presence may be a consequence of either internalization or local synthesis. In the latter case, they should not play an important role in hormone action. They have been less well studied than the intracellular non-receptor binding proteins for thyroid hormones and retinoids. It has been shown that these plasma proteins bind to membrane 'receptors' of target cells (Rosner 1991).

The concept of allosteric modulation may unify most aspects of steroid action according to current data. This is the case for intracellular receptors. Binding of the hormone to its ligand binding domain induces a profound change designated as 'transformation' or 'activation', promotes interaction of the DNA-binding domain with DNA, and triggers transcriptional activity (via transcription-activating functions, TAFs). Several membrane receptors for neurotransmitters are also allosterically modulated by steroids.

Examples of steroid action at the plasma membrane level

The progesterone receptor of Xenopus laevis oocytes:
decreased adenylate cyclase activity, increased Ca^{2+} uptake

Mature *Xenopus laevis* oocytes (stage VI of Dumont) resume meiosis when incubated in a solution containing micromolar concentrations of progesterone, or of several other steroids with or without affinity for the classical intracellular progesterone receptor. The possibility of a primary site of steroid hormone action at the plasma membrane level had already been considered by several scientists in different systems, when Sabine Schorderet-Slatkine asked one of us to look for a progesterone receptor in amphibian oocytes. Smith, Schutz and colleagues (reviewed in Baulieu et al 1978) had already provided indirect evidence for an action of progesterone at the external side of the plasma membrane, since progesterone was active when added to the incubation medium, but not when injected into the cytoplasm. It was also known that enucleated oocytes exposed to progesterone generated the meiosis-promoting factor (MPF), thus excluding a genomic effect of the hormone.

We did not detect an intracellular progesterone receptor. We also failed to detect a membrane receptor with conventional equilibrium binding assays using radioactive hormone. An interaction of the hormone with membrane lipids, approximately following the polarity rule (the most hydrophobic steroids should be the most active), did not explain our data: some steroids with the same partition coefficient as progesterone were inactive, and most importantly several steroids with no or weak meiosis-promoting activity inhibited progesterone action, implying a limited number of specific sites. The failure of binding experiments was then attributed to the high lipid content of the plasma membrane. We were encouraged in our search for a membrane receptor by the finding that meiosis was induced by incubation of *Xenopus* oocytes with a 'macromolecular progesterone' (progesterone covalently linked to poly[ethylene oxide]), that could not enter oocytes. Moreover, progesterone inhibited the adenylate cyclase activity of an oocyte membrane preparation, thus reproducing the decrease in cAMP levels observed in intact oocytes exposed to progesterone, a result in accordance with the well-documented antimeiotic action of cAMP (Finidori-Lepicard et al 1981).

Finally, photoaffinity labelling of oocyte membranes with an active progestin, [^{3}H]R5020, allowed us to detect a molecule of approximately 30 kDa, definitely smaller than adenylate cyclase (about 120 kDa). This protein showed the required features of a receptor for meiosis-inducing steroids (Blondeau & Baulieu 1984). Unfortunately, this putative receptor was not further studied. Thus, the apparent discrepancy between these results and another report which claims the presence of an approximately 120 kDa protein that also binds RU486, indeed a weak agonist of progesterone in this system, has not been resolved (Sadler et al 1985). The inhibition of adenylate cyclase activity suggests that the progesterone receptor is coupled to G protein, although pertussis toxin does not block progesterone action (Goodhart et al 1984). It is likely that the progesterone receptor of *Xenopus* oocytes belongs to a family including receptors for the steroids that trigger meiosis in other amphibian and in fish oocytes (Jalabert 1976, Patino & Thomas 1990), and even possibly the adenine receptor of starfish oocytes.

Concurrently with the decrease in cAMP, progesterone increases the concentration of intracellular free Ca^{2+} in *Xenopus* oocytes. Both uptake of extracellular Ca^{2+} and mobilization of intracellular Ca^{2+} stores might be involved. Gammexane, an inhibitor of phosphoinositide generation, blocks reinitiation of meiosis.

Finally, insulin, operating through the receptor for insulin-like growth factor 1 (IGF-1), is also active, and the effects of steroid and insulin potentiate each other (Baulieu & Schorderet-Slatkine 1983). The two types of hormones act via different membrane receptors and intracellular cascades. The transduction mechanisms that may explain their cooperation are still unknown.

In conclusion, *Xenopus laevis* oocytes constitute a well-documented case of steroid action at the level of the membrane. The putative progesterone receptor definitely differs from the classical intracellular one by its ligand binding specificity, its size and its transduction mechanism. This is a remarkable example of cooperation between a lipophilic hormone and a water-soluble peptidic growth factor at the plasma membrane level. Cooperation between a membrane (dopamine) receptor and an intracellular (progesterone) receptor is another, more recent and conceptually quite different example, since it implies a transcriptional effect via an intracellular steroid receptor in the absence of hormone (Power et al 1991).

Steroid receptors as integral components of ligand-gated ion channels

The type A γ-aminobutyric acid (GABA$_A$) receptor is an oligomeric protein complex that, when activated by agonists, produces an increase in neuronal membrane conductance to Cl$^-$ ions, resulting in membrane hyperpolarization and reduced neuronal excitability. A number of centrally active drugs, including convulsants, anticonvulsants, anaesthetics and anxiolytics, bind to distinct but interacting domains of this receptor complex, to modulate Cl$^-$ conductance (Olsen & Tobin 1990).

There are both inhibitory and excitatory steroids which, respectively, excite or inhibit GABA activity in acting on GABA$_A$ receptors in nerve cell membranes (Majewska 1992, Paul & Purdy 1992, Gee 1988). Some steroid metabolites, such as the reduced derivative of progesterone, allopregnanolone (3α-hydroxy-5α-pregnan-20-one), both mimic and enhance the effects of GABA. These steroids potentiate both muscimol and benzodiazepine binding, whereas they inhibit the binding of the convulsant t-butylbicyclo-phosphorothionate (TBPS). Pharmacological evidence indicates that steroid interaction site(s) is (are) distinct from those of both benzodiazepines and barbiturates. The most active molecules operate in the low nanomolar range. The 3α-hydroxyl group is an absolute structural requirement. A planar conformation of the A/B ring junction (5α-reduced) is preferred; however, 5β-hydroxy derivatives still display significant activity.

Pregnenolone sulphate and DHEA sulphate are prototypic naturally excitatory neurosteroids (Majewska 1992). At low micromolar concentrations, they antagonize GABA$_A$ receptor-mediated ^{36}Cl$^-$ uptake into synaptoneuro-somes and Cl$^-$ conductance in cultured neurons. Pregnenolone sulphate bimodally modulates [^{3}H]muscimol binding to synaptosomal membranes, slightly potentiates benzodiazepine binding, and inhibits the binding of the convulsant [^{35}S]TBPS to the GABA$_A$ receptor Cl$^-$ channel. Conversely, pregnenolone sulphate is displaced from its synaptosomal binding site by barbiturates at millimolar concentrations. Moreover, despite mutual com-petition in these membranes, DHEA sulphate and pregnenolone sulphate behave differently in their capability to displace [^{35}S]TBPS and 1-[^{3}H]-phenyl-4-t-butyl-2,6,7-trioxabicyclo[2-2-2]octane and to enhance [^{3}H]benzodiazepine binding. Furthermore, unlike the sulphate esters of Δ5-3β-hydroxysteroids, those of 3α-hydroxysteroids behave as their unconjugated counterparts, potentiating GABA-induced Cl^- transport in synaptoneurosomes (El-Etr et al 1992). Therefore, distinct sites for neurosteroids, mediating distinct allosteric modes of interaction, seem to exist on the $GABA_A$ receptor or in its membrane vicinity. Nevertheless, the complexity of $GABA_A$ receptors must be kept in mind when interpreting these results—transfection experiments with defined wild-type or mutated $GABA_A$ receptor subunits may provide more direct evidence for their sites of interaction with neurosteroids.

In addition to its effects on $GABA_A$ receptors, pregnenolone sulphate positively modulates N-methyl-D-aspartate receptors, and negatively regulates glycine receptors in spinal cord neurons (Wu et al 1991). Progesterone (free and bovine serum albumin [BSA]-linked forms) modulates a neuronal nicotinic acetylcholine receptor, assembled from two subunits α4 and nα1 expressed in *Xenopus* oocytes (Valera et al 1992), and progesterone decreases the binding of muscarinic cholinergic agents in hypothalamic and pituitary membranes (Klangkalya & Chan 1988).

Membrane receptors for oestrogens

While intracellular receptors for oestrogens were the first steroid receptors to be detected, specific oestrogen binding sites on plasma membranes of endometrial and liver cells were detected early on, thanks to the use of an affinity material made up of oestrogen immobilized by covalent linkage to an inert support (Pietras & Szego 1977, 1980). Oestradiol binding to plasma membranes was also detected, by equilibrium binding assays, on rat synaptosomal membranes (Towle & Sze 1983) and pituitary cells (Bression et al 1986), or by immunocytochemical methods in nerve endings of rat hypothalamic slices (Blaustein et al 1992), and plasma membranes of human breast cancer cells (Nenci et al 1980) or prolactin-secreting GH3 cells (T. C. Pappas, personal communication 1994).

Short-term effects of oestradiol are consistent with specific interactions of the hormone with plasma membrane components. This is the case with septopreoptic neurons (Kelly et al 1977), pituitary cells in culture (Dufy et al 1979), cerebellar Purkinje cells (Smith et al 1988), hippocampal CA1 neurons (Wong & Moss 1992) and medial amygdala neurons (Nabekura et al 1986).

In cultured rat osteoblasts, a cell surface receptor for oestrogens, possibly coupled to phospholipase C via a G protein, has been described (Lieberherr et al 1993). The rapid release of intracellular Ca^{2+} by oestrogen in chicken granulosa cells may also be due to the function of a cell surface receptor (Morley et al 1992). The same cells may have both membrane and intracellular oestradiol receptors (Blaustein et al 1992).

In summary, it is still impossible to connect precisely the binding characteristics of the postulated receptors, with corresponding electro-physiological and biochemical responses to oestrogens. In some cases, steroid specificity resembles that of the intracellular oestradiol receptor (i.e. 17β-oestradiol binds but 17α-oestradiol does not), but not completely (i.e. no binding of hydroxytamoxifen reported in Bression et al 1986).

Other membrane receptors for progesterone

The mechanism by which progesterone modulates gonadotropin-releasing hormone (GnRH) release is poorly understood. The effect of progesterone may involve regulation of GnRH gene expression, although indirectly via action on other cells than GnRH neurons, since progesterone receptors are not present in the latter cells; it was assumed that the effect of progesterone ions was exerted through neurotransmitter/neuropeptide-containing neurons in the hypothalamus that surround GnRH neurons. Progesterone coupled to BSA via the 3-ketone or a 11α-hydroxyl group was used to uncover a membrane-binding component that might be responsible for the stimulation of GnRH release by superfused rat hypothalamus (Tischkau & Ramirez 1993). However, the site of progesterone action was not determined. GnRH neurons possess postsynaptic $GABA_A$ receptors, they convert progesterone to its neuroactive metabolite allopregnanolone, which, in contrast to progesterone itself, potentiates the GABA-induced release from an established cell line of GnRH neurons (El-Etr et al 1995). Thus, progesterone may modulate GnRH synthesis and release by indirect, genomic and membrane mechanisms, and by a direct, allopregnanolone-mediated membrane action. Progesterone was identified as a major component of the follicular fluid for inducing acrosome reaction preceded by an immediate, transient Ca^{2+} influx into spermatozoa (Thomas & Meizel 1988). These rapid biological effects have been attributed to a sperm cell-surface receptor (Blackmore et al 1991). Whether or not this receptor is analogous to the one found in Xenopus laevis oocytes is unknown. In contrast, RU486 induced an immediate dose-dependent decrease in intracellular free Ca^{2+} in human sperm, and it was suggested that RU486 interacts with a membrane site distinct from the putative progesterone receptor (Yang et al 1994).

Glucocorticosteroid membrane receptors

A glucocorticosteroid receptor-like antigen, recognized by specific antibodies to the intracellular glucocorticoid receptor, has been detected in the plasma membranes of S49 mouse lymphoma cells. It has been claimed to be responsible for the lympholytic response to corticosteroids (Gametchu 1987). This membrane glucocorticoid receptor, which is larger than its intracellular counterpart, undergoes patching and capping after hormone binding and can be removed from the cell surface by trypsin (Gametchu et al 1991), resulting in the loss of lympholytic activity. Although the membrane component is antigenically related to the intracellular glucocorticoid receptor, it is unknown whether the latter is inserted in the membrane, possibly after a post-translational modification, or whether a membrane-specific isoform is present. Binding sites for corticosteroids have also been found in liver plasma membranes and/or neurons.

A glucocorticosteroid receptor has been observed in the synaptosomal membranes of an amphibian (*Taricha granulosa*), and it is involved in the rapid suppression of male reproductive behaviour. The affinities of corticosteroids and analogues are linearly related to their potencies (Orchinik et al 1991). Its steroid-binding specificity is different from that of the intracellular gluco-corticoid receptor. Whether intracellular and membrane glucocorticoid receptors coexist in the same cell is unknown. Guanyl nucleotides negatively modulate the binding of corticosterone to the membrane receptor, suggesting coupling to G proteins (Orchinik et al 1992).

A preliminary report (Wehling et al 1993) has described a membrane aldosterone receptor in human mononuclear leukocytes which modulates their Na^+-H^+ exchanger.

Membrane receptors for androgens

In confluent male rat osteoblasts, testosterone (either free or coupled to BSA) increased intracellular Ca^{2+} flux and increased inositol 1,4,5-trisphosphate and diacylglycerol formation within seconds, thus suggesting a cell-surface mechanism (Lieberherr & Grosse 1994).

A special case: steroidal pheromones

A pheromone is a diffusible external chemical messenger secreted by an individual and received by another individual of the same species in which it induces a response (Luscher & Karlson 1959, Monti-Bloch et al 1994). In mammals, pheromones are present in several secretions and excretions from various organs, including the skin, and may reach receptors located in the vomeronasal organ.

The vomeronasal organ is atrophic in adult primates, and functions only very early in life as a means of maternal recognition by the newborn. 'Vomeropherins' are not 'odorants' and sex specificity has been described by electrophysiological methods. Volatile Δ16-androstene derivatives are candidates for steroid pheromonal functions (Gower & Ruparelia 1993).

An example of a water-diffusible steroid pheromone has been described in some fishes. Progesterone derivatives, such as 17,20α-dihydroxy-pregn-4-en-3-one, have two functions. In the female, the steroid, made in the ovaries, triggers oocyte meiotic maturation, similarly to the action of progesterone in amphibian oocytes (Sorensen et al 1989). The steroid is also excreted into the water and reaches the nasal epithelium of males in amounts sufficient to stimulate heterosexual behaviour and sperm release, probably via the hypothalamopituitary gonadal axis. Whether the nasal sensory receptor engaged in this response is related to the Δ16-androstene steroid receptor(s) previously postulated is unknown. Another, possibly related, observation is that the mRNAs of some odorant receptors cloned from nasal epithelium have been found on dog sperm (Vanderhaeghen et al 1993).

Summary and conclusions

The classical mechanism of steroid hormone action, which implies the binding of activated hormone–receptor complexes to regulatory elements of target genes, cannot account for a variety of rapid (within seconds or minutes) effects of steroids. Membrane mechanisms have therefore been postulated and documented. They are the main topic of the present review. Although none of these receptors has been duly isolated and purified, they seem generally distinct from the intracellular receptors. This conclusion is based, despite limited analogies, on the fact that ligand affinities and specificities are markedly different between the two types of receptors. Since some intracellular and membrane receptors are antigenically related, the existence of either classical receptors integrated in the plasma membrane (eventually after post-translational modification) or of membrane-specific isoforms remain possibilities. In most cases, relatedness of membrane receptors to their intracellular counterparts seems excluded, and even their coexistence in the same cells awaits conclusive demonstration.

Distinct categories of membrane steroid receptors have been distinguished. Steroids can be allosteric modulators of neurotransmitter receptors. The best known example is the large family of $GABA_A$ receptor isoforms, but steroids can also modulate glutamate, glycine, acetylcholine, opioid and σ receptors. The physiological significance of almost all these modulations is uncertain. In other instances, steroids appear as the primary ligands of membrane receptors: this is the case for progesterone receptors in oocytes and sperm, and for corticosteroid receptors in the CNS and in lymphocytes. Several attempts have been made to define the transduction mechanisms triggered by the binding of steroids to membrane receptors, but the results obtained are very preliminary, and further work is needed to unravel the molecular mechanism, the physiological significance, and the eventual pathological implications of membrane steroid receptors.

References

Baulieu EE, Alberga A, Jung I 1967 Récepteurs hormonaux. Liaison spécifique de l'oestradiol à des protéines utérines. C R Acad Sci Ser III Sci Vie 265:354–357

Baulieu EE 1978 Cell membrane, a target for steroid hormones. Mol Cell Endocrinol 12:247–254

Baulieu EE, Godeau F, Schorderet M, Schorderet-Slatkine S 1978 Steroid-induced meiotic division in *Xenopus laevis* oocytes: surface and calcium. Nature 275: 593–598

Baulieu EE, Schorderet-Slatkine S 1983 Steroid and peptide control mechanisms in membrane of *Xenopus laevis* oocytes resuming meiotic division. In: Molecular biology of egg maturation. Pitman, London (Ciba Found Symp 98) p 137–158

Beato M 1989 Gene regulation by steroid hormones. Cell 56:335–344

Blackmore PF, Neulen J, Lattenzio F, Beebe SJ 1991 Cell surface-binding sites for progesterone mediate calcium uptake in human sperm. J Biol Chem 266: 18655–18659

Blaustein JD, Lehman MN, Turcotte JC, Greene G 1992 Estrogen receptors in dendrites and axon terminals in the guinea pig hypothalamus. Endocrinology 131:281–290

Blondeau JP, Baulieu EE 1984 Progesterone receptor characterized by photo-affinity labelling in the plasma membrane of *Xenopus laevis* oocytes. Biochem J 219: 785–792

Bression D, Michard M, Le Dafniet M, Pagesy P, Peillon F 1986 Evidence for a specific estradiol binding site on rat pituitary membranes. Endocrinology 119:1048–1051

DeSombre ER, Mohla S, Jensen EV 1975 Receptor transformation, the key to estrogen action. J Steroid Biochem 6:469–480

Dufy B, Vincent JD, Fleury H, Du Pasquier P, Gourdji D, Tixier-Vidal A 1979 Membrane effects of thyrotropin-releasing hormone and estrogen shown by intracellular recording from pituitary cells. Science 204:509–511

Duval D, Durant S, Homo-Delarche F 1983 Non-genomic effects of steroids. Interactions of steroid molecules with membrane structures and functions. Biochim Biophys Acta 737:409–442

El-Etr M, Corpechot C, Young J, Akwa Y, Robel P, Baulieu EE 1992 Modulating effect of steroid sulfates on muscimol-stimulated ^{36}Cl uptake. Eur J Neurosci 4:128(abstr)

El-Etr M, Akwa Y, Fiddes R, Robel P, Baulieu EE 1995 A progesterone metabolite stimulates the release of gonadotropin-releasing hormone (GnRH) from GT1-1 hypothalamic neurons via the γ-aminobutyric acid type A receptor. Proc Natl Acad Sci USA, in press

Evans RM 1988 The steroid and thyroid hormone receptor family. Science 240:889–895

Finidori-Lepicard J, Schorderet-Slatkine S, Hanoune J, Baulieu EE 1981 Progesterone inhibits membrane-bound adenylate cyclase in *Xenopus laevis* oocytes. Nature 292:255–256

Gametchu B 1987 Glucocorticoid receptor-like antigen in lymphoma cell membranes: correlation to cell lysis. Science 236:456–461

Gametchu B, Watson CS, Pasko D 1991 Size and steroid-binding characterization of membrane-associated glucocorticoid receptor in S-49 lymphoma cells. Steroids 56:402–410

Gee KW 1988 Steroid modulation of the GABA/benzodiazepine receptor-linked ionophore. Mol Neurobiol 2:291–317

Goodhart M, Ferry N, Buscaglia M, Baulieu EE, Hanoune J 1984 Does the guanine regulatory protein Ni mediate the progesterone inhibition of *Xenopus* oocyte adenylate cyclase? EMBO J 3:2653–2657

Gower DB, Ruparelia BA 1993 Olfaction in humans with special reference to odorous 16-androstenes: their occurrence, perception and possible social, psychological and sexual impact. J Endocrinol 137:167–187

Green S, Chambon P 1988 Nuclear receptors enhance our understanding of transcription regulation. Trends Genet 4:309–314

Jalabert B 1976 *In vitro* oocyte maturation and ovulation in rainbow trout (*Salmo gairdneri*), northern pike (*Esox lucius*), and goldfish (*Carassius auratus*). J Fish Res Board Can 33:974–988

Jensen EV, Jacobson HI 1962 Basic guides to the mechanism of estrogen action. Rec Prog Horm Res 18:387–414

Kelly MJ, Moss RL, Dudley CA, Fawcett CP 1977 The specificity of the response of preoptic–septal area neurons to estrogen: 17α-estradiol versus 17β-estradiol and the response of extrahypothalamic neurons. Exp Brain Res 30:43–52

Klangkalya B, Chan A 1988 Inhibition of hypothalamic and pituitary muscarinic receptor binding by progesterone. Neuroendocrinol 47:294–302

Lieberherr M, Grosse B 1994 Androgens increase intracellular calcium concentration and inositol 1,4,5-trisphosphate and diacylglycerol formation via a pertussis toxin-sensitive G-protein. J Biol Chem 269:7217–7223

Lieberherr M, Grosse B, Kachkache M, Balsan S 1993 Cell signaling and estrogens in female rat osteoblasts: a possible involvement of unconventional nonnuclear receptors. J Bone Miner Res 8:1365–1376

Liao S, Kokontis J, Sai T, Hiipakka RA 1989 Androgen receptors: structures, mutations, antibodies and cellular dynamics. J Steroid Biochem 34:41–51

Luscher M, Karlson P 1959 Pheromones: a new term for a class of biologically active substances. Nature 183:55–56

Majewska MD 1992 Neurosteroids: endogenous bimodal modulators of the GABA$_A$ receptor. Mechanism of action and physiological significance. Prog Neurobiol 38:379–395

Milgrom E, Atger M, Baulieu EE 1973 Studies on estrogen entry into uterine cells and on oestradiol–receptor complex attachment to the nucleus. Is the entry of estrogen into uterine cells a protein-mediated process? Biochim Biophys Acta 320:267–283

Monti-Bloch L, Jeennings-White C, Dolberg DS, Berliner DL 1994 The human vomeronasal system. Psychoneuroendocrinology 19:673–686

Morley P, Whitfield JF, Wanderhyden BC, Tsang BK, Schwartz J-L 1992 A new, nongenomic estrogen action: the rapid release of intracellular calcium. Endocrinology 131:1305–1312

Mueller GG, Herranen AM, Jervell KF 1958 Studies on the mechanism of action of estrogens. Rec Prog Horm Res 14:95–139

Munck A 1957 The interfacial activity of steroid hormones and synthetic estrogens. Biochim Biophys Acta 24:507–514

Nabekura J, Oomura Y, Minami T, Mizuno Y, Fukuda A 1986 Mechanism of the rapid effect of 17β-estradiol on medial amygdala neurons. Science 233:226–228

Nenci I, Fabris G, Marchetti E, Marzola A 1980 Cytochemical evidence for steroid binding sites in the plasma membrane of target cells. In: Bresciani (ed) Perspectives in steroid receptor research. Raven Press, New York, p 61–72

Olsen RW, Tobin AJ 1990 Molecular biology of GABA$_A$ receptors. FASEB J 4:1469–1480

O'Malley BW, Tsai SY, Bagchi M, Weigel NL, Schrader WT, Tsai MJ 1991 Molecular mechanism of action of a sex steroid receptor. Rec Progr Horm Res 47:1–26

Orchinik M, Murray TF, Moore FL 1991 A corticosteroid receptor in neuronal membranes. Science 252:1848–1851

Orchinik M, Murray TF, Franklin PH, Moore FL 1992 Guanyl nucleotides modulate binding to steroid receptors in neuronal membranes. Proc Natl Acad Sci USA 89:3830–3834

Patino R, Thomas P 1990 Characterization of membrane receptor activity for 17α,20α,21-trihydroxy-4-pregnen-3-one in ovaries of spotted seatrout (*Cynoscion nebulosus*). Gen Comp Endocrinol 78:204–217

Paul SM, Purdy RH 1992 Neuroactive steroids. FASEB J 6:2311–2322

Pietras RJ, Szego CM 1977 Specific binding sites for oestrogen at the outer surfaces of isolated endometrial cells. Nature 265:69–72

Pietras RJ, Szego CM 1980 Partial purification and characterization of oestrogen receptors in subfractions of hepatocyte plasma membranes. Biochem J 191:743–760

Power RF, Mani SK, Codina J, Conneely OM, O'Malley BW 1991 Dopaminergic and ligand-independent activation of steroid hormone receptors. Science 254:1636–1639

Raynaud-Jammet C, Baulieu EE 1969 Action de l'oestradiol *in vitro*: augmentation de la biosynthèse d'ARN dans les noyaux utérins. C R Acad Sci Ser III Sci Vie 268: 3211–3214

Rosner W 1991 Plasma steroid-binding proteins. Endocrinol Metab Clin North Am 20:697–719

Sadler SE, Bower MA, Maller JL 1985 Studies of a plasma membrane steroid receptor in *Xenopus* oocytes using the synthetic progestin RU486. J Steroid Biochem 22: 419–426

Smith SS, Waterhouse BD, Woodward DJ 1988 Locally applied estrogens potentiate glutamate-evoked excitation of cerebellar Purkinje cells. Brain Res 475:272–282

Sorensen PW, Hara TJ, Stacey NE, Dulka JG 1989 Extreme olfactory specificity of male goldfish to the preovulatory steroidal pheromone 17α,20β-dihydroxy-4-pregnen-3-one. J Comp Physiol A Sens Neural Behav Physiol 166:373–383

Talalay P, Williams-Ashman HG 1960 Mechanisms of hormone action. I. Participation of steroid hormones in the enzymatic transfer of hydrogen. Rec Prog Horm Res 16: 1–47

Thomas P, Meizel S 1988 An influx of extracellular calcium is required for initiation of the human sperm acrosome reaction induced by human follicular fluid. Gamete Res 20:397–411

Tischkau SA, Ramirez VD 1993 A specific membrane-binding protein for progesterone in rat brain: sex differences and induction by estrogen. Proc Natl Acad Sci USA 90:1285–1289

Toft D, Gorski J 1966 A receptor molecule for estrogens: isolation from the rat uterus and preliminary characterization. Proc Natl Acad Sci USA 55:1574–1581

Towle AC, Sze PY 1983 Steroid binding to synaptic plasma membrane: differential binding of glucocorticosteroids and gonadal steroids. J Steroid Biochem 18:135–143

Truss M, Candau R, Chávez S, Beato M 1995 Transcriptional control by steroid hormones: the role of chromatin. In: The non-reproductive actions of sex steroids. Wiley, Chichester (Ciba Found Symp 191) p 7–23

Ts'o PO, Lu P 1964 Interaction of nucleic acids. I. Physical binding of thymine, adenine, steroids, and aromatic hydrocarbons to nucleic acids. Proc Natl Acad Sci USA 51: 17–24

Valera S, Ballivet M, Bertrand D 1992 Progesterone modulates a neuronal nicotinic acetylcholine receptor. Proc Natl Acad Sci USA 89:9949–9953

Vanderhaeghen P, Schurmans S, Vassart G, Parmentier M 1993 Olfactory receptors are displayed on dog mature sperm cells. J Cell Biol 123:1441–1452

Vig E, Barrett TJ, Vedeckis WV 1994 Coordinate regulation of glucocorticoid receptor and c-jun RNA levels: evidence for cross-talk between two signaling pathways at the transcriptional level. Mol Endocrinol 8:1336–1346

Wehling M, Eisen C, Christ M 1993 Membrane receptors for aldosterone: a new concept of non-genomic mineralocorticoid action. News Physiol Sci 8:241–244

Willmer EN 1961 Steroids and cell surfaces. Biol Rev 36:368–398

Wong M, Moss RL 1992 Long-term and short-term electrophysiological effects of estrogen on the synaptic properties of hippocampal CA1 neurons. J Neurosci 12: 3217–3225

Wu S, Gibbs TT, Farb DH 1991 Pregnenolone sulfate: a positive allosteric modulator at the *N*-methyl-D-aspartate receptor. Mol Pharmacol 40:333–336

Yamamoto KR 1985 Steroid receptor regulated transcription of specific genes and gene networks. Annu Rev Genet 19:209–252

Yang J, Serres C, Philibert D, Robel P, Baulieu EE, Jouannet P 1994 Progesterone and RU486: opposing effects on human sperm. Proc Natl Acad Sci USA 91:529–533

DISCUSSION

McEwen: From the effects of RU486 on sperm acrosomal Ca^{2+} mobilization, it would appear to be an antagonist for the action of progesterone on this membrane effect.

Baulieu: We observed opposite effects when we compared the activities of progesterone with those of RU486. If we put the two steroids together in the incubation medium, one cancels the other out.

McEwen: RU486 is an antagonist which most of us presumed is antagonizing an intracellular progesterone or glucocorticoid receptor. With your results you can no longer assume that the RU486 effect is exclusively via an intracellular receptor, unless you analyse it very carefully.

Baulieu: When progesterone binds to the intracellular progesterone receptor it is an agonist in classical target cells, such as uterine cells, as is the synthetic analogue R5020. In human sperm, progesterone promotes Ca^{2+} uptake but R5020 doesn't. RU486 has the opposite action to progesterone in human sperm, and is an antagonist at the intracellular receptor level. We are looking for analogues of RU486 that do not bind to intracellular steroid receptors but still have the RU486 effect at the sperm membrane. There is a compound with a side chain at the 7α position (instead of 11β) in RU486 which has no intracellular receptor binding at all, but which has the same effect as RU486 on human sperm. We believe that we have been able to dissociate the intracellular receptor and membrane-mediated effects.

Horwitz: Concerning the difference between the effects of progesterone and RU486 on the Ca^{2+} flux, I thought you said that this was a non-competitive inhibition. You implied that progesterone and RU486 bind to different membrane proteins and that by photoaffinity labelling R5020 binds to a 30 kDa protein. If, indeed, RU486 binds to a different site, you should also be able to identify this alternative protein by photoaffinity labelling. Are there two proteins and do both occur in oocytes and sperm?

Baulieu: This is very difficult to do. In the amphibian oocyte system, RU486 is a weak agonist. Even though we haven't worked things out in detail, I believe that RU486 is a progesterone agonist interacting with the same receptor as progesterone itself. In the sperm, we believe there are two separate receptors. So we cannot generalize even between one germ cell and another.

Horwitz: Can R5020 be photoaffinity labelled to a protein in the sperm membrane?

Baulieu: R5020 doesn't stimulate Ca^{2+} flux in the sperm, so we cannot use it. There are other progesterone analogues that work which you could use.

Manolagas: In oesteoblasts, vitamin D and androgens—two hormones with entirely different spectra of actions—both mediate Ca^{2+} entry. How can this be reconciled with the distinct effects of these steroids?

Related to this, the effects on Ca^{2+} entry are seen in a matter of seconds, whereas the biological changes in response to these hormones takes hours or days. Why would such a rapid effect be important for the biological action of a steroid when it takes as long as a day for the system to be adapted?

Baulieu: I don't remember if vitamin D and androgens have exactly the same pattern of Ca^{2+} increase. In addition, I should say that oestrogens do this too (Lieberherr et al 1993). Like you, I can't see why these complications are necessary if it is only to have a long-term response to a membrane mechanism. Although people haven't realized it before, it is clear that an effect initiated at the membrane level may last a long time.

Manolagas: Are you proposing that the Ca^{2+} effects, rather than underlying the specific actions of the steroids, may represent a tonic control for the actions of other hormones? If so, one should not expect to find a steroid-specific effect but rather a general effect of other, say systemic hormones.

Baulieu: We all live with the beautiful work of D. Ingle (Chicago) who described the 'permissive' action of glucocorticosteroids. He introduced the concept that most steroids (at least glucocorticosteroids) establish a sort of general background or a stage permitting specific signals to act. It is now clear that any single physiological response of a cell is dependent on many inputs. This is why it's so difficult to do physiology and, eventually, medicine with hormones.

Jensen: In the case of vitamin D, one used to hear that its action on Ca^{2+} uptake is due to stimulation of transcription for a specific Ca^{2+} transport protein. Is that still the case?

Manolagas: Not any more. The mechanism of the phenomenon that was described initially by Tony Norman from Riverside is non-genomic: it is not related to the Ca^{2+}-binding proteins you are referring to. Hence, there is a physiological equivalent to what Etienne Baulieu described, i.e. fast Ca^{2+} entry into the intestine that is affected by 1,25-dihydroxyvitamin D_3.

Oelkers: Cortisol suppresses adrenocorticotrophic hormone (ACTH); this effect is very rapid. It also has a long-term effect. Do you know the mechanism by which the rapid effect of cortisol (within minutes) is mediated at the pituitary level?

Baulieu: I don't know. Exactly how rapid is this phenomenon? Whether it is 5 s or 5 min can be significant, because there are rather rapidly produced genomic effects observable in a few minutes.

Oelkers: There are at least two components of this action. One is compatible with a nuclear effect, but the other one lasts only 5 min.

What is the minimum time required for a change in cell function that is mediated by nuclear effects of steroid hormones? You mentioned that if rapid turnover proteins are involved it doesn't have to be 1 h, but is it really as short as 5 min?

Baulieu: Yes, you can observe signs of transcription activation within a couple of minutes with appropriate technology.

Oelkers: Including changes in protein synthesis?

Baulieu: I don't know.

Rochefort: I am not working in this field, but I believe that ACTH is coming from pro-opio melanocortin (POMC). This is a classical model for negative regulation at the gene level by glucocorticoids. Dr Drouin's laboratory in Canada is working on this type of regulation. There is a negative glucocorticoid response element which mediates the effects of the glucocorticoid receptor on transcription of the *POMC* gene.

Baulieu: The problem is, if you are looking for an effect dependent on the synthesis of a constitutive protein (such as a component of the cytoskeleton or something relatively abundant), then it takes some time for enough product to accumulate for a response to occur. If you are investigating an enzyme, a few molecules of which are enough to trigger a reaction, it may only take a couple of minutes for enough to be produced to promote the events it triggers. It's difficult to generalize. I don't see any reason for not believing that in a few minutes you can have a critical synthesis of a newly transcribed protein.

Oelkers: Martin Wehling (Munich) has discovered a new (probably membrane-mediated) effect of aldosterone in human lymphocytes (Wehling et al 1992). Although lymphocytes also harbour the classical mineralocorticoid receptor, the effect of aldosterone (0.1 nmol/l) on cellular Na^+ content and cell volume is demonstrable within 10 min. It is probably mediated by inositol 1,4,5-trisphosphate.

Baulieu: There is evidence from some studies that intracellular hormone receptors may be attached to the membrane: this has been claimed by B. Gametchu (1987) who studied lymphocytolysis. From his immunocyto-chemical experiments, he believes that the glucocorticosteroid receptor can be detected in the membrane. When this does not happen (as occurs in several mutants), he finds that corticosteroids do not have their lymphocytolytic activity. So he believes that part of the mechanism of action of the intracellular receptor occurs at the membrane.

Jensen: Using a sensitive immunohistochemical technique, Milgrom found that antibodies to the intracellular progesterone receptor did not detect any antigen in the membrane (Perrot-Applanat et al 1986). So whatever is in the membrane must be different from the intracellular receptor.

Baulieu: The diversity of systems is such that it is very difficult to generalize the concept that part of the receptor activity occurs at the membrane level. From the literature and from our own experience I do not believe it.

Muramatsu: As a membrane mechanism of steroid action, Matsuda et al (1993) found that oestrogen can bind to c-ErbB2 with very high affinity— almost as high affinity as the binding of oestrogen to oestrogen receptor. Oestrogen can elicit its transduction signals by ErbB2, which used to be an orphan receptor, with a tyrosine kinase domain similar to the epidermal growth factor receptor.

Horwitz: After the previous paper (Truss et al 1995, this volume) we discussed the differences in results that one can obtain using *in vitro* and *in vivo* methods. I would like to get back to that issue with a question about the membrane effects that you are describing. One of the things that I have noted in reading the literature is the relative non-specificity of these effects; namely, that the same effect is observed with many steroids. Another is that the effects are rapid. A third is that relatively high concentrations of steroid hormones are required. Since most of these studies have been done *in vitro*, I wonder what happens if you add the serum-binding proteins for these hormones, so that their effective concentration is lowered to physiological levels. Can one then still see some of these membrane effects? In other words, can one begin to mimic the *in vivo* system by adding serum-binding proteins that may bind up some of these steroids so as to target the responses or enhance their specificity?

Baulieu: I think that's a very important question. The *in vivo* situation is different from *in vitro*. For example, in the amphibian oocyte, testosterone is as effective as progesterone at reinitiating meiosis. *In vivo*, the granular cells which surround the oocytes present progesterone directly to the oocyte, and not testosterone. The receptor would be able to respond to testosterone if it was present, but it is not. In other words, the membrane receptor specificity, which we do not find to be as narrow as that of the intracellular receptor, should be considered differently, because here we are dealing with a 'paracrine' system (in contrast to an endocrine system where the receptor is exposed to all hormones present in the blood). Here, the source of steroid is very close, and it provides it directly and at a concentration that may be very high locally. Therefore, there is *in vivo* only one given steroid presented to the target cell, and not all those which we can add in the test-tube. I would like to propose that here there is 'physiological specificity' rather than molecular specificity. We should not be misled by sophisticated experiments which do not mean too much in relation to physiology. A relatively low affinity of the membrane receptor may well be adjusted to the relatively high local concentration of hormone (in contrast to the concentration of blood-borne steroids).

To give an example, I would like to mention the $GABA_A$ receptor, which reacts to allosteric ligands that are metabolites of progesterone. The first thing to be discovered was a potentiation effect of GABA at a relatively low concentration of the $3\alpha,5\alpha$-reduced steroids in the nanomolar range. When GABA is absent and only the steroid is present, there is no GABA-like effect; however, at a much higher steroid concentration, the progesterone metabolites become active on their own. The concentrations of steroids in the brain are high enough to explain many of these effects described *in vitro*.

Bäckström: The steroid effects on the $GABA_A$ receptor are very specific: even slight changes of the hydroxyl group at the 3 position from α to β will demolish the effects. They are seen in physiological situations. For instance, we have measured the maximal saccadic velocity (the eye movement), which is controlled by the $GABA_A$ receptor. Changes in the saccadic velocity are seen during the menstrual cycle (Sundström & Bäckström 1993).

Beato: Do you know anything about the effects of steroid hormones on the mitochondrial genome?

Baulieu: I found nothing in the literature on this.

Bonewald: In the same vein, what about the effects of sex steroids on matrix vesicles produced by chondrocytes and osteoblasts? These are membrane organelles that are found in the osteoid in bone. They play a role in the mineralization of the osteoid (Anderson 1989). It has been shown that vitamin D has very specific non-genomic effects on these. 1,25-dihydroxyvitamin D_3 leads to the activation of certain proteases or enzymes, whereas 24,25-dihydroxyvitamin D_3 has little or no effect (Boyan et al 1994). Oestrogen also has a specific effect: it has been shown that oestrogen elicits an increase in membrane fluidity by matrix vesicles from female rats, whereas oestrogen has no effect on matrix vesicles isolated from male rats (Gates et al 1994). So this is a very good system for examining totally non-genomic effects of hormones.

Baulieu: When I reviewed the literature, I did not find a single example of a steroid being an allosteric ligand for an enzyme. Some people have suggested activation of protease activity in several aspects of cellular action of steroids, including in chromatin.

Beato: A number of serine proteases are supposed to be modulated by steroid binding, for instance transcortin. This has been claimed to be one mechanism by which glucocorticoids influence inflammation.

Castagnetta: We know that there is redundant expression of steroid response elements. We also have clear evidence of the activity of several enzymes of steroid metabolism and this agrees well with the different effects of several steroid derivatives. For instance, oestrogen metabolites may have distinct, even opposing, biological effects. Do you have any evidence that different forms of the oestrogen response element may have a role in this respect?

Beato: The different hormone-responsive elements respond differently—this has been shown experimentally. If a different hormone-responsive element has a high affinity for the receptor, you need less receptor or less hormone to generate induction. Some of the elements are supposed to act as negative elements (this idea has been mainly put forward by Keith Yamamoto) that do not induce the same type of conformational change in the receptor as is seen with the positive binding sites. Binding to these negative binding sites could lead to what is supposed to be an inhibitory conformation of the receptor. If this is true, it is not only the ligand that modulates the structure of the receptor, but also the DNA binding site.

In some cells, several mRNAs for receptors and several receptor proteins are found. Some of them lack important domains of the protein and you would expect them to behave in a different way. In fact, they have been shown to behave in a different way *in vitro*. Whether or not they have a very specific function *in vivo* is an open question, because most transcription factors are members of large families with many structural variants and very small differences in behaviour. I think we have to get used to this apparent redundancy in regulatory elements and their degenerated response elements. It's making our life very difficult. In knockout experiments we can inactivate one gene and see no phenotype because there are other genes that can take over.

References

Anderson HC 1989 Mechanism of mineral formation in bone. Lab Invest 60:320–330
Boyan BD, Schwartz Z, Park-Snyder S et al 1994 Latent transforming growth factor-β is produced by chondrocytes and activated by extracellular matrix vesicles upon exposure to 1,25-(OH)$_2$D$_3$. J Biol Chem, in press
Gametchu B 1987 Glucocorticoid receptor-like antigen in lymphoma cell membranes: correlation to cell lysis. Science 236:456–461
Gates PA, Mendez J, Schwartz Z et al 1994 Sex-dependent effects of 17β-estradiol on resting zone chondrocyte membrane fluidity. J Dent Res 73(suppl 1):376(abstr)
Lieberherr M, Grosse B, Kachkache M, Balsan S 1993 Cell signalling and estrogens in female rat osteoblasts: a possible involvement of unconventional nonnuclear receptors. J Bone Miner Res 8:1365–1376
Matsuda S, Kadowaki Y, Ichino M, Akiyama T, Toyoshima K, Yamamoto T 1993 17β-Estradiol mimics ligand activity of the c-*erb*B2 protooncogene product. Proc Natl Acad Sci USA 90:10803–10807
Perrot-Applanat M, Groyer-Picard M-T, Logeat F, Milgrom E 1986 Ultrastructural localization of the progesterone receptor by an immunogold method: effect of hormone administration. J Cell Biol 102:1191–1199
Sundström I, Bäckström T 1993 Studies of GABA$_A$ receptor sensitivity variations during the menstrual cycle. Neuropsychopharmacology 9:925(abstr)
Truss M, Candau R, Chávez S, Beato M 1995 Transcriptional control by steroid hormones: the role of chromatin. In: Non-reproductive actions of sex steroids. Wiley, Chichester (Ciba Found Symp 191) p 7–23
Wehling M, Eisen C, Christ M 1992 Aldosterone-specific membrane receptors and rapid non-genomic actions of mineralocorticoids. Mol Cell Endocrinol 90:C5–C9

General discussion I

Oestrogen-responsive finger protein suggests
a transcriptional cascade of steroid hormone regulation

Muramatsu: I am going to describe work I have carried out with Satoshi Inoue, Akira Orimo and Hajime Orimo.

The oestrogen receptor, a member of the steroid/thyroid hormone receptor superfamily (Green & Chambon 1988, Evans 1988), mediates oestrogen action by binding to the oestrogen-responsive element that exists in the enhancer region of target genes, regulating their transcription directly. However, in contrast to the diverse effects of oestrogen on a variety of organs, tissues and cells, relatively few genes are known that are responsive to the oestrogen receptor. The few that are include those that encode vitellogenin, prolactin, ovalbumin, progesterone receptor and pS2. The genes important in regulating the growth and differentiation of female organs are not known. The targets of oestrogen receptor found in the CNS have yet to be identified.

We have recently developed a new method to isolate oestrogen receptor-binding fragments from human genomic DNA (Inoue et al 1991). This procedure makes use of the DNA-binding domain of the oestrogen receptor that is expressed in *Escherichia coli* to bind and isolate human genomic DNA fragments that have the oestrogen-responsive element (Fig. 1). Using one of these oestrogen receptor-binding fragments (E0, Fig. 2A) as a probe, we detected positive signals by Northern blot analysis of human placental RNA (Fig. 2B). This suggested the existence of a transcribed region adjacent to the genomic DNA fragment. We then screened human placental cDNA libraries and found positive clones with the E0 probe. As shown in Fig. 2A, λC1 had the largest insert containing a polyadenylated tail and the λC3 had the longest open reading frame (see below), a part of which overlapped with 5' regions of λC1. λC1 contained the entire E0 fragment and λC3 contained the *PstI–Eco*RI fragment of E0 as an exon. λC3 encodes a protein containing a new zinc finger motif called RING finger (Freemont et al 1991, Reddy et al 1992) which we have named the oestrogen-responsive finger protein (Efp).

The E0 fragment that is located in the 3' untranslated region of the *efp* gene contains an oestrogen response element sequence called ERE0, which is compared with the *Xenopus* vitellogenin oestrogen response element sequence in Fig. 2A (Klein-Hitpass et al 1986). The filter binding assay showed that the E0 fragments bind to the oestrogen receptor DNA-binding domain tightly. The E0 fragment was inserted into a reporter vector and transfected into COS-7

43

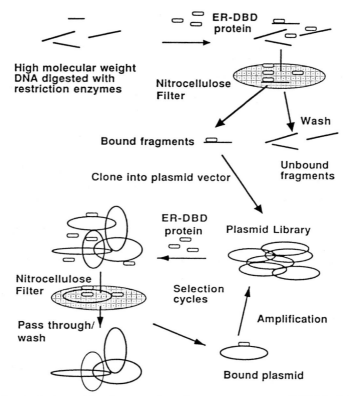

FIG. 1. The strategy used for the screening of oestrogen receptor (ER) binding elements in genomic DNA.

cells. In the presence of both the oestrogen receptor expression vector and 17β-oestradiol, the chloramphenicol acetyltransferase (CAT) activity was stimulated significantly (Fig. 2C). The results show that the 3′ oestrogen receptor-binding region of the *efp* gene can act as a downstream oestrogen-dependent enhancer. It has been reported that some enhancers exist in 3′ untranslated regions of genes such as K-*fgf* (Curatola & Basilico 1990).

The Efp protein predicted from the cDNA sequence consists of 630 amino acids, with a calculated M_r of 73 166. By immunoblotting with an anti-Efp antibody we detected a specific band in HBL-100 cells, which are derived from the human mammary gland (Gaffney 1982). The size of the natural product agreed with the predicted M_r of Efp and also with that expressed in COS-7 cells transfected with the λC3 expression vector. We therefore concluded that the λC3 clone possessed the full-length open reading frame of Efp. Immunostaining of both COS-7 cells transfected with the Efp expression vector and the HBL-100 cells demonstrated the nuclear localization of Efp.

Efp belongs to a new family of nuclear proteins which contain a RING finger motif (Fig. 3A). These proteins not only have a common variant zinc finger structure, but also appear to be related to cell proliferation and differentiation (Freemont et al 1991, Reddy et al 1992). They are assumed to bind with Zn^{2+} and then to DNA, using the zinc finger domains (Reddy et al 1992). Some members of this family possess a second, novel CH domain, B box domain (Reddy & Etkin 1991, Reddy et al 1992), downstream of the RING finger. Interestingly, Efp, PML (a putative transcription factor found fused to the retinoic acid receptor α in promyelocytic leukaemia translocations: Kakizuka et al 1991, de The et al 1991, Goddard et al 1991) and T18 (Miki et al 1991) contain two B box motifs and appear to form a subgroup, as shown in Fig. 3B. Furthermore, all members of the B box-containing family possess a predicted coiled-coil domain downstream of the B box (Fig. 3B). Three of the family members, PML, T18 and Rfp (Takahashi et al 1988), show transforming potential when fused with another protein by chromosomal translocation. In each of these translocations, RING finger, B box and coiled-coil domain are retained and fused to other genes, suggesting that these domains have an important role in transformation potential. Four of these six—PML, Rfp, Rpt-1 (Patarca et al 1988) and Xnf-7 (Reddy & Etkin 1991)—are thought to be transcription factors or transcriptional regulators. The existence of the two characteristic domains, zinc finger and coiled-coil, and the nuclear localization, strongly suggest that Efp is also a transcription factor.

Northern blot analysis of HBL-100 cells probed by the *efp* cDNA indicates that the *efp* gene is regulated by oestrogen at the mRNA level (Fig. 4A). The steady-state level of mRNA in HBL-100 cells was elevated from 2 h after oestrogen treatment, reached the highest level (3.5 times) at 10 h, and then returned to the initial level by 20 h, whereas expression of the oestrogen receptor was not changed. Immunoblotting showed that Efp protein is also regulated by oestrogen in HBL-100 cells (Fig. 4B). Similar tendencies were also observed in MCF-7 cells derived from human breast cancer (data not shown).

Oestrogen has a wide variety of effects on different organs, but the oestrogen receptor, the putative sole mediator of oestrogen action, was found as a single molecular species (Green & Chambon 1988, Evans 1988). To achieve this diversity of oestrogen action, we postulate a second mediator of oestrogen action, the oestrogen receptor-regulated transcription factor. In this model (Inoue et al 1993), the oestrogen-responsive transcription factor mediates and amplifies the action of oestrogen, forming a cascade of gene regulation to provide diverse and specific pathways in each target organ. Efp is a good candidate as a mediator in this model (Fig. 5). We have recently examined the tissue distribution of *efp* by *in situ* hybridization and found that its mRNA is expressed in endometrial cells, testicular cells and in the brain, including the

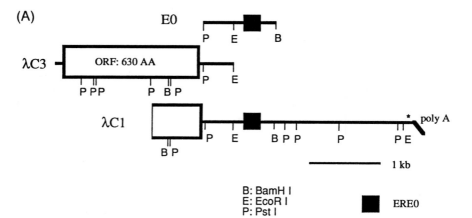

(A)

ERE0   TTCAG**GGTCA**TGG**TGACC**CTGAT
vitERE AGTCA**GGTCA**CAG**TGACC**TGATC
conERE   **GGTCA**NNN**TGACC**

B: BamH I
E: EcoR I
P: Pst I

ERE0 ■

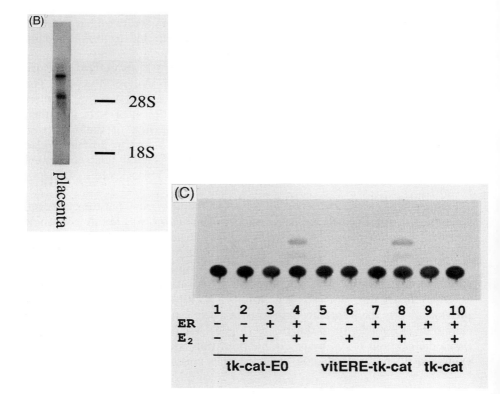

(B)

— 28S

— 18S

placenta

(C)

	1	2	3	4	5	6	7	8	9	10
ER	−	−	+	+	−	−	+	+	+	+
E_2	−	+	−	+	−	+	−	+	−	+

tk-cat-E0   vitERE-tk-cat   tk-cat

nuclei in the hypothalamus, in accordance with the hypothesis outlined above (A. Orimo, S. Inoue, M. Muramatsu, unpublished results). To further explore the functions of *efp* we are now looking at its transgenic expression and doing gene targeting experiments.

Beato: Have you any evidence that *efp* is involved in growth regulation?

Muramatsu: At present, we are investigating the expression of *efp* in breast cancers as compared with normal mammary glands. In some cases, we find elevation in mammary carcinomas, but not always. We are going to have to examine more samples.

Manolagas: Have you found *efp* in bone cells and/or the endometrium?

Muramatsu: Yes, it is expressed in endometrial cells. We have not yet checked bone cells, but an osteoclastic cell line does express *efp* abundantly.

Manolagas: Have you found it in any other tissues?

Muramatsu: We have detected it in other organs, including the brain. We found rather small amounts in the uterus by Northern blots. This may be because only a fraction of uterine cells express *efp*. Therefore, we are now doing *in situ* hybridization to see exactly where it is expressed.

Baulieu: In tissues and cells where oestrogen receptor is present, is *efp* always present?

Muramatsu: So far we have checked only a few tissues. In some tissues, both mRNAs appear to co-localize, but in others we couldn't detect oestrogen receptor where *efp* mRNA was present. In fact, we found that *in situ* hybridization with oestrogen receptor was quite tricky. Sometimes we could demonstrate it, but other times we couldn't. As far as we can tell, there are some places where they co-localize, but in other places they don't. But at the same time, we have to realize that *efp* is not always regulated by

---

FIG. 2. (A) Restriction map of the E0 fragment and the cDNAs (λC1, λC3) cloned for use as probes. The oestrogen-responsive element sequence (ERE0) is depicted as a black box and compared with the *Xenopus* vitellogenin oestrogen-responsive element (vitERE) and consensus oestrogen-responsive element sequences (conERE). The open reading frame is indicated by a white box and the potential polyadenylation site is denoted by an asterisk. (B) Northern blot analysis of human placental RNA with the E0 probe. The lower band (6 kb) corresponds to the size of the cDNA picked up by λC1 and λC3. Migration positions of ribosomal RNA markers are shown on the right. (C) Oestrogen-dependent enhancer activity of the E0 fragment. The E0 fragment was inserted downstream of the reporter vector pBLCAT2 (Luckow & Schutz 1987), with a herpes simplex virus tk promoter to produce the tk-CAT-E0 construct. The reporter plasmid (2 μg) was co-transfected with (+) or without (−) 0.1 μg of the oestrogen receptor expression vector pSVcER (ER) (Koike et al 1987) into COS-7 cells. As controls, the reporter plamids without the insert (tk-CAT) and with the oestrogen-responsive element of the vitellogenin enhancer in the upstream position (vitERE-tk-cat) were assayed simultaneously. They were incubated in the presence (+) or absence (−) of 10 nM 17β-oestradiol (E$_2$) and CAT activities were assayed.

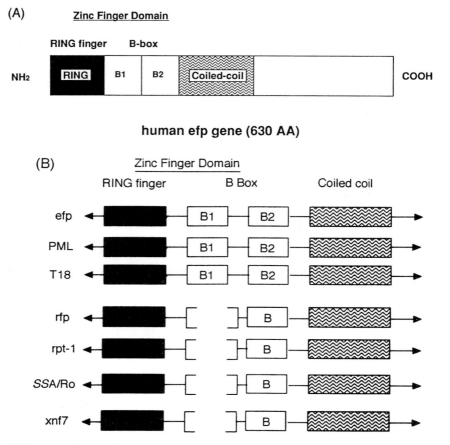

FIG. 3. (A) Schematic structure of the *efp* gene. (B) Schematic representation of the B box-containing RING finger proteins. The black boxes represent the RING finger, the white boxes represent the B boxes and the wavy-lined boxes represent the coiled-coil domain. The square brackets represent a gap.

oestrogen receptor—the oestrogen receptor is one of the regulating factors for the *efp* gene, which may also be activated by other factors, depending on the tissue. In some places, like the hippocampus, we don't know if oestrogen receptor is present, but *efp* is expressed. Perhaps other gene products and transcription factors are regulating *efp* together with the oestrogen receptor.

Rochefort: *efp* appears to be one of the first genes described in mammals to have a consensus oestrogen-responsive element. Most of them have no consensus: they have at least one or two mismatched sequences compared with

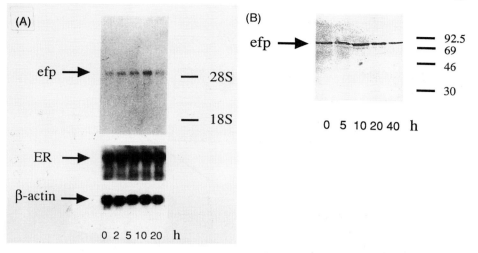

FIG. 4. (A) The *efp* gene is regulated by oestrogen. Poly(A)+ RNA (5 μg) was prepared from the HBL-100 cells at the indicated times after 17β-oestradiol treatment and analysed by Northern blot analysis using *efp*, oestrogen receptor (ER) (Koike et al 1987) and β-actin cDNA probes. Migration positions of ribosomal RNA markers are shown on the right. (B) The Efp protein is regulated by oestrogen. Nuclear extracts (25 μg) were prepared from the HBL-100 cells at the indicated times after 17β-oestradiol treatment and analysed by immunoblotting using anti-Efp antibody. Migration positions of markers (in kDa) are shown on the right.

the vitellogenin oestrogen-responsive element. Have you looked in breast cancer cell lines to see how rapid the induction of *efp* by oestrogen is?

Muramatsu: MCF-7 is the only breast cancer cell line we have looked at. On the addition of oestrogen to the culture, the *efp* mRNA goes up from two hours, reaches a maximum at 10 h, and comes down by 20 h.

Horwitz: We have worked with a large number of different breast cancer cell lines, some of which are oestrogen resistant and appear to have no normal oestrogen receptors. But if we do a gel mobility shift assay using an oestrogen response element, we see proteins that bind beautifully to this DNA. These are not oestrogen receptors because you cannot supershift the bands with an anti-oestrogen receptor antibody. Thus they appear to be other oestrogen response element-binding proteins.

Baulieu: Do you have any idea of their molecular weight?

Horwitz: No, we have only done gel mobility shift assays.

Sutherland: Sometimes oestrogen effects can be mimicked by growth factors. I wonder whether you have any evidence for effects of EGF, TGF-α or bFGF on the expression of *efp*?

Muramatsu: No, we have not done such an experiment.

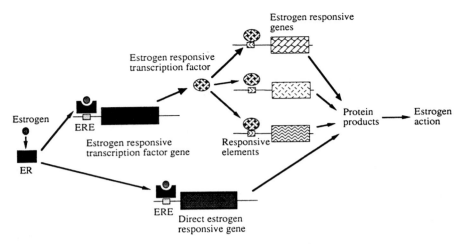

FIG. 5. A model for oestrogen action through an oestrogen-responsive transcription factor. In this model, an oestrogen receptor (ER)-regulated transcription factor, such as Efp, mediates and possibly amplifies the oestrogen effect. ERE, oestrogen-responsive element.

References

Curatola AM, Basilico C 1990 Expression of the K-*fgf* proto-oncogene is controlled by 3′ regulatory elements which are specific for embryonal carcinoma cells. Mol Cell Biol 10:2475–2484

de The H, Lavau C, Marchio A, Chomienne C, Degos L, Dejean A 1991 The PML–RAR-α fusion mRNA generated by the t(15–17) translocation in acute promyelocytic leukemia encodes a functionally altered RAR. Cell 66:675–684

Evans RM 1988 The steroid and thyroid hormone superfamily. Science 240:309–314

Freemont PS, Hanson IM, Trowsdale J 1991 A novel cysteine-rich sequence motif. Cell 64:483–484

Gaffney EV 1982 A cell line (HBL-100) established from human breast milk. Cell Tissue Res 227:563–568

Goddard AD, Borrow J, Freemont PS, Solomon E 1991 Characterization of a zinc finger gene disrupted by the t(15–17) in acute promyelocytic leukemia. Science 254:1371–1374

Green S, Chambon P 1988 Nuclear receptors enhance our understanding of transcription regulation. Trends Genet 4:309–314

Inoue S, Kondo S, Hashimoto M, Kondo T, Muramatsu M 1991 Isolation of estrogen receptor binding sites in human genomic DNA. Nucleic Acids Res 19:4091–4096

Inoue S, Orimo A, Hosoi T et al 1993 Genomic binding site cloning reveals an estrogen-responsive gene that encodes a RING finger protein. Proc Natl Acad Sci USA 90:11117–11121

Kakizuka A, Miller WH, Umesono K et al 1991 Chromosomal translocation t(15–17) in human acute promyelocytic leukemia fuses RAR-α with a novel putative transcription factor. Cell 66:663–674

Klein-Hitpass L, Schorpp M, Wagner U, Ryffel GU 1986 An estrogen-responsive element derived from the 5′ flanking region of the *Xenopus* vitellogenin A2 gene functions in transfected human cells. Cell 46:1053–1061

Koike S, Sakai M, Muramatsu M 1987 Molecular cloning and characterization of rat estrogen receptor cDNA. Nucleic Acids Res 15:2499–2513

Luckow B, Schütz G 1987 CAT constructions with multiple unique restriction sites for the functional analysis of eukaryotic promoters and regulatory elements. Nucleic Acids Res 15:5490

Miki T, Fleming TP, Crescenzi M et al 1991 Development of a highly efficient expression cDNA cloning system: application to oncogene isolation. Proc Natl Acad Sci USA 88:5167–5171

Patarca R, Schwartz J, Singh RP et al 1988 Rpt-1, an intercellular protein from helper inducer T-cells that regulates gene expression of interleukin 2 receptor and human immunodeficiency virus type 1. Proc Natl Acad Sci USA 85:2733–2737

Reddy BA, Etkin LD 1991 A unique bipartite cysteine–histidine motif defines a subfamily of potential zinc finger proteins. Nucleic Acids Res 19:6330

Reddy BA, Etkin LD, Freemont PS 1992 A novel zinc finger coiled-coil domain in a family of nuclear proteins. Trends Biochem 17:344–345

Takahashi M, Inaguma Y, Hiai H, Hirose F 1988 Developmentally resulated expression of a human finger-containing gene encodes by the 5′ half of the *ret* transforming oncogene. Mol Cell Biol 8:1853–1856

Oestrogens and the structural and functional plasticity of neurons: implications for memory, ageing and neurodegenerative processes

Bruce S. McEwen, Elizabeth Gould, Miles Orchinik, Nancy G. Weiland and Catherine S. Woolley*

Laboratory of Neuroendocrinology, The Rockefeller University, 1230 York Avenue, New York, NY 10021, USA

Abstract. Oestrogens have numerous effects on the brain, beginning during gestation and continuing on into adulthood. Many of these actions involve areas of the brain that are not primarily involved in reproduction, such as the basal forebrain, hippocampus, caudate putamen, midbrain raphe and brainstem locus coeruleus. This paper describes three actions of oestrogens that are especially relevant to brain mechanisms involved in memory processes and their alterations during ageing and neurodegenerative diseases: (1) the regulation of cholinergic neurons by oestradiol in the rat basal forebrain, involving induction of choline acetyltransferase and acetylcholinesterase according to a sexually dimorphic pattern; (2) the regulation of synaptogenesis in the CA1 region of the hippocampus by oestrogens and progestins during the four- to five-day oestrus cycle of the female rat. Formation of new excitatory synapses is induced by oestradiol and involves N-methyl-D-aspartate receptors; removal of these synapses involves intracellular progestin receptors; (3) sex differences in hippocampal structure, which may help to explain differences in the strategies that male and female rats use to solve spatial navigation problems. During the period of development when testosterone is elevated in the male, aromatase and oestrogen receptors are also elevated, making it likely that this pathway is involved in the masculinization of hippocampal structure.

1995 Non-reproductive actions of sex steroids. Wiley, Chichester (Ciba Foundation Symposium 191) p 52–73

Gonadal hormones have many effects on the nervous system that extend beyond their very important actions of regulating gonadotrophin and prolactin

---

*Present address: Department of Neurosurgery, University of Washington, Seattle, WA 98195, USA.

secretion and modulating sexual behaviour. Perhaps the most prominent examples are the effects of oestrogens and androgens on verbal fluency and performance of spatial tasks, verbal memory tests and fine-motor skills (Hampson 1990, Phillips & Sherwin 1992a,b). Similarly important are the actions of oestrogens on locomotory activity in animals (Smith 1991) and in Parkinson's disease and tardive dyskinesia in humans (Bedard et al 1977). The hormonal influences on memory processes appear to involve actions on brain structures such as the hippocampus and basal forebrain, whereas the effects on normal and abnormal motor activity involve brain structures such as the caudate putamen, nucleus accumbens and substantia nigra and ventral tegmental (A9 and A10, respectively) dopaminergic nuclei of the midbrain.

Many of these oestrogen effects differ qualitatively or quantitatively between the sexes, suggesting that they may be influenced by sexual differentiation during pre- or postnatal development; alternatively, circulating hormones may have different effects in adult males and females. Sex differences in brain function also include differences in the incidence of psychopathologies such as depressive illness (more common in women), substance abuse and antisocial behaviour (more common in men) as well as pain sensitivity. Some clinically relevant states that are influenced by ovarian hormones and also show sex differences are summarized below (for review see McEwen et al 1994).

Motor disturbances. Oestrogens modulate activity of the nigostriatal and mesolimbic dopaminergic systems and have effects on normal and abnormal locomotory activity. High levels of oestrogens are recognized to exacerbate symptoms of Parkinson's disease.

Epilepsy. Catamenial epilepsy varies according to the menstrual cycle, with the peak frequency of occurrence corresponding to the lowest ratio of progesterone : oestradiol during the cycle. There are at least three potential mechanisms: (1) induction by oestrogen of excitatory synapses in the hippocampus, leading to decreased seizure thresholds (see **Oestrogens and synaptogenesis**, p 58); (2) progesterone actions via the steroid metabolites which act via the type A γ-aminobutyric acid ($GABA_A$) receptor to decrease excitability; and (3) actions of hormones on the liver to increase clearance rates of antiseizure medication.

Premenstrual syndrome. Premenstrual tension is a cyclic mood disorder which, in its most severe form, is referred to as premenstrual syndrome (PMS). Its symptoms are eliminated by arresting the menstrual cycle, although specific hormonal causes are unknown. It remains to be determined whether or not PMS involves any of the specific morphological or neurochemical effects of ovarian steroids described in this paper.

Depressive illness.   One-month prevalence studies of mental disorders have revealed that males suffer more frequently from substance abuse and hostility/ conduct disorders, whereas women are more prone to develop anxiety disorders and depressive illness (Regier et al 1988). Early experience and social and cultural factors also affect sex differences. Although it is unclear whether these differences may be attributed to circulating gonadal hormones or to the actions of hormones during sexual differentiation, high-dose oestrogen replacement produces a significant reduction in anxiety and depression.

Pain.   Recent studies in mice indicate that males and females use functionally distinct pain pathways, and that gonadal steroids, particularly oestrogens, play a major role in regulating these pathways. Thus, basic research on pain mechanisms, as well as clinical practice in treating pain must now take into consideration the sex differences in the neural mechanisms mediating pain.

Cognitive function.   Oestrogens influence short-term verbal memory as well as performance on tests of fine-motor skills and spatial ability; sex differences exist in humans and in animals among strategies used in solving spatial navigation problems (see *Cognitive function*, p 63).

Dementia.   Oestrogen therapy in open trials has been reported to improve cognitive function in patients with Alzheimer's disease, and there is a reportedly lower fatal prevalence of Alzheimer's disease in elderly women who receive oestrogen replacement therapy postmenopausally. Thus, there is reason to study the benefits of oestrogen replacement for brain function in the ageing population (see *Dementia*, p 63).

The diversity of these effects implies the involvement of regions of the brain outside of the hypothalamus. Indeed, mapping of intracellular receptors, which modulate genomic actions, revealed the presence of oestrogen and progestin receptors in regions such as the amygdala, hippocampus, cingulate cortex, locus coeruleus and midbrain raphe nuclei and central grey matter. Although the density of such receptors is far lower in many of these brain areas compared with the hypothalamus and amygdala, the existence of prominent oestrogen and progestin effects in many of these regions requires a careful examination of the role of the cells in these brain areas that do express intracellular receptors, as well as consideration of possible alternative mechanisms of steroid action. There are also indications that the expression of steroid receptors is developmentally regulated, and that oestrogen and progestin receptor expression is transient in some brain regions and stable in others.

This paper will review the neural actions of gonadal hormones on non-reproductive brain structures and processes, placing particular emphasis on the

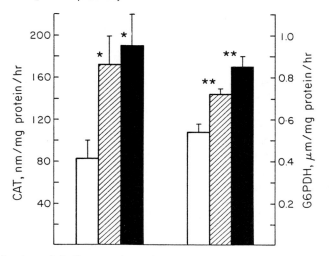

FIG 1.   Induction of choline acetyltransferase (CAT) activity in the horizontal limb of the diagonal band of Broca of rats is compared with induction of glucose-6-phosphate dehydrogenase (G6PDH) activity in the pituitary. Horizontal bars represent means ± SEM (*n* = 5 per group). Oestrogen administration was for 6 h (▨) and 24 h (■) in 1 cm Silastic capsules. Significant differences from ovariectomized control rats (□) are indicated by *$P < 0.05$ and **$P < 0.01$. (Reproduced with permission from Luine & McEwen 1983.)

effects of oestrogen and progestin in the basal forebrain and hippocampal formation of the rat, because these brain structures are prominent in learning and memory and also are sites of neural degeneration in dementia, including Alzheimer's disease.

Oestrogens and the cholinergic system

We begin the story with a discussion of the basal forebrain of the rat and the cholinergic neurons that project to cerebral cortex and hippocampus, where they play an important role in cognitive function. Studies of the effects of oestrogen on the expression of cholinergic enzymes were among the first that indicated non-reproductive actions of gonadal steroids (for review see McEwen et al 1987). Experiments with ovariectomy and oestrogen replacement therapy revealed an induction of choline acetyltransferase (ChAT), the rate-limiting enzyme for acetylcholine formation, within 6–24 h in the basal forebrain of female rats (Fig. 1). In addition, there was evidence from measurements of increased ChAT activity in projection areas, namely, the CA1 region of hippocampus and cerebral cortex, 10 days after hormone injection that oestrogen-induced ChAT was transported from cell bodies to nerve endings.

TABLE 1 Effect of three days of oestradiol treatment on activity of choline acetyltransferase and acetylcholinesterase in the horizontal limb of the diagonal band of Broca in gonadectomized male and female rats

Treatment group	Choline acetyltransferase activity (nmol/mg protein per h)	Acetylcholinesterase activity (nmol/mg protein per h)
Gonadectomized male	79 ± 12	7.4 ± 1.1
Gonadectomized male + oestradiol	109 ± 14	7.6 ± 0.8
Ovariectomized female	260 ± 28	11.6 ± 1.5
Ovariectomized female + oestradiol	491 ± 50**	17.1 ± 2.0*

Values represent means ± SEM; $n = 6-8$ females and 10–12 males for each treatment. Sprague–Dawley rats of 200–220 g body weight were gonadectomized for 3–5 d and then given oestradiol in 1 cm Silastic capsules for 3 d.
*$P < 0.05$, **$P < 0.01$ when compared with gonadectomized control. (Adapted from Luine & McEwen 1983.)

Acetylcholinesterase activity was also induced by oestradiol, suggesting that a general trophic effect on the cholinergic neurons might occur (Table 1).

A recent investigation of long-term (five- to 28-week) ovariectomy revealed a decline in high-affinity choline uptake and ChAT activity in the frontal cortex and hippocampus that was at least partly prevented by oestrogen replacement therapy (Singh et al 1994). Also, long-term ovariectomy caused a decline in learned performance of active avoidance behaviour, which was prevented by oestrogen replacement therapy (Singh et al 1994). One potential candidate as a regulator of the cholinergic system of the basal forebrain is nerve growth factor (NGF), which is produced by the hippocampus and retrogradely transported to neurons in the basal forebrain to produce trophic effects. Although the effects of oestrogen treatment on NGF levels in the hippocampus remain to be investigated, Toran-Allerand et al (1992) reported the co-localization of oestrogen receptors with low-affinity NGF receptors in cholinergic neurons of the basal forebrain of the rat.

The basal forebrain of male rats failed to show the same response to oestradiol treatment as that of females (Table 1) and postnatal oestrogen treatment of female rats or blockade of aromatization in males failed to change this sex difference (Luine & McEwen 1983, McEwen et al 1987). The basal forebrain cholinergic system differs between male and female rats, with females having smaller and more densely packed cholinergic neurons compared with untreated males (Westlind-Danielsson et al 1991). Moreover, application of triiodothyronine (T3) to newborn male and female rats, creating transient hyperthyroidism during the first postnatal week, revealed further indications of sexual differentiation of the basal forebrain cholinergic system

(Westlind-Danielsson et al 1991). For example, treatment with T3 increased cholinergic cell density, ChAT activity and muscarinic receptor binding in the septum/diagonal band region of male rats, whereas female rats did not respond to T3, except in the medial septum where they showed the opposite effect to males; namely, an increased cholinergic cell body area (Westlind-Danielsson et al 1991).

On the other hand, in another study of sex differences in the cholinergic system, female rats showed larger effects than males to the cholinergic lesions produced in the hippocampus by the specific cholinergic neurotoxin, AF64A. Female rats were particularly sensitive when the toxin was administered into the lateral ventricles on the day of pro-oestrus (Hortnagl et al 1993). These results suggest a sexually dimorphic organization of the basal forebrain cholinergic system in the rat, involving a prenatally programmed difference in the neuroanatomical organization, as well as sex differences in cholinergic enzyme induction in response to oestradiol and the effects of T3 treatment within the first week of postnatal life. These differences may underlie, at least in part, the sex differences in spatial learning that are discussed below.

Developmentally regulated sex differences in the hippocampus

The hippocampus is another brain structure that shows subtle sex differences. For example, there are sex differences in the density of apical dendritic excrescences and branching of dendrites of CA3 pyramidal neurons. Treatment with T3 during the first week of postnatal life enhanced these differences (Gould et al 1990; see Fig. 2). Excrescences on the proximal region of apical dendrites receive input from mossy fibre synapses from granule neurons of the dentate gyrus. Therefore, the greater density of excrescences in males is consistent with a report that male rats have a greater number of mossy fibre synapses than female rats have (see Parducz & Garcia-Segura 1993). Other studies have indicated sex differences in hippocampal morphology that are dependent on the rearing environment (see Juraska 1991).

In mice and rats the dentate gyrus also shows sex differences. In mice there are strain-dependent sex differences—in strains with large numbers of granule neurons, males have more neurons than females, whereas in strains with fewer granule neurons, both sexes have the same number (Wimer & Wimer 1989). Male rats have a larger and more asymmetric dentate gyrus than female rats have, and neonatal testosterone treatment caused the genetically female dentate gyrus to appear masculine (Roof & Havens 1992). Neonatal testosterone treatment in female rats also improved spatial learning ability in a Morris water maze (Roof & Havens 1992).

How do these sex differences come about during development? Similarly to the cerebral cortex, the rat hippocampus expresses oestrogen receptors transiently during perinatal development (O'Keefe & Handa 1990). The

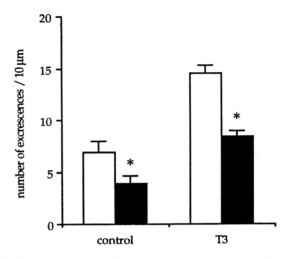

FIG. 2.   Sex differences in number of excrescences per 10 μm length on apical dendrites of CA3 pyramidal neurons in euthyroid and neonatal rats of both sexes treated with triiodothyronine (T3). Horizontal bars represent means ± SEM (*n* = 4 per group). *P < 0.01, compared with male rats in the same treatment group. There are also strong T3 treatment effects for both sexes. (Reproduced with permission from Gould et al 1990.)

presence of oestrogen receptors in the hippocampus coincides with the transient expression of the aromatizing enzyme system that converts testosterone to oestradiol (MacLusky et al 1987); as a result, oestrogen receptors in male rats would be exposed to locally generated oestradiol and this could lead to sexual differentiation of hippocampal structure and function. Consistent with this scenario are data showing that although neonatal castration of male rats produced feminine learning curves in a Morris water maze, administration of oestradiol to newborn female rats produced a masculine learning curve (Williams & Meck 1991).

Oestrogens and synaptogenesis

Even though the hippocampus loses most of its oestrogen receptors as postnatal development proceeds, this structure displays a robust response to oestrogen and progestin treatment and to endogenous ovarian steroids during the natural oestrus cycle. This was first revealed in terms of cyclic variations in the threshold of the dorsal hippocampus to elicitation of seizures, with the greatest sensitivity occurring on prooestrus (Terasawa & Timiras 1968).

Mapping studies of [³H]oestradiol uptake in the hippocampus showed a sparse distribution of oestrophilic neurons containing intracellular oestrogen receptors in what appear to be interneurons in the CA1 region and in other regions of Ammon's horn (Loy et al 1988). Similar results were reported for the guinea pig using immunocytochemistry to detect the oestrogen receptor (Don Carlos et al 1991). There was also an indication that the CA1 region of the hippocampus contains some oestrogen-inducible progesterone receptors, albeit at much lower levels than in the hypothalamus (Parsons et al 1982). Complementary data from *in situ* hybridization studies revealed the presence of low levels of progestin receptor mRNA in both the CA1 and CA3 regions of Ammon's horn (Hagihara et al 1992).

None of these data suggest that the hippocampus is a major target for oestrogen action, compared with brain regions such as the hypothalamus. The picture changed when morphological studies demonstrated that oestrogen induces dendritic spines not only in the ventromedial hypothalamus of the female rat (where it also induces new synapses), but also on pyramidal neurons in the hippocampus (for review see McEwen & Woolley 1994). Oestrogen effects on dendritic spines were found only in the CA1 region and not in the CA3 region or in the dentate gyrus. Moreover, spine density changed cyclically during the oestrus cycle of the female rat. There were also parallel changes in synapse density on dendritic spines revealed by electron microscopy, strongly supporting the notion that new synapses are induced by oestradiol. Taken together, these morphological studies indicate that synapses are formed and broken down rapidly during the natural reproductive cycle of the rat.

The critical involvement of progesterone was indicated by the fact that progesterone administration rapidly potentiated oestrogen-induced spine formation, but then triggered the down-regulation of spines on CA1 neurons. The down-regulation of dendritic spines occurred slowly when oestrogen was withdrawn, but took place within 8–12 h when progesterone was administered; moreover, the natural down-regulation of dendritic spines between the pro-oestrus peak and the trough on the day of oestrus was blocked by the progesterone antagonist RU486 (Fig. 3; Woolley & McEwen 1993).

Antagonism of this effect of progesterone by RU486 is consistent with the involvement of intracellular progestin receptors, and is compatible with the finding, noted above, of oestrogen-inducible progestin receptors in the CA1 region of the hippocampus (Parsons et al 1982). However, we know little about the specificity of the progesterone recognition sites in neuronal membranes, and the intracellular progestin receptors have not been localized by immunocytochemistry. In addition, as noted above, the only detectable intracellular oestrogen receptors appear to be in interneurons, not in the CA1 pyramidal neurons themselves.

There is yet another piece to the puzzle, namely, the fact that the induction of new spine synapses on CA1 pyramidal neurons by oestrogen is blocked by concurrent administration of the *N*-methyl-D-aspartate (NMDA) receptor antagonists, MK801 or CGP43487 (Woolley & McEwen 1994). Scopolamine

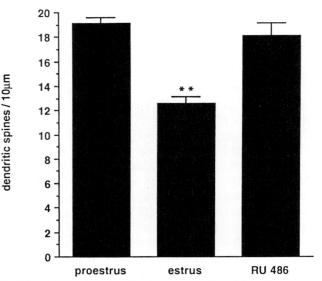

FIG. 3. Apical dendritic spine density (per 10 μm length) in intact rats in the pro-oestrus and oestrus phases of the oestrus cycle, and in rats given RU486 during the pro-oestrus phase of the oestrus cycle. Horizontal bars represent means $\pm$ SEM ($n=7$ per group). Note that the spine density is decreased in the oestrus compared with the pro-oestrus phase of the cycle, but it remains elevated in rats given RU486 during pro-oestrus. $**P<0.01$, compared with both pro-oestrus and RU486. (Reproduced with permission from Woolley & McEwen 1993.)

and NBQX (2,3-dihydro-6-nitro-7-sulphanoylbenzo(F)quinoxaline), antagonists of muscarinic and non-NMDA excitatory amino acid receptors, respectively, were ineffective at doses that interfere with cognition and block ischaemic damage, respectively (Woolley & McEwen 1994). Spine synapses are excitatory and it is likely that NMDA receptors occur on them; therefore, it is of particular interest that one of the long-term effects of oestradiol is to induce NMDA receptor binding sites in the CA1 region of the hippocampus (Fig. 4; see Weiland 1992a).

Therefore, it is relevant to ask whether activation of NMDA receptors themselves could lead to induction of new synapses, in which case oestrogen induction of NMDA receptors would then become a primary event leading to synapse formation. As discussed by Woolley & McEwen (1994), NMDA receptors gate calcium ions and this may be an important factor in the extension and retraction of dendritic spines. It was found, for example, that NMDA receptor activation promotes dephosphorylation of MAP2 (microtubule-associated protein 2) and alters the interaction of this cytoskeletal protein with actin and tubulin. There are, of course, many questions to be answered regarding the necessary and sufficient conditions for

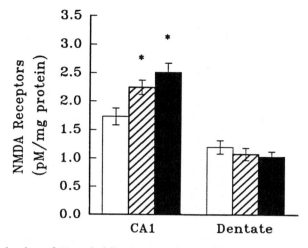

FIG. 4.   Induction of *N*-methyl-D-aspartate (NMDA) receptors by oestradiol (▧) and oestradiol plus progesterone (■) treatment in ovariectomized female rats. Oestradiol was administered to ovariectomized rats in Silastic capsules for two days, and progesterone was given at 10 mg/kg intraperitoneally for 4.5 h to oestrogen-primed rats. Horizontal bars represent means ± SEM ($n = 10$ per group). *$P < 0.05$ compared with ovariectomized control rats (□). (Reproduced with permission from Weiland 1992a.)

NMDA receptor activation to induce synapse formation and whether the cascade of events triggered by NMDA receptor activation involves induction of cytoskeletal proteins or of enzymes that modify their interactions (for example, by phosphorylation and dephosphorylation).

We now return to the possible significance of the presence of oestrogen receptors in inhibitory interneurons and not in CA1 pyramidal neurons in which spine formation takes place. Oestradiol treatment induced glutamic acid decarboxylase mRNA in inhibitory interneurons within the CA1 pyramidal cell layer (Weiland 1992b), thus potentially increasing inhibitory activity within those neurons, although it is not clear where they exert their inhibitory effect (on the pyramidal neurons themselves, or on other inhibitory interneurons; see below). In this connection, there are studies indicating effects of oestrogens that excite hippocampal CA1 pyramidal neurons, possibly through disinhibition. Wong & Moss (1992) have shown that two-day oestradiol treatment prolongs the excitatory postsynaptic potential and increases the probability of repetitive firing in some CA1 neurons in response to Schaffer collateral (glutamatergic) stimulation. One explanation is that actions of oestrogen on inhibitory interneurons might, in some manner, disinhibit the pyramidal neurons, allowing removal of the magnesium blockade and activation of NMDA receptors (for discussion, see Woolley & McEwen 1994). This would have to occur by negation of the inhibitory influence of other interneurons on pyramidal cell firing, and the neuroanatomical pathway for such an effect remains to be worked out. Moreover, it is not clear how such a sequence of events would lead to induction of NMDA receptors, or, for that matter, to induction of $GABA_A$ receptors or, finally, to the induction of new excitatory synapses on dendritic spines.

Other brain areas and systems affected by oestradiol

There are a number of other brain regions that show oestrogen effects on neural systems that are not specifically associated with reproductive functions and in which the classical oestrogen receptors have either been found in small numbers or have not yet been detected. First, brainstem cate-cholaminergic neurons contain small numbers of oestrogen receptors (Heritage et al 1980), and oestrogen treatment after gonadectomy exerts complex, time-dependent effects on the level of tyrosine hydroxylase mRNA (Liaw et al 1992). Second, there are actions of oestrogens and progestins on the dopaminergic system, involving both pro- and antidopaminergic effects, that depend on the dose and time course of oestrogen administration and are manifested in both the nigrostriatal and mesolimbic dopaminergic systems; one of the most puzzling features of these actions of ovarian hormones is manifested in the absence of detectable classical oestrogen receptors (for review, see DiPaolo 1994).

Also, there are sex differences in the synthesis and turnover of 5-hydroxytryptamine in the rat forebrain and, particularly, in the hippocampus, in which female rats have significantly higher 5-hydroxytryptamine metabolism than males (reviewed in Haleem et al 1990). There are several potential mechanisms for ovarian steroid influences on the serotonergic system. First, in the hippocampus, oestrogen treatment increases the efficacy of 5-hydroxy-tryptamine$_{1A}$ receptors to inhibit cAMP formation in isolated membrane fractions (Clarke & Miani 1990). This might at least partly explain how females respond more to the actions of the 5-hydroxytryptamine$_{1A}$ agonist 8-hydroxy-di-N-propylaminotetraline to decrease hippocampal 5-hydroxytrypamine metabolism (Haleem et al 1990). Another facet of oestrogen action in the raphe itself is the induction of progesterone receptors (Bethea 1994). So far, these findings apply only to the macaque, and there has been no direct demonstration of classical oestrogen receptors in the raphe nuclei, raising many of the same questions about detectability and distribution of oestrogen receptors that apply to the hippocampus, nigrostriatal and mesolimbic dopamine and basal forebrain cholinergic systems.

Therapeutic potential of oestrogens

As noted in the introduction, ovarian steroids produce a variety of actions on non-reproductive functions in the human brain. Besides the examples mentioned above, there are results demonstrating actions of ovarian steroids on cognitive function, as well as implications that oestrogens may be beneficial in dementia.

Cognitive function. In women, cyclic variations in oestradiol affect performance of spatial tasks and fine-motor skills (Hampson 1990), as well as paired-associate learning (Phillips & Sherwin 1992b). At present, it is not known whether or not the oestrogen-induced changes in human performance are mediated by changes in synapse density. Nevertheless, these effects are specific to verbal memory and spatial learning, which suggests the involvement of the hippocampus and temporal lobe.

Dementia. Early reports indicated that therapy of elderly women with oestradiol, progesterone and testosterone improved some measures of intellectual functioning (Caldwell 1954). More recent work has demonstrated beneficial effects on verbal memory tests of oestrogen replacement therapy in surgically menopausal women (Phillips & Sherwin 1992a). There is a higher frequency of Alzheimer's disease in elderly women than in elderly men (Henderson et al 1994), and two types of studies indicate a possible link between oestrogen deficiency and dementia in postmenopausal women. First, women with a history of cerebrovascular disease showed improved cognitive function on oestrogen replacement therapy (Funk et al 1991). Second, two other studies on elderly female Alzheimer's patients revealed improvements in tests of mental status over a six-week period of oestrogen replacement therapy (Fillit et al 1986, Honjo et al 1989).

Furthermore, there are indications that women given oestrogen replacement therapy are less likely to develop dementia. A recent study (Henderson et al 1994) revealed a significant reduction in Alzheimer's disease as a cause of death in women who had taken oestrogens postmenopausally, with a greater apparent protective effect resulting from higher dosage and longer exposure. Henderson et al (1994) are very cautious in pointing out the need for prospective, randomized treatment trials to substantiate any beneficial effects of oestrogen treatment. Nevertheless, these findings warrant further investigation of the relationship between oestrogens and both normal cognitive function and dementia.

Conclusions

Besides the potentially important therapeutic implications of oestrogen effects on cognitive processes, and the indications that cognitive function, mood disorders, pain pathways and motor activity are modulated by oestrogens, the studies of the actions of oestrogen on extra-hypothalamic areas of the brain have highlighted our ignorance about basic cellular and molecular mechanisms. These considerations apply to the serotonergic system and the nigrostriatal and mesolimbic dopaminergic systems, and they have been illustrated in this paper by our studies on the hippocampus and basal forebrain. Future work must attempt to identify hitherto unrecognized oestrogen receptors in cells that show oestrogen effects but do not appear to have classical receptors, or, alternatively, proceed to elucidate the types of trans-synaptic interactions that lead to oestrogen effects in cells that do not express oestrogen receptors. The suggestion that NMDA receptors are involved in oestrogen induction of new excitatory synapses on CA1 pyramidal neurons is one indication of a possible trans-synaptic action in which excitatory amino acid neurotransmitters play a pivotal role. It seems very likely that there will be similar interactions operating in other neural systems that respond to ovarian steroids.

Acknowledgements

Research support was obtained from National Institutes of Health grant NS 07080 to B. S. M., NS 09129 to M. O. and NS 30105 to N. W. M. O. was also the recipient of a Pharmaceutical Manufacturers Association Fellowship.

References

Bedard P, Langelier P, Villeneuve A 1977 Oestrogens and the extrapyramidal system. Lancet II 31:1367–1368

Bethea CL 1994 Regulation of progestin receptors in raphe neurons of steroid-treated monkeys. Neuroendocrinology 60:50–61

Caldwell BM 1954 An evaluation of psychological effects of sex hormone administration in aged women. J Gerontol 9:168–174

Clarke WP, Miani S 1990 Oestrogen effects on $5\text{-}HT_{1A}$ receptors in hippocampal membranes from ovariectomized rats: functional and binding studies. Brain Res 518:287–291

DiPaolo T 1994 Modulation of brain dopamine transmission by sex steroids. Rev Neurosci 5:27–42

DonCarlos LL, Monroy E, Morrell JI 1991 Distribution of oestrogen receptor immunoreactive cells in the forebrain of the female guinea pig. J Comp Neurol 305:591–612

Fillit H, Weinreb H, Cholst I et al 1986 Observations in a preliminary open trial of estradiol therapy for senile dementia—Alzheimer's type. Psychoneuroendocrinology 11:337–345

Funk J, Mortel K, Meyer J 1991 Effects of oestrogen replacement therapy on cerebral perfusion and cognition among postmenopausal women. Dementia 2:268–272

Gould E, Westlind-Danielsson A, Frankfurt M, McEwen BS 1990 Sex differences and thyroid hormone sensitivity of hippocampal pyramidal cells. J Neurosci 10: 996–1003

Hagihara K, Hirata S, Osada T, Kato J 1992 Distribution of cells containing progesterone receptor mRNA in the female rat di- and telencephalon: an *in situ* hybridization study. Mol Brain Res 14:239–249

Haleem DJ, Kennett GA, Curzon G 1989 Hippocampal 5-hydroxytryptamine synthesis is greater in female rats than in males and more decreased by the $5-HT_{1A}$ agonist, 8-OH-DPAT. J Neural Transm 79:93–101

Hampson E 1990 Oestrogen-related variations in human spatial and articulatory motor skills. Psychoneuroendocrinology 15:97–111

Henderson VW, Paganini-Hill A, Emanuel CK, Dunn ME, Buckwalter JG 1994 Oestrogen replacement therapy in older women: comparisons between Alzheimer's disease and nondemented controls. Arch Neurol 51:896–900

Heritage AS, Stumpf W, Sar M, Grant LD 1980 Brainstem catecholamine neurons are target sites for sex steroid hormones. Science 207:1377–1380

Honjo H, Ogino Y, Naitoh K et al 1989 *In vivo* effects by estrone sulfate on the central nervous system senile dementia (Alzheimer's type). J Steroid Biochem 34: 521–525

Hortnagl H, Hansen L, Kindel G, Schneider B, Tamer AE, Hanin I 1993 Sex differences and oestrus cycle-variation in the AF64A-induced cholinergic deficit in the rat hippocampus. Brain Res Bull 31:129–134

Jursaka J 1991 Sex differences in 'cognitive' regions of the rat brain. Psychoneuroendocrinology 16:105–119

Liaw J-J, He J-R, Hartman RD, Barracolough CA 1992 Changes in tyrosine hydroxylase mRNA levels in medullary A1 and A2 neurons and locus coeruleus following castration and oestrogen replacement in rats. Brain Res 613:213–238

Loy R, Gerlach JL, McEwen BS 1988 Autoradiographic localization of oestradiol binding neurons in the hippocampal formation and the entorhinal cortex. Dev Brain Res 39:245–251

Luine VN, McEwen BS 1983 Sex differences in cholinergic enzymes of diagonal band nuclei in the rat preoptic area. Neuroendocrinology 36:475–481

MacLusky N, Clark AS, Naftolin F, Goldman-Rakic PS 1987 Oestrogen formation in the mammalian brain: possible role of aromatase in sexual differentiation of the hippocampus and neocortex. Steroids 50:459–474

McEwen BS 1994 Ovarian steroids have diverse effects on brain structure and function. In: Berg G, Hammar M (eds) The modern management of the menopause. Parthenon Publishing, Park Ridge, NJ, p 269–278

McEwen BS, Woolley CS 1994 Oestradiol and progesterone regulate neuronal structure and synapatic connectivity in adult as well as developing brain. Exp Gerontol 29: 431–436

McEwen BS, Luine V, Fischette C 1987 Developmental actions of hormones: from receptors to function. In: Easter S, Barald K, Carlson B (eds) From message to mind. Sinauer Associates, Sunderland, MA, p 272–287

O'Keefe JA, Handa RJ 1990 Transient elevation of oestrogen receptors in the neonatal rat hippocampus. Dev Brain Res 57:119–127

Parducz A, Garcia-Segura LM 1993 Sexual differences in the synaptic connectivity in the rat dentate gyrus. Neurosci Lett 161:53–56

Parsons B, Rainbow TC, MacLusky N, McEwen BS 1982 Progestin receptor levels in rat hypothalamic and limbic nuclei. J Neurosci 2:1446–1452

Phillips S, Sherwin B 1992a Effects of oestrogen on memory function in surgically menopausal women. Psychoneuroendocrinology 17:485–495

Phillips S, Sherwin B 1992b Variations in memory function and sex steroid hormones across the menstrual cycle. Psychoneuroendocrinology 17:497–506

Regier DA, Boyd JH, Burke JD et al 1988 One-month prevalence of mental disorders in the United States. Arch Gen Psychiatry 45:977–986

Roof RL, Havens MD 1992 Testosterone improves maze performance and induces development of a male hippocampus in females. Brain Res 572:310–313

Singh M, Meyer EM, Millard WJ, Simpkins JW 1994 Ovarian steroid deprivation results in a reversible learning impairment and compromised cholinergic function in female Sprague–Dawley rats. Brain Res 644:305–312

Smith S 1991 The effects of oestrogen and progesterone on GABA and glutamate responses at extrahypothalamic sites. In: Costa E, Paul SM (eds) Neurosteroids and brain function. Thieme Medical, New York (Fidia Res Found Symp Ser) 8:87–94

Terasawa E, Timiras P 1968 Electrical activity during the oestrus cycle of the rat: cyclic changes in limbic structures. Endocrinology 83:207–216

Toran-Allerand CD, Miranda RC, Bentham WDL et al 1992 Estrogen receptors colocalize with low-affinity nerve growth factor receptors in cholinergic neurons of the basal forebrain. Proc Natl Acad Sci USA 89:4668–4672

Weiland NG 1992a Estradiol selectively regulates agonist binding sites on the N-methyl-D-aspartate receptor complex in the CA1 region of the hippocampus. Endocrinology 131:662–668

Weiland NG 1992b Glutamic acid decarboxylase messenger ribonucleic acid is regulated by estradiol and progesterone in the hippocampus. Endocrinology 131:2697–2702

Westlind-Danielsson A, Gould E, McEwen BS 1991 Thyroid hormone causes sexually distinct neurochemical and morphological alterations in rat septal-diagonal band neurons. J Neurochem 56:119–128

Williams CL, Meck W 1991 The organizational effects of gonadal steroids on sexually dimorphic spatial ability. Psychoneuroendocrinology 16:155–176

Wimer CC, Wimer RE 1989 On the sources of strain and sex differences in granule cell number in the dentate area of house mice. Dev Brain Res 48:167–176

Woolley C, McEwen BS 1993 Roles of estradiol and progesterone in regulation of hippocampal dendritic spine density during the oestrus cycles in the rat. J Comp Neurol 336:293–306

Woolley C, McEwen BS 1994 Estradiol regulates hippocampal dendritic spine density via an NMDA receptor-dependent mechanism. J Neurosci 14:7680–7687

Wong M, Moss RL 1992 Long-term and short-term electrophysiological effects of estrogen on the synaptic properties of hippocampal CA1 neurons. J Neurosci 12:3217–3225

DISCUSSION

Thijssen: One of the issues that has been discussed in recent years has been the effect of hormone replacement therapy on psychological, cognitive and other functions of the brain in women. How solid is the evidence for the influence of oestrogens on brain function?

McEwen: The person whose work is most often cited is Barbara Sherwin at McGill University. She has studied a series of women, many of whom have undergone a surgical menopause, who have experienced cognitive difficulty such as short-term memory problems. When she treated them with oestrogen she found specific improvements in areas of cognitive function that seem most closely linked to the hippocampus and temporal lobe, such as various kinds of short-term verbal memory (Phillips & Sherwin 1992). The problem with these kind of data is that if you look at women after menopause, not all women show these kinds of cognitive deficits. If you treat women with oestrogens, some will improve and some won't. The difficulty is how to distinguish between them, and whether there's a meaningful difference between them. I trust Barbara Sherwin's work—she's a very careful investigator who has spent a lot of time on this—but even she would admit that she's puzzled at why there are responders and non-responders.

Thijssen: In addition to this, there's been a recent report on the psychological behaviour of transsexuals. L. Gooren (1994) from the Amsterdam Free University Hospital, has more than a thousand transsexuals under his treatment. He has found that male-to-female transsexuals improved on their verbal capability and had reduced spatial capability. Female-to-male transsexuals went in the other direction. This fits in with what you said.

McEwen: Elizabeth Hampson has reported oestrus cycle variations in fine-motor skills and spatial capability (Hampson 1990). She hasn't really looked, until recently, at verbal memory tests. She has found, very reliably, these relationships between high oestrogen levels and good fine-motor coordination and somewhat depressed spatial ability. Her colleague, Doreen Kimura, is also beginning to see improved short-term verbal memory and protection against intellectual decline after the menopause (Kimura 1994).

Uvnäs-Moberg: Could the positive effects of oestradiol on memory simply be a response to a certain dosage level? It is well established that during pregnancy, when the levels of steroids are very high, memory is not very good (Silber et al 1990).

McEwen: You have to bear in mind the level of progesterone. Progesterone down-regulates the synapses in the hippocampus. We do not know yet whether or not progesterone also down-regulates the cholinergic enzymes of the basal forebrain that are induced by oestradiol. Progesterone often modulates the actions of oestrogen, potentiating them when given sequentially with oestradiol and inhibiting them when given concurrently. I understand that concurrent oestradiol and progesterone therapy may not be as beneficial as unopposed oestrogen therapy as far as cardiovascular and bone measures are concerned.

Uvnäs-Moberg: Most studies on the steroid dependency of female skills compare the midluteal and menstrual phases of the oestrus cycle. That is, a period with high levels of both oestrogen and progesterone is compared with a period with low levels. An exception is the study by Hampson (1990) in which the follicular phase is compared with the menstrual phase. Is there a difference between the effects of oestrogen alone and the two hormones together?

McEwen: One has to bear in mind the lag period between the hormone rise and the effect you're measuring. If you don't know what this is, you may make the wrong inference about the hormone level at that time. (This may seem like an evasion, but it's true.) Until we know a little bit more about the timing, we can't really tell what correlates with what.

Bäckström: Smith (1989) has shown that the effect of oestradiol on the NMDA receptor in the cerebellum is very rapid—it arises within three minutes. Couldn't the effect of oestrogen on the NMDA receptor be similar to the effect of progesterone metabolites on the GABA receptor in that both affect the receptor directly?

McEwen: I can't answer that question directly, but Bob Moss has done studies on rapid oestrogen effects on electrical activity in the CA1 region of the hippocampus, the same one that shows the synaptic changes (Wong & Moss 1992). He finds that there are rapid non-genomic oestrogen effects via a non-NMDA receptor. The NMDA-sensitive response is up-regulated slowly, which fits very well with the induction of the new spine synapses, which we presume have NMDA receptors on them. So there's no indication, at least from his studies, that there is a rapid modulation of the NMDA receptor itself. Rather, there seems to be rapid modulation of a non-NMDA receptor.

Jensen: I think the effect of oestrogen on Alzheimer's disease is very exciting, because there is no effective clinical management for these patients. But from what I have read, this effect is preventive—women who have had replacement therapy are less likely to develop Alzheimer's. Is there evidence that oestrogen treatment is beneficial for people who already have Alzheimer's?

McEwen: This was the basis of the intervention studies that Howard Fillit did, and that have now been replicated on small groups of Alzheimer's patients (Fillit et al 1986). Another group has replicated this finding (Honjo et al 1989). These studies showed that oestrogen does have a fairly long-term effect to reduce the dementia. There hasn't yet been a large-scale trial, but the studies so far are remarkably consistent in showing an improvement that outlasts the withdrawal of oestrogen therapy.

Uvnäs-Moberg: Oestrogen in animals has been shown to be anti-aggressive: is there any risk that if you treat women with oestrogens, they will become less assertive?

Pfaff: We have some very preliminary data with oestrogen receptor knockout mice, in collaboration with Ken Korach in North Carolina, who created these mice (S. Ogawa, K. Korach & D. Pfaff, unpublished results). We've been working with the males for several months now. The results have been very interesting: in the absence of the oestrogen receptor during development, as well as during adulthood, the male mice still have sexual motivation, but they have another deficit: that is, there is something wrong with their peripheral apparatus or with the neural substrate governing it, such that they don't achieve proper penetration and so they are rarely fertile. In standard aggressive behaviour tests with an olfactory bulbectomized male as the target the oestrogen receptor knockout male mice have absolutely no aggression. But, if they get frustrated when they're supposed to be mating with females, they may show aggression towards the female. So the effect of the gene on a given social behaviour depends on the target of that social behaviour. The females simply don't show lordosis behaviour.

Manolagas: I have been wondering whether there is a common link between the protective effects of oestrogen in osteoporosis, Alzheimer's disease and the cardiovascular system. I am aware of literature linking Alzheimer's progression and cytokines. To your knowledge, how convincing is the evidence that cytokines are involved in Alzheimer's?

McEwen: There is debate as to which mechanism is more important. Certainly, inflammatory mechanisms are among the hot topics of late. I am not able to talk about how oestrogens might interact with those systems, especially the ones in the hippocampus and basal forebrain: I don't think it has been very well studied. But just to show you how varied the actions of oestrogen are, there's a recent paper from Paul Greengard's lab (at the Rockefeller University, NY) on the production of the amyloid precursor protein (Jaffe et al 1994). Apparently, all cells produce this protein, which is secreted, and the abnormal form that's truncated by a protease is the one that seems to be toxic. The normal secreted form, if anything, has a beneficial effect. The Greengard lab has shown that oestrogen treatment increases the secretory pathway and decreases the abnormal pathway. Results from a number of labs (including Don Pfaff's; Chung et al 1988) indicate that long-term oestrogen treatment has beneficial effects, or at least synapse-enhancing effects, in a number of brain areas. Matsumoto & Arai (1979) showed that if you damage the hypothalamus, oestrogen treatment actually increases colateral sprouting in synapse formation. So there's reason to believe that oestrogens have a beneficial trophic effect. For reasons of time, in my paper I passed over the fact that the serotonergic, noradrenergic and dopaminergic neurons tend to die as an animal ages; these die out particularly rapidly in Alzheimer's disease (Wallin et al 1991). If oestrogens have some kind of trophic action on those neurons, as well as on the cholinergic neurons, then you can imagine a rather broad-based effect of oestrogens throughout the entire nervous system.

Baulieu: Are the Greengard experiments you mentioned studies with oestrogen-receptor-containing cells?

McEwen: I don't think they transfected oestrogen receptors into the cells, so only endogenous oestrogen receptors were present. These were ZR75-1 human breast carcinoma cells, which have the highest oestrogen receptor concentration of any known oestrogen-responsive cell line.

Pfaff: We found that oestrogens had massive morphological effects on neurons. We observed changes in the endoplasmic reticulum that suggested that the cells had come out of G0 and were in G1 and would divide; we wanted to find out why those neurons don't go ahead and divide. This would be quite important for the issues that we're talking about. So Bob Gibbs, in my lab, did a rather unusual experiment, to transplant the uterus into the ventricle of the rat brain, and to ask whether the brain's environment could inhibit the uterine cells from dividing, or vice versa. We made a lot of discoveries, but unfortunately the answer, in both directions, was 'no'. The brain's environment could not keep the uterine cells from dividing, and uterine fluids, at least in so far as they could diffuse, could not make the brain cells divide.

Thijssen: You mentioned the role of aromatase in the hippocampus. The androgens you have been using are supposed to work after aromatization on that particular area of the brain. Has anyone done experiments using androgens which are not aromatizable? Would they give different effects on testosterone?

McEwen: That's a very good question. No one has looked at the hippocampal system in terms of spatial ability with a non-aromatizable androgen. It would be an interesting experiment.

Parker: There was a report a year or so ago that the main oestrogen receptor in the brain is a variant that lacks exon 4 (Skipper et al 1993).

McEwen: I don't think it's the main one, although it has caused problems, because by using the wrong probe for the oestrogen receptor (that is, one that doesn't specifically look at the part that's missing, which happens to be part of the oestrogen DNA-binding domain) some workers have identified 'oestrogen receptor message' in the adult hippocampus and the basal forebrain, where there is very little receptor binding. So it has created a nuisance that has to be taken into account. I understand that this receptor lacking exon 4 turns out not to have much function at all. We don't know what it is doing in any cell.

Parker: But if you use the right probe, is there such a thing as an exon 4 variant in the brain?

McEwen: That was the claim from David Crews (Skipper et al 1993).

Thijssen: We have confirmed this in meningioma (Koehorst et al 1993). We also demonstrated that the receptor lacking exon 4 is not active—it does not inhibit or support oestrogen function (Koehorst et al 1994).

Baulieu: Has this receptor lacking exon 4 been cloned, or has it been identified by *in situ* hybridization?

Parker: Its presence is argued on the basis of polymerase chain reaction analyses where you look for a version of mRNA that lacks the exon 4 sequences.

Pfaff: Does it make a protein?

Parker: In theory, it can make a protein.

Castagnetta: It is a debated question. We have isolated this 323 bp oestrogen receptor mRNA from MCF-7 and ZR75-1 mammary carcinoma cell lines, suggesting that it is a product of alternative splicing (Pfeffer et al 1993). However, it remains to be proved that it is translated into a protein with a functional role. Coming to a different point, high tissue levels of catecholoestrogens have been found in different areas of the CNS (Paul et al 1977) and there is also sparse evidence that these peculiar oestrogen metabolites have specific binding sites (Vandewalle et al 1985). In particular, 4-hydroxy oestradiol seems to be able to bind both macromolecules and DNA; it has been proposed that this unique feature could be relevant to the carcinogenic potential of these compounds (Weisz 1991).

McEwen: As far as catecholoestrogens and the brain are concerned, there was initial excitement because there were indications that they would interact with catecholamine receptors, noradrenaline uptake and tyrosine hydroxylase (Lloyd & Weisz 1978, Paden et al 1982, Ghraf et al 1983). However, it was found that these effects occurred at micromolar concentrations or even higher, making a physiological role unlikely. Moreover, careful analysis of receptor affinities and *in vivo* efficacies of catchecoloestrogens, as well as the conversion of oestradiol to catchecoloestrogens, led to the conclusion that they are not major metabolites of oestrogen, nor are they obligatory intermediates of oestrogen action on reproductive endpoints in the nervous system (Pfeiffer et al 1986, MacLusky et al 1986). These findings caused interest in catchecol-oestrogens to wane, at least in studies of the nervous system.

Castagnetta: We have previously observed that tissue levels of catecholoes-trogens in rat CNS were far greater than those of oestradiol (L. Castagnetta et al, unpublished results). This was true in rat pituitary, hypothalamus and brain cortex.

McEwen: Regarding protein binding of modified oestrogens, there is a paper by Robert Lustig, with Don Pfaff and I, building on findings by Jack Fishman, that 16α-hydroxyoestrone binds very strongly to oestrogen receptors (Lustig et al 1989).

Thijssen: 16α-hydroxyoestrone is also supposed to bind to DNA, according to Jack Fishman and Leon Bradlow (Bradlow et al 1985, Fishman & Martucci 1980, Swaneck & Fishman 1988).

References

Bradlow HL, Herschopf RJ, Martucci CP, Fishman J 1985 Estradiol 16α-hydroxylation in the mouse correlates with mammary tumor incidence and presence of murine mammary tumor virus: a possible model for the etiology of breast cancer in humans. Proc Natl Acad Sci USA 82:6295–6299

Chung SK, Pfaff DW, Cohen RS 1988 Estrogen-induced alterations in synaptic morphology in the midbrain central gray. Exp Brain Res 69:522–530

Fillit H, Weinreb H, Cholst I et al 1986 Observations in a preliminary open trial of estradiol therapy for senile dementia—Alzheimer's type. Psychoneuroendocrinology 11:337–345

Fishman J, Martucci C 1980 Biological properties of 16α-hydroxyestrone: implications in estrogen physiology and pathophysiology. J Clin Endocrinol & Metab 51:611–615

Ghraf R, Micel M, Hiemke C, Knuppen R 1983 Competition by monophenolic oestrogens and catecholoestrogens for high-affinity uptake of [³H](−)norepinephrine into synaptosomes from rat cerebral cortex and hypothalamus. Brain Res 277:163–168

Gooren L 1994 Effect of androgens on brain function in males and females. Eur J Endocrinol 130(suppl 2):77–78

Hampson E 1990 Estrogen-related variations in human spatial and articulatory-motor skills. Psychoneuroendocrinology 15:97–111

Honjo H, Ogino Y, Naitoh K et al 1989 In vivo effects by estrone sulfate on the central nervous system senile dementia (Alzheimer's type). J Steroid Biochem 34:521–525

Jaffe AB, Toran-Allerand CD, Greengard P, Gandy SE 1994 Estrogen regulates metabolism of Alzheimer amyloid β precursor protein. J Biol Chem 269: 13065–13068

Kimura D 1994 Estrogen replacement therapy protects against intellectual decline in post-menopausal women. Univ Ontario Res Bull 724 (September 1994)

Koehorst SGA, Jacobs HM, Thijssen JHH, Blankenstein MA 1993 Wild-type and alternatively spliced estrogen receptor messenger RNA in human meningioma tissue and MCF-7 breast cancer cells. J Steroid Biochem Mol Biol 45:227–233

Koehorst SGA, Cox JJ, Donker GH et al 1994 Functional analysis of an alternatively spliced estrogen receptor lacking exon 4 isolated from MCF-7 human breast cancer cells and meningioma tissue. Mol Cell Endocrinol 101:237–245

Lloyd T, Weisz J 1978 Direct inhibition of tyrosine hydroxylase activity by catechol estrogens. J Biol Chem 253:4841–4843

Lustig R, Mobbs C, Bradlow L, McEwen BS, Pfaff DW 1989 Differential effects of estradiol and 16α-hydroxyoestrone on pituitary and preoptic estrogen receptor regulation. Endocrinology 125:2701–2709

MacLusky NJ, Krey LC, Parsons B et al 1986 Are catechol oestrogens obligatory mediators of oestrogen action in the central nervous system? II. Potencies of natural and synthetic oestrogens for induction of gonadotrophin release and female sexual behavior. J Endocrinol 110:499–505

Matsumoto A, Arai Y 1979 Synaptogenic effect of estrogen on the hypothalamic arcuate nucleus of the adult female rat. Cell Tissue Res 198:427–433

Paden C, McEwen BS, Fishman J, Snyder L, DeGroff V 1982 Competition by estrogens for catecholamine receptor binding in vitro. J Neurochem 39:512–520

Paul SM, Axelrod J, Diliberto J Jr 1977 Catechol estrogen-forming enzyme of brain: demonstration of a cytochrome P450 monooxygenase. Endocrinology 101:1604–1610

Pfeffer U, Fecarotta E, Castagnetta L, Vidali G 1993 Estrogen receptor variant messenger RNA lacking exon 4 in estrogen-responsive human breast cancer cell lines. Cancer Res 53:741–743

Pfeiffer DG, MacLuskey NJ, Barnea E et al 1986 Are catechol oestrogens obligatory mediators of oestrogen action in the central nervous system? I. Characterization of pharmacological probes with different receptor binding affinities and catechol oestrogen formation in rats. J Endocrinol 110:489–497

Phillips S, Sherwin B 1992 Effects of oestrogen on memory function in surgically menopausal women. Pshychoneuroendocrinology 17:485–495

Silber M, Almkvist O, Larsson B, Uvnäs-Moberg K 1990 Temporary peripartal impairment in memory and attention and its possible relation to oxytocin concentration. Life Sci 47:57–65

Skipper JK, Young LJ, Bergeron JM, Tetzlaff MT, Osborn CT, Crews D 1993 Identification of an isoform of the estrogen receptor messenger RNA lacking exon four and present in the brain. Proc Natl Acad Sci USA 90:7172–7175

Smith SS 1989 Estrogen produces long-term increases in excitatory neural responses to NMDA and quisqualate. Brain Res 503:354–357

Swaneck GE, Fishman J 1988 Covalent binding of the endogenous estrogen 16α-hydroxyestrone to the estradiol receptor in human breast cancer cells: characterization and intranuclear localization. Proc Natl Acad Sci USA 85:7831–7835

Vandewalle B, Peyrat J-P, Bonneterre J, Lefebvre J 1985 Catecholestrogen binding sites in breast cancer. J Steroid Biochem 23:603–610

Wallin A, Carlsson A, Ekman R et al 1991 Hypothalamic monoamines and neuropeptides in dementia. Eur Neuropsychopharmacol 1:165–168

Weisz J 1991 Metabolism of estrogens by target cells: diversification and amplification of hormone action and the catecholestrogen hypothesis. In: Hochberg RG, Naftolin F (eds) The new biology of steroid hormones (Sereno Symp 74). Raven Press, NY, p 101–112

Wong M, Moss RL 1992 Long-term and short-term electrophysiological effects of estrogen on the synaptic properties of hippocampal CA1 neurons. J Neurosci 12:3217–3225

Actions of sex steroids on behaviours beyond reproductive reflexes

Catherine A. Priest and Donald W. Pfaff

Neurobiology and Behavior Laboratory, The Rockefeller University, 1230 York Avenue, New York, NY 10021, USA

Abstract. The actions of sex steroids in the brain have been shown, from molecular to systems levels, to control reproductive behaviour in a wide range of vertebrates. It has become increasingly clear that gonadal steroid hormones have regulatory functions which extend far beyond the direct coordination of an animal's physiological state and its display of sexual behaviour. While some of these actions may include changes in mood or other behavioural measures, such as exploration or excitability, sex steroid hormones also influence neural plasticity, neuronal activity and, possibly, learning and memory, as reflected by long-term potentiation or age-related deficits. Here we describe two systems that have been used to explore the non-reproductive roles of gonadal steroid hormones. The first of these is to examine the oestrogen-sensitive opioid peptide gene expression in the hypothalamus. Currently, we are attempting to identify the types of behaviour which may be altered consequent to the oestrogenic induction of the preproenkephalin gene. The second approach involves studying the effects of progesterone at the neuronal cell membrane and characterizing the metabolites of progesterone which have benzodiazepine-like actions in the brain. A number of studies suggest that this may provide an alternative mechanism through which progesterone can influence mood or behaviour.

1995 Non-reproductive actions of sex steroids. Wiley, Chichester (Ciba Foundation Symposium 191) p 74–89

Traditionally, studies to determine the physiological roles of gonadal steroid hormones have focused primarily on their regulation of behaviour associated with reproduction. Not only are reproductive behaviours such as lordosis, the female-typical reproductive reflex of many quadruped species, of obvious biological importance, but also they are composed of reflexive motor responses to defined, hormonally dependent stimuli and are elicited, observed and quantified easily in a laboratory setting (Pfaff et al 1973). These advantages have led to the elucidation of neural circuitry which is sufficient to integrate environmental and sensory signals with the circulating steroid milieu for the production of lordosis behaviour (Pfaff 1980). The integration of these internal and external signals has been examined further, at the level of individual hypothalamic neurons, using electrophysiological (Kow et al 1994) and immunocytochemical (Pfaus et al 1993) techniques. Additionally, transcriptional systems have been identified in which gonadal steroid hormones regulate gene expression within these cells to modify production of neuropeptides and receptors involved in reproductive behaviour. Reproductively relevant genes sensitive to oestrogen include those encoding the progesterone receptor, muscarinic receptors, adrenergic α_1 receptors, preproenkephalin, oxytocin and its receptor, and gonadotropin-releasing hormone (GnRH/LHRH; for review, see Pfaff et al 1994).

Recently, however, it has become clear that sex steroids have a much wider involvement in developmental, behavioural and cellular processes. While the 'organizing' effects on neural development and sexual behaviour of neonatal exposure to androgens are well known (for review, see Micevych & Ulibarri 1992, Pfaff et al 1994), measures of exploratory behaviour also differ between the sexes and can be altered by manipulations of the circulatory gonadal steroid environment during development (Pfaff & Zigmond 1971). Similarly, adult rats differ in the degree to which a novel environment determines their social interactions. While males tend to interact less with an unknown conspecific in a neutral environment than in one which is familiar, this distinction is lost in post-pubertal females (Etzel & Kellog 1993). In humans, sex differences have been reported in the incidence of depression-based illnesses (Rice et al 1984), although this also may reflect undefined social influences. Additionally, there has been a great deal of interest in identifying changes in emotionality and task performance during the menstrual cycle. While a subset of women appear to experience mood changes, the exact nature of the endocrine involvement remains unclear (for review, see Endicott 1993). Roles for progesterone (Freeman et al 1992) and its metabolites (Freeman et al 1993) have been suggested on the basis of correlations between circulating plasma levels of these steroids and measures of fatigue, immediate recall and confusion. Cyclical fluctuations in oestrogen levels also may also influence performance and memory. Alterations in neuronal morphology have been noted following oestrogen treatment of ovariectomized rats (Cohen & Pfaff 1981) and ovariectomy is followed by a deterioration in performance in the Morris water maze, which is reversible by oestrogen replacement treatment (Singh et al 1994). Because this effect can be correlated with changes in cholinergic neurons, it may be related to synthetic changes in the oestrogen-sensitive cells which produce nerve growth factor (Gibbs & Pfaff 1994).

Indeed, it is no longer sufficient to consider only the genomic effects of sex steroid hormones. A number of neurosteroids (Baulieu et al 1987) have been identified that can elicit almost immediate effects at the level of the cell membrane (for review, see Paul & Purdy 1992). The best-studied of these have been the 3α-hydroxy ring A-reduced pregnane steroids, allopregnanolone and pregnenolone, which have sedative/hypnotic effects. These progesterone metabolites do not interact with classical nuclear progesterone receptors; instead, they augment Cl^- conductance across the cell membrane via their facilitatory interactions with type A γ-aminobutyric acid $(GABA_A)$/benzo-diazepine binding sites. Conversely, other steroidal metabolites, such as pregnenolone sulphate or the synthetic RU5135 (3α-hydroxy-16-imino-5β-17-azaandrostan-11-one) have been shown to antagonize $GABA_A$/benzodiazepine-mediated Cl^- conductance. The actions of these steroids can be altered pharmacologically by benzodiazepine-like agonists and antagonists and, while their primary site of synthesis is the adrenal gland, they can be measured in the brains of adrenalectomized animals at pharmacologically relevant levels (Medina et al 1993). Furthermore, increased steroidogenesis in peripheral tissues may produce higher levels of $GABA_A$/benzodiazepine receptor-active steroid metabolites in the brain. Indeed, allopregnanolone and pregnenolone are induced by stress (Purdy et al 1991; for review, see Myslobodsky 1993) and fluctuate, in parallel with progesterone, across the oestrous and menstrual cycles (Purdy et al 1990). Interestingly, circulating levels of allopregnanolone drop during pregnancy and begin to rise again at parturition (Ichikawa et al 1974). Concurrent with these changes is an increase in maternal pain threshold during pregnancy (Gintzler 1980). On the basis of these observations, it has been hypothesized (Schwartz-Giblin & Pfaff 1987) that allopregnanolone contributes to the maintenance of a general tonic inhibition of antinociceptive mechanisms in non-pregnant animals, although this association remains to be proven. Interestingly, *in vivo* administration of progesterone increases the density of $GABA_A$/benzodiazepine binding sites for tritiated flunitrazepam in the substantia gelatinosa, an area of the spinal cord that is involved in nociceptive processing (Schwartz-Giblin et al 1988). Given the high degree of bioactivity of these steroid metabolites, it is imperative that their actions also be considered as alternative mechanisms through which gonadal steroid hormones can affect behaviour.

Thus, current molecular analyses of sex steroid actions in the brain have revealed a much broader scope of relationships between hormones and behaviour than previously was anticipated. In this paper, we describe two model systems which illustrate in greater detail some of the other ways in which gonadal steroid hormones can affect brain function. In the first section, the genomic effects of oestrogen and progesterone on preproenkephalin gene expression will be discussed; in the second section, we will present information regarding the genomic and cell-surface actions of the progesterone receptor.

Opioid peptide genes in female rat brain

Throughout the brain, the distribution of opioid peptide mRNAs (Harlan et al 1987) overlaps extensively with the distribution of cells that concentrate tritiated oestradiol, as demonstrated by autoradiographic techniques (Morrell et al 1986), a hallmark of cells that contain oestrogen receptors. In some areas of the hypothalamus and limbic system, cells have been shown to co-localize markers for both opioid peptides and oestrogen receptors (Akesson & Micevych 1991). These findings have been followed by molecular determinations of sex hormone effects on opioid peptide gene expression. Briefly, oestradiol treatments have been found to increase dramatically the expression of the preproenkephalin gene, as measured by filter hybridization or by *in situ* hybridization techniques (Romano et al 1988, Priest et al 1995). The induction of preproenkephalin by oestrogen occurs within one hour of steroid exposure (Romano et al 1989a, Zhu & Pfaff 1994) and is potentiated by subsequent administration of progesterone (Romano et al 1989a). Following acute oestrogen treatment, hypothalamic preproenkephalin is expressed in a biphasic pattern across time, with peaks in expression occurring one and 48 h after oestrogen exposure (Priest et al 1995). In Sprague–Dawley rats, the induction of preproenkephalin by oestrogen can be detected in females but not in males (Romano et al 1990); however, the hormonal induction also can be seen in male Fisher 344 rats, a strain that is very sensitive to oestrogen (Hammer et al 1993). DNase hypersensitivity studies have been used to determine which areas of the rat preproenkephalin gene promoter and its first intron may be involved in the effects of hormone-influenced transcription factors (Funabashi et al 1993). The potential importance of the first intron of the rat preproenkephalin gene has been highlighted further by *in situ* hybridization demonstrations of an unusual hormone-sensitive nuclear transcript which it encodes (Brooks et al 1993). The preproenkephalin promoter demonstrates oestrogen receptor-like binding to DNA, as revealed by gel retardation assays (Zhu et al 1994) and contains *cis*-regulatory elements sufficient for a transcriptional effect, as revealed by *in vivo* promoter analysis using neurotropic viral vectors (Yin et al 1994). Thus, the preproenkephalin gene is a primary candidate for direct involvement in behaviour regulated by oestrogen-sensitive circuitry in the hypothalamus.

Although a role for preproenkephalin in the regulation of reproductive behaviour is possible, and the induction of preproenkephalin mRNA by oestrogen is, indeed, positively correlated with the display of lordosis behaviour (Lauber et al 1990), a causal relationship between the two factors has not been shown. The ways in which opioid peptides are involved in the modulation of reproductive behaviour are characteristically several and complex, with the result that a variety of opioid effects on sexual behaviour have been reported, some of which are conflicting (for review, see Pfaus & Pfaff 1992). In relation to the effects of preproenkephalin gene products, it is interesting that intracerebroventricular infusion of peptides that are active at δ opioid receptors (but not at μ opioid receptors) can increase significantly the performance of lordosis behaviour by ovariectomized, oestrogen-treated female rats (Pfaus & Gorzalka 1987, Pfaus & Pfaff 1992). However, the modest, yet statistically significant, effects on lordosis behaviour contrast with the dramatic, oestrogen-dependent changes in preproenkephalin gene expression. Moreover, we have not yet defined the exact neural sites of action of the δ receptor peptides following intraventricular administration. It is possible that the behavioural effects of these peptides are indirect or reflect the altered activities of distant target sites. Notably, during sexual behaviour tests, infusions of the δ-active peptide DPDPE (D-Phe–D-Phe–enkephalin) tend to reduce the female rat's rejection behaviour towards a male rat (Pfaus & Pfaff 1992). A possible explanation for this observation involves the analgesic effects of opioid peptides. It may be that a primary functional role of the endogenous, oestrogen-induced preproenkephalin is to alter the disposition of the female towards somatosensory input and to reduce the aversiveness of cutaneous stimuli. A second possible role for the oestrogen induction of the preproenkephalin gene products involves the anatomical projections from the hypothalamus to the limbic system and basal forebrain. Under the influence of gonadal oestrogens, the female rat may experience changes in disposition towards olfactory inputs, especially those which could signal potentially aversive events. On a broader scope, endogenous opioid peptides may be regulated by gonadal steroid hormones to influence the overall emotional state of an animal, beyond specific modalities such as somatosensation or olfaction. Currently, we are investigating some of these behavioural possibilities and are attempting to determine the neural sites of oestrogen-sensitive opioid peptide action (C. A. Priest, L.-M. Kow & D. W. Pfaff, unpublished results). Relevant projections of hypothalamic neurons that express enkephalin may extend from the basal medial hypothalamus towards the forebrain or midbrain (for review, see Priest & Pfaff 1994), although the local actions of opioid interneurons within the hypothalamus also must be investigated, as suggested by the high proportion of intrinsic hypothalamic synapses (Nishizuka & Pfaff 1989). Thus, while molecular analyses of the effects of gonadal steroid hormones on preproenkephalin gene expression illustrate an exquisitely sensitive regulatory system, further studies are needed to determine the underlying range of the behavioural effects of endogenous opioid peptides.

Progesterone actions: nuclear receptors and neurosteroids

Most investigations into the effects of progestins on behaviour have examined nuclear progesterone receptors and their genomic actions as transcription factors. In the hypothalamus, levels of nuclear progesterone receptors show a high positive correlation with the production of female reproductive behaviour (Parsons & Pfaff 1985). In the hypothalamus of females, previous oestradiol priming is required for the display of lordosis and nuclear progesterone receptors are regulated by oestradiol; oestrogen does not appear to induce expression of the progesterone receptor gene in males (Lauber et al 1991). In fact, the requisite coupling between oestradiol induction of progesterone receptor gene expression and lordosis behaviour provides the first example of the synthesis of a transcription factor which is related directly to the performance of a specific behaviour (Romano et al 1989b). The behavioural importance of the nuclear progesterone receptor is demonstrated further by the ability of either progestin receptor antagonists (Brown & Blaustein 1984, Vathy et al 1989) or antisense DNA directed against the mRNA for progesterone receptor (Ogawa et al 1994) to block the effectiveness of progesterone on steroid-dependent copulatory and courtship behaviours. We have used gel retardation techniques to illustrate differential nuclear protein/DNA binding patterns at compound oestrogen response elements in the rat progesterone receptor gene (Lauber et al 1993, Wu-Peng & Pfaff 1994, Scott et al 1994) and we are currently expanding the results from these studies through transcriptional analyses using neurotropic viral vectors.

In dramatic contrast to the genomic effects of the progesterone receptor are the neurosteroidal actions of the progestin metabolites, allopregnanolone and pregnenolone (Baulieu et al 1987). Through interactions with the $GABA_A$/ benzodiazepine receptor, these compounds have been reported to affect neural transmission, behaviour and mood (for review, see Paul & Purdy 1992) by augmented Cl^- conductance across the cell membrane (Majewska et al 1986). Among their best-established functions are their anxiolytic effects. When rats are administered progesterone at doses that do not affect overall locomotion (Bitran et al 1993a), its anxiolytic effect is revealed by increased exploratory behaviour in an elevated plus-maze (Pellow et al 1985), and is correlated with increased $GABA_A$/benzodiazepine receptor function (Bitran et al 1993a). Similar anxiolytic results are seen when intact males are treated with testosterone propionate, an androgenic compound that is reduced to the positive $GABA_A$/benzodiazepine receptor modulators androstanediol and androsterone (Bitran et al 1993b). Further, the endocrine status of a female rat influences the anxiolytic effects mediated by the $GABA_A$/benzodiazepine receptor. Diazepam has decreased anxiolytic effects in ovariectomized females, as compared with its effects in pro-oestrous females, and this is correlated with a reduced efficacy of $GABA_A$/benzodiazepine-stimulated Cl^- transport (Bitran et al 1991). Importantly, these effects seem to *require* the bioconversion of progesterone to allopregnanolone, as blocking the enzymic conversion prevents the anxiolytic effects of the administered progesterone (Bitran et al 1993c).

Recent developments in molecular biology and pharmacology have opened new avenues for examining the effects of progesterone metabolites. For example, control over the rate of steroidogenesis in mitochondria, including those in glial cells, may be a regulatory step of potential behavioural importance (Romeo et al 1993). Compounds that increase mitochondrial steroidogenesis can increase significantly the number of entries into the open arms of an elevated plus-maze and increase the amount of time spent there (Romeo et al 1992), indicating anxiolytic effects. Furthermore, the rate of mitochondrial steroidogenesis has been shown to be an important factor in at least two animal models of anxiety, neophobia and response to conflict. These effects are demonstrated, for example, by the ability of PK1195 to block the anticonflict activity of FGIN1-27 (Auta et al 1993). In addition, pharmacological treatments to increase the rate of steroidogenesis have anticonvulsant actions (tables II and III in Auta et al 1993). This may be relevant to the finding that seizures caused by picrotoxin, a $GABA_A$/benzodiazepine channel antagonist, can be reduced by high doses of oestradiol (Schwartz-Giblin et al 1989).

Importantly, the two model systems discussed in this chapter may interact. In the local circuitry within individual cell groups of the hypothalamus, basal forebrain or midbrain central grey, the actions of progesterone metabolites through $GABA_A$/benzodiazepine receptors may be related to the actions of opioid peptides. Namely, in proposed mechanisms of opioid-induced antinociception (from the work of Moreau & Fields 1986, Depaulis et al 1987, as reviewed in Schwartz-Giblin & Pfaff 1987), opioid-containing projections may synapse on local, intranuclear, GABA neurons to inhibit them, thus acting to disinhibit the 'output cells' from a nuclear cell group. The proposed opioid-initiated, GABA-mediated disinhibition could be important not only for reproductive behaviour (for example, in the ventromedial hypothalamus), but also in the control of pain mechanisms (for example, in the midbrain central grey) or for a variety of emotional and mood-related behaviours mediated by limbic–hypothalamic systems.

References

Akesson TR, Micevych PE 1991 Endogenous opioid-immunoreactive neurons of the ventromedial nucleus of the hypothalamus concentrate estrogen in male and female rats. J Neurosci Res 28:359–366

Auta J, Romeo E, Kozikowski A, Ma D, Costa E, Guidotti A 1993 Participation of mitochondrial diazepam binding inhibitor receptors in the anticonflict, antineophobic and anticonvulsant action of 2-aryl-3-indoleacetamide and imidazopyridine derivatives. J Pharmacol Exp Ther 265:649–656

Baulieu EE, Robel P, Vatier O, Haug A, Le Goascogne C, Bourreau E 1987 Neurosteroids: pregnenolone and dehydroepiandrosterone in the rat brain. In: Fuxe K, Agnati LF (eds) Receptor–receptor interaction: a new intramembrane integrative mechanism. Macmillan, Basingstoke, p 89–104

Bitran D, Hilvers RJ, Kellogg CK 1991 Ovarian endocrine status modulates the anxiolytic potency of diazepam and the efficacy of γ-aminobutyric acid–benzodiazepine receptor-mediated chloride ion transport. Behav Neurosci 105:653–662

Bitran D, Purdy RH, Kellogg CK 1993a Anxiolytic effect of progesterone is associated with increases in cortical allopregnanolone and $GABA_A$ receptor function. Pharmacol Biochem Behav 45:423–428

Bitran D, Kellogg CK, Hilvers RJ 1993b Treatment with an anabolic–androgenic steroid affects anxiety-related behavior and alters the sensitivity of cortical $GABA_A$ receptors in the rat. Horm Behav 27:568–583

Bitran D, McLead M , Shiekh M 1993c Blockade of the bioconversion of progesterone to allopregnanolone prevents the anxiolytic effect and potentiation of cortical $GABA_A$ receptor function observed in progesterone-treated ovariectomized rats. Soc Neurosci Abstr 19:373

Brooks PJ, Funabashi T, Kleopoulos SP, Mobbs CV, Pfaff DW 1993 Cell-specific expression of preproenkephalin intronic heteronuclear RNA in the rat forebrain. Mol Brain Res 19:22–30

Brown TJ, Blaustein JD 1984 Inhibition of sexual behavior in female guinea pigs by a progestin receptor antagonist. Brain Res 301:343–349

Cohen R, Pfaff DW 1981 Ultrastructure of neurons in the ventromedial nucleus of the hypothalamus in ovariectomized rats with or without estrogen treatment. Cell Tissue Res 217:451–470

Depaulis A, Morgan MM, Liebeskind JC 1987 GABAergic modulation of the analgesic effects of morphine microinjected in the ventral periaqueductal grey matter of the rat. Brain Res 436:223–228

Endicott J 1993 The menstrual cycle and mood disorders. J Affect Dis 29:193–200

Etzel BA, Kellogg CK 1993 Hormonal influence on the development of gender-specific social interaction in female rats. Soc Neurosci Abstr 19:1191

Freeman EW, Weinstock L, Rickels K, Sondheimer SJ, Coutifaris C 1992 A placebo-controlled study of effects of oral progesterone on performance and mood. Br J Clin Pharmacol 33:293–298

Freeman EW, Purdy RH, Coutifaris C, Rickels K, Paul SM 1993 Anxiolytic metabolites of progesterone: correlation with mood and performance measures following oral progesterone administration to healthy female volunteers. Clin Neuroendocrinol 58:478–484

Funabashi T, Brooks PJ, Mobbs CV, Pfaff DW 1993 DNA methylation and DNase-hypersensitive sites in the 5′ flanking and transcribed regions of the rat preproenkephalin gene: studies of mediobasal hypothalamus. Mol Cell Neurosci 4:499–509

Gibbs RB, Pfaff DW 1994 In situ hybridization detection of trkA mRNA in brain: distribution, co-localization with p75[NGFR] and up-regulation by nerve growth factor. J Comp Neurol 341:324–339

Gintzler AR 1980 Endorphin-mediated increases in pain threshold during pregnancy. Science 210:193–195

Hammer RP, Bogic L, Handa RJ 1993 Estrogenic regulation of proenkephalin messenger RNA expression in the ventromedial hypothalamus of the adult male rat. Mol Brain Res 19:129–134

Harlan RE, Shivers BD, Romano GJ, Howells RD, Pfaff DW 1987 Localization of preproenkephalin messenger RNA in the rat brain and spinal cord by in situ hybridization. J Comp Neurol 258:159–184

Ichikawa S, Sawada T, Nakamura Y, Morioka H 1974 Ovarian secretion of pregnane compounds during the estrous cycle and pregnancy in rats. Endocrinology 94:1615–1620

Kow LM, Mobbs CV, Pfaff DW 1994 Roles of second-messenger systems and neuronal activity in the regulation of lordosis by neurotransmitters, neuropeptides and estrogen: a review. Neurosci Biobehav Rev 18:251–268

Lauber AH, Romano GJ, Pfaff DW 1991 Sex difference in estradiol regulation of progestin receptor mRNA in rat mediobasal hypothalamus as demonstrated by *in situ* hybridization. Neuroendocrinology 53:608–613

Lauber AH, Romano GJ, Mobbs CV, Howells RD, Pfaff DW 1990 Estradiol induction of proenkephalin messenger RNA in hypothalamus: dose–response and relation to reproductive behavior in the female rat. Mol Brain Res 8:47–54

Lauber AH, Pfaff DW, Alroy I, Freedman LP 1993 Hypothalamic estrogen receptor binding to a consensus ERE and a putative ERE from a progesterone receptor gene. Soc Neurosci Abstr 19:698

Majewska MD, Harrison NL, Schwartz RD, Barker JL, Paul SM 1986 Steroid hormone metabolites are barbiturate-like modulators of the GABA receptor. Science 232:1004–1007

Medina JH, Paladini AC, Izquierdo I 1993 Naturally occurring benzodiazepines and benzodiazepine-like molecules in brain. Behav Brain Res 58:1–8

Micevych PE, Ulibarri C 1992 Development of the limbic–hypothalamic cholecystokinin circuit: a model of sexual differentiation. Dev Neurosci 14:11–34

Moreau J-L, Fields HL 1984 Evidence for GABA involvement in midbrain control of medullary neurons that modulate nociceptive transmission. Brain Res 397:37–46

Morrell JI, Krieger MS, Pfaff DW 1986 Quantitative autoradiographic analysis of estradiol retention by cells in the preoptic area, hypothalamus and amygdala. Exp Brain Res 62:343–354

Myslobodsky MS 1993 Pro- and anti-convulsant effects of stress: the role of neuroactive steroids. Neurosci Biobehav Rev 17:129–139

Nishizuka M, Pfaff DW 1989 Intrinsic synapses in the ventromedial nucleus of the hypothalamus: an ultrastructural study. J Comp Neurol 286:260–268

Ogawa S, Olazabal UE, Parhar IS, Pfaff DW 1994 Effects of intrahypothalamic administration of antisense DNA for progesterone receptor mRNA on reproductive behavior and progesterone receptor immunoreactivity in female rat. J Neurosci 14:1766–1774

Paul SM, Purdy RH 1992 Neuroactive steroids. FASEB J 6:2311–2322

Parsons B, Pfaff DW 1985 Progesterone receptors in CNS correlated with reproductive behavior. In: Ganten D, Pfaff DW (eds) Current topics in neuroendocrinology, vol 5: Actions of progesterone on the brain. Springer-Verlag, Heidelberg, p 103–140

Pellow S, Chopin P, File SE, Briley M 1985 Validation of open : closed arm entries in an elevated plus-maze as a measure of anxiety in the rat. J Neurosci Methods 14:149–167

Pfaff DW 1973 Luteinizing hormone-releasing factor potentiates lordosis behavior in hypophysectomized ovariectomized female rats. Science 182:1148–1149

Pfaff DW 1980 Estrogens and brain functions: neural analysis of a hormone-controlled mammalian reproductive behavior. Springer-Verlag, New York

Pfaff DW, Zigmond RE 1971 Neonatal androgen effects on sexual and nonsexual behavior of adult rats tested under various hormone regimes. Neuroendocrinology 7:129–145

Pfaff DW, Lewis C, Diakow C, Keiner M 1973 Neurophysiological analysis of mating behaviour responses as hormone-sensitive reflexes. In: Steller E, Sprague JM (eds) Progress in physiological psychology. Academic Press, San Diego, CA

Pfaff DW, Schwartz-Giblin S, McCarthy MM, Kow L-M 1994 Cellular and molecular mechanisms of female reproductive behaviors. In: Knobil E, Neill J (eds) The physiology of reproduction (edn 2). Raven Press, New York, p 107–220

Pfaus JG, Gorzalka BB 1987 Selective activation of opioid receptors differentially affects lordosis behavior in female rats. Peptides 8:309–317

Pfaus JG, Pfaff DW 1992 Mu, delta, and kappa opioid receptor agonists selectively modulate sexual behaviors in the female rat: differential dependence on progesterone. Horm Behav 26:457–473

Pfaus JG, Kleopoulos SP, Mobbs CV, Gibbs RB, Pfaff DW 1993 Sexual stimulation activates c-*fos* within estrogen concentrating regions of the female rat forebrain. Brain Res 624:253–267

Priest CA, Pfaff DW 1994 Functional considerations of enkephalinergic projections from the hypothalamic ventromedial nucleus of the rat. Regul Pept 54:231–232

Priest CA, Eckersell CB, Micevych PE 1995 Temporal regulation by estrogen of preproenkephalin-A mRNA expression in the rat ventromedial nucleus of the hypothalamus. Mol Brain Res 28:251–262

Purdy RH, Moore PH, Rao PM et al 1990 Radioimmunoassay of 3α-hydroxy-5α-pregnane-20-one in rat and human plasma. Steroids 55:290–296

Purdy RH, Morrow AL, Moore PH, Paul SM 1991 Stress-induced elevations of γ-aminobutyric acid type A receptor-active steroids in the rat brain. Proc Natl Acad Sci USA 88:4553–4557

Rice J, Reich T, Andreasen NC et al 1984 Sex-related differences in depression: familial evidence. J Affect Dis 71:199–210

Romano GJ, Harlan RE, Shivers BD, Howells RD, Pfaff DW 1988 Estrogen increases proenkephalin messenger ribonucleic acid levels in the ventromedial hypothalamus of the rat. Mol Endocrinol 2:1320–1328

Romano GJ, Mobbs CV, Howells RD, Pfaff DW 1989a Estrogen regulation of proenkephalin gene expression in the ventromedial hypothalamus of the rat: temporal qualities and synergism with progesterone. Mol Brain Res 5:51–58

Romano GJ, Krust A, Pfaff DW 1989b Expression and estrogen regulation of progesterone receptor mRNA in neurons of the mediobasal hypothalamus: an *in situ* hybridization study. Mol Endocrinol 3:1295–1300

Romano GJ, Mobbs CV, Lauber A, Howells RD, Pfaff DW 1990 Differential regulation of proenkephalin gene expression by estrogen in the ventromedial hypothalamus of male and female rats: implications for the molecular basis of a sexually differentiated behavior. Brain Res 536:63–68

Romeo E, Auta J, Kozikowski AP et al 1992 2-Aryl-3-indolacetamides (FGIN-1): a new class of potent and specific ligands for the mitochondrial DBI receptor (MDR). J Pharmacol Exp Ther 262:971–978

Romeo E, Cavallaro S, Korneyev A et al 1993 Stimulation of brain steroidogenesis by 2-aryl-indole-3-acetamide derivatives acting at the mitochondrial diazepam-binding inhibitor receptor complex. J Pharmacol Exp Ther 267:462–471

Schwartz-Giblin S, Pfaff DW 1987 Actions of steroids on neurons: role in personality, mood, stress, and disease. Integr Psychiatry 5:258–273

Schwartz-Giblin S, Canonaco M, McEwen BS, Pfaff DW 1988 Effects of *in vivo* estradiol and progesterone on tritiated flunitrazepam binding in rat spinal cord. Neuroscience 25:249–257

Schwartz-Giblin S, Korotzer A, Pfaff DW 1989 Steroid hormone effects on picrotoxin-induced seizures in female and male rats. Brain Res 476:240–247

Scott REM, Wu-Peng S, Yen PM, Chin WW, Pfaff DW 1994 Estrogen receptor and thyroid receptor in ventromedial hypothalamus bind to the progesterone receptor gene sequence. Soc Neurosci Abstr 20:57

Singh M, Meyer EM, Millard WJ, Simpkins JW 1994 Ovarian steroid deprivation results in a reversible learning impairment and compromised cholinergic function in female Sprague–Dawley rats. Brain Res 644:305–312

Vathy IU, Etgen AM, Barfield RJ 1989 Actions of RU 38486 on progesterone facilitation and sequential inhibition of rat estrous behavior: correlation with neural progestin receptor levels. Horm Behav 23:43–56

Wu-Peng XS, Pfaff DW 1994 Estrogen regulation of the rat progesterone receptor (PR) gene. Soc Neurosci Abstr 20:56

Yin J, Kaplitt MG, Kwong AD, Pfaff DW 1994 *In vivo* promoter analysis for detecting an estrogen effect on preproenkephalin (PPE) transcription in hypothalamic neurons. Endocr Soc Abstr 76:318

Zhu YS, Pfaff DW 1994 Protein–DNA binding assay for analysis of steroid-sensitive neurons in the mammalian brain. In: De Kloet ER, Sutano W (eds) Neurobiology of steroids: methods in neurosciences. Academic Press, San Diego, CA, in press

Zhu YS, Frieden M, Pfaff DW 1994 DNA binding of hypothalamic and pituitary nuclear proteins on ERE and proenkephalin (Penk) promoter. Soc Neurosci Abstr 20:53

DISCUSSION

Thijssen: Other studies you have been doing on prolactin in the brain (e.g. reviewed in Dutt et al 1994) have all been conducted in rats. Are you comfortable about extrapolating from prolactin in rats to humans? There are big differences in secretion of prolactin in rats with ageing compared with humans (Meites 1982). In humans, the level of prolactin in peripheral blood decreases with age, or at least stays similar: it does not increase like it does in rats.

Pfaff: Firstly, we have to be careful to remember that rats are nocturnal and humans are diurnal: there could be some differences because of this. Secondly, the circulating prolactin comes from the transcription of pituitary prolactin. You could ask, straight away, how can prolactin be expressed in the brain at all if there's a relative absence of Pit-1? So I would like to hold open the possibility, first, that the transcriptional control of prolactin in the brain is different from the pituitary and, secondly, that in so far as the oestrogen really drives it, the decline of gonadal steroids during ageing could be biologically important, at least at a speculative level.

Oelkers: Is dopamine involved in the regulation of the expression of prolactin in the anterior hypothalamus, as it is in the pituitary?

Pfaff: Nobody knows about it. The expression in hypothalamus is so low that regulation studies are hard to do. It was difficult to prove that prolactin is expressed in hypothalamus. For example, in our lab we used 13 control conditions in order to see what would affect the immunocytochemistry for prolactin in the brain (Harlan et al 1989). Out of the 13, 12 were perfect— that is, those that should not disturb the immunocytochemistry didn't and those that should, did. But the 13th control was to use the N-terminus of pro-opiomelanocortin, which for some reason blocked prolactin immuno-cytochemistry in the brain. We still don't know why this happened.

Oelkers: When you stimulated the animals with oestrogen, prolactin expression was rather strong (Shivers et al 1989), so one could look at whether this is inhibited by dopaminergic agonists in the brain.

Pfaff: I hasten to say that all Dr Shivers did was to show that oestrogen receptors are present in the nuclei of prolactin cells. Others, using quantitative polymerase chain reaction analysis, demonstrated the oestrogen stimulation. Those studies need to be confirmed before they are used as the basis for further neurochemical work.

Baulieu: Is there a relationship between the prolactin-making cells and the distribution of prolactin-receptor-bearing cells?

Pfaff: No, because the prolactin receptor is much more broadly spread than the cell bodies of the prolactin immunoreactive neurons. The receptors are really dense throughout the hypothalamus.

Baulieu: So how do you envisage the distribution between cells?

Pfaff: Paracrine mechanisms are one possibility.

We now have a lot of examples of neuropeptides which don't operate strictly on a point-to-point, conventionally synaptic basis, but which may diffuse to reach a cluster of cells.

Baulieu: Is there any evidence for this for prolactin?

Pfaff: There's no direct evidence for it. There are simply some precedents for it in other systems. For example, some GnRH neurons don't actually have normal synaptic morphology. The GnRH may 'leak' out in the manner of an adrenergic system, potentially affecting a mesencephalic periventricular system. We would like to open the speculative possibility that prolactin in the brain could do a similar thing.

Baulieu: Is there any evidence for this, for example, from prolactin injections into the brain?

Pfaff: Behaviourally, yes, but in terms of neuronal growth, no.

If you put prolactin into the midbrain central grey, where there are prolactinergic synapses, you can turn on reproductive behaviour. If you put in the NIH antibody to prolactin, you can significantly reduce reproductive behaviour. So there is some functional consequence.

Furthermore, in birds there's a tremendous amount of evidence for prolactin affecting the brain to control parental behaviour.

James: Do patients with prolactinomas show behavioural changes?

Pfaff: Prolactinoma patients will have high peripheral levels, so this could have a behavioural effect which is entirely different, simply because of its distribution throughout the brain and the body. Secondly, if it gets large enough, a prolactinoma can actually press on the hypothalamus, so it could have neural and behavioural effects for other reasons.

Muramatsu: What is the nature of the prolactin receptor? What family does it belong to?

Pfaff: It's closely linked to the growth hormone receptor.

Muramatsu: Is it highly specific for prolactin, or does it accept other growth hormones?

Pfaff: As I recall, it is quite specific, but that's a nice thought because we could be talking about growth hormone as well.

Baulieu: You have excluded that, haven't you?

Pfaff: I haven't worked on prolactin binding at all.

Uvnäs-Moberg: It is important to remember that in humans the role of prolactin is complex. Prolactin levels go up during pregnancy and lactation. During these periods, prolactin levels correlate with personality traits such as being pleasing (the psychological term is social desirability) (Uvnäs-Moberg et al 1990) and prolactin levels are decreased by stress. In contrast, stress increases prolactin levels in men and women in stressful situations, in particular if the stress is related to loss of control or ability to cope (Jeffcoate et al 1986).

In primates, testosterone levels are linked to dominance, and prolactin levels to subordination (Rose et al 1971). It has been suggested that stress-linked prolactin secretion in humans is related to this basal psychophysiological endocrine pattern. There are some data that indicate that prolactin may indeed influence behaviour in humans. Prolactinoma patients, even if the tumours are small, may gain weight and become depressed and lose vitality. These symptoms disappear if the tumour is taken away. But this is of course pathology and not physiology.

Thijssen: You are generalizing—there are many different kinds of prolactinoma patients. The majority have just a slight elevation of prolactin, which means the only affect you will see is on fertility.

Pfaff: A further caution is that there might be indirect reactions behaviourally to the metabolic consequences of peripheral actions. As far as brain prolactin is concerned, it could be under quite different transcriptional control.

Uvnäs-Moberg: Absolutely, anything is possible.

Oelkers: While we are discussing the effects of peripheral prolactin on the brain, we should consider that the protein hormone has access to the brain itself only in parts where the blood–brain barrier is not functioning.

Pfaff: Yes, prolactin can come from the circulation through the chorioid plexus into the cerebrospinal fluid, and so have access to some parts of the periventricular system that way. But it is restricted; you are right.

Beato: I was also very impressed with your lab's use of a neurotropic herpes virus to study promoters *in situ* (Kaplitt et al 1991, 1993, 1994)—I think this is potentially a very interesting method. How efficiently does it infect your cells?

Pfaff: It's in the neighbourhood of 1%. That is, if you divide the number of cells expressing the reporter gene by the number of viral particles that you put into that brain region, it is in the neighbourhood of 1%—par for the course.

Beato: Do you know whether there is more than one viral genome in each cell?

Pfaff: We've never been able to work that out quantitatively. We believe this vector remains episomal, but we have not yet done a Southern blot to prove this.

Beato: Do you have any preference for a particular type of cell or do you have the impression that this is a general finding?

Pfaff: If you use a non-specific cytomegalovirus promoter, then it's very clear that neurons will both take up the viral particle and use its DNA to make a reporter gene. But we haven't ruled out the possibility that glial cells in the brain can also do it. That is, we have not yet combined β-galactosidase histochemistry with GFAP immunocytochemistry.

Horwitz: I have also been very impressed with your lab's results using viral vectors. From the amount of *lacZ* expression that you showed (Kaplitt et al 1994), it seems that the preproenkephalin promoter is constitutively expressed at very high levels. I was surprised that you were able to superimpose oestrogen regulation onto that high expression level.

Pfaff: Remember, we simply measured the number of cells expressing *lacZ*, not the optical density.

Horwitz: Were there some cells in which *lacZ* expression was zero until you added oestrogen, and some cells in which *lacZ* was constitutively expressed and not regulated by oestrogen?

Pfaff: At least as measured by β-galactosidase, yes, because the numbers of cells expressing *lacZ* went up by a factor of two and a half or three.

Horwitz: So one might conclude that the promoter is turned on in some cells, but absolutely regulated by oestradiol in other cells?

Pfaff: Yes, that is possible. We are now studying this with histochemical and neuroanatomical techniques. To have the cellular resolution gives many opportunities. On the other hand, the studies are very laborious.

Beato: This is the normal behaviour of all inducible genes, including the MMTV. They are either on or off, but there are no intermediate levels of activity. What changes when you induce gene expression is not the intensity of the expression of individual promoters, but the number of cells expressing a particular promoter, because the promoter can only be on or off.

Pfaff: A possible exception to that rule is with ribosomal RNA (which may be a special case), where we can see an oestrogen induction in the brain; obviously it's not totally absent in oestrogen-free neurons.

Beato: In your case, of course, you don't have just one promoter in the cells. When a gene is episomal and present in multiple copies, you can have a situation where a fraction of the promoters are turned on. But if you have one gene per cell, it's on or off.

Pfaff: That's a very attractive formulation.

Beato: These studies were done with exactly the same results in the *lacZ* operon of *Escherichia coli*. In animal cells with a single copy of MMTV/β-galactosidase, induction with different concentrations of hormone leads to the classical dose–response curve when the population of cells is studied as a whole. But if you look at the individual cells, there are only two stages. What changes is the proportion of cells that are recruited into the active state. What the activated receptors are doing, in this case, is to increase the probability that a promoter is activated by recruitment of transcription factors.

Baulieu: Is such a concept valid for oestrogen?

Beato: This is for glucocorticoid.

McEwen: I was musing about sex, stress and enkephalin. The paraventricular nucleus has enkephalin-expressing cells in which the mRNA is induced by certain stressors. Do you know whether or not these cells are oestrogen-sensitive?

Pfaff: Not yet. We're addressing this question using transgenic mice.

McEwen: Obviously, it would be very interesting to find out what effect the combination of the presence or absence of oestrogens, and the application of one of the stressors, would have upon the expression of those genes. This could help to unravel some of the sex differences in stress responsiveness.

Pfaff: Miguel Beato, what do you think about the enkephalin gene's intronic sequence? In the first intron there may be an alternative start site which is oestrogen sensitive.

Beato: All the indications of hormone-responsive elements in introns are inconclusive, particularly in the first intron of the growth hormone gene. Nobody has really shown that these elements are functional *in situ*. It has been shown that these elements, when placed in front of a tk promoter, can activate it. No one has done careful studies with mutations *in situ*. I think this is the way to do the experiment. Now that you have the virus, you should make a mutant of this intronic element and see what happens to the induction.

Rochefort: You showed that oestrogen induces preproenkephalin. This would have a beneficial effect in treating stress. Do you know whether tamoxifen, which is used to treat breast cancer, would behave as a partial oestrogen agonist on the preproenkephalin gene? Have you done this experiment?

Pfaff: We haven't done it yet.

Rochefort: That would be of relevance to the treatment of breast cancer patients.

Pfaff: Yes, absolutely.

References

Dutt A, Kaplitt MG, Kow L-M, Pfaff DW 1994 Prolactin, central nervous system and behaviour: a critical review. Neuroendocrinology 59:413–419

Harlan RE, Shivers BD, Fox SR, Kaplove KA, Schachter BS, Pfaff DW 1989 Distribution and partial characterization of immunoreactive prolactin in the rat brain. Neuroendocrinology 49:10–22

Jeffcoate WJ, Lincoln NB, Selby C, Herbert M 1986 Correlation between anxiety and serum prolactin in humans. J Psychosom Med 30:217–222

Kaplitt MG, Pfaus JG, Kleopoulos SP, Hanlon BA, Rabkin SD, Pfaff DW 1991 Expression of a functional foreign gene in adult mammalian brain following *in vivo* transfer via a herpes simplex virus type I defective viral vector. Mol Cell Neurosci 2:320–330

Kaplitt MG, Rabkin S, Pfaff DW 1993 Molecular alterations in nerve cells: direct manipulation and physiological mediation. Curr Top Neuroendocrinol 11:169–191

Kaplitt MG, Kwong AD, Kleopoulos SP, Mobbs CV, Rabkin SD, Pfaff DW 1994 Preproenkephalin promoter yields region-specific and long-term expression in adult brain following direct *in vivo* gene transfer via a defective herpes simplex viral vector. Proc Natl Acad Sci USA 91:8979–8983

Meites J 1982 Changes in neuroendocrine control of anterior pituitary function during aging. Neuroendocrinology 34:151–156

Rose RM, Holady JW, Berstein IS 1971 Plasma testosterone, dominance rank and aggressive behaviour in male rhesus monkeys. Nature 231:366–368

Shivers BD, Harlan RE, Pfaff DW 1989 A subset of neurons containing immunoreactive prolactin is a target for estrogen regulation of gene expression in rat hypothalamus. Neuroendocrinology 49:23–27

Uvnäs-Moberg K, Widström AM, Nissen E, Björvell H 1990 Personality traits in women 4 days post partum and their correlation with plasma levels of oxytocin and prolactin. J Psychosom Obstet Gynaecol 11:261–273

Neurosteroids: synthesis and functions in the central and peripheral nervous systems

Michael Schumacher* and Etienne-Emile Baulieu*†

*Lab. Hormones, INSERM U33, 80 rue de Général Leclerc, 94276 Bicêtre, France and †Collège de France

Abstract. Some steroids are synthesized within the central and peripheral nervous systems, mostly by glial cells. These are known as neurosteroids. In the brain, neurosteroids have been shown to act directly on membrane receptors for neurotransmitters. For example, progesterone inhibits the neuronal nicotinic acetylcholine receptor, whereas its 3α,5α-reduced metabolite 3α,5α-tetrahydroprogesterone (allopregnanolone) activates the type A γ-aminobutyric acid receptor complex. Besides these effects, neurosteroids also regulate important glial functions, such as the synthesis of myelin proteins. Thus, in cultures of glial cells prepared from neonatal rat brain, progesterone increases the number of oligodendrocytes expressing the myelin basic protein (MBP) and the 2′,3′-cyclic nucleotide-3′-phophodiesterase (CNPase). An important role for neurosteroids in myelin repair has been demonstrated in the rodent sciatic nerve, where progesterone and its direct precursor pregnenolone are synthesized by Schwann cells. After cryolesion of the male mouse sciatic nerve, blocking the local synthesis or action of progesterone impairs remyelination of the regenerating axons, whereas administration of progesterone to the lesion site promotes the formation of new myelin sheaths.

1995 Non-reproductive actions of sex steroids. Wiley, Chichester (Ciba Foundation Symposium 191) p 90–112

Sex steroids: effects on the nervous system

The nervous system is a target for sex steroids (McEwen et al 1995, Priest & Pfaff 1995, this volume). During sensitive periods in late fetal and early postnatal life, gonadal steroids influence the survival, the differentiation and the connectivity of specific neuronal populations in both the brain and spinal cord (De Vries et al 1984). At this stage of development, the nervous tissue is highly plastic and some of the hormone effects are permanent. In the adult, sex steroids still influence neuronal functions, mainly by regulating synaptic transmission. They do so either by increasing the transcription of specific genes after binding to intracellular receptors (McEwen 1991, Truss et al 1995, this volume), or by acting directly on the neuronal membrane, most likely by binding to membrane receptors for neurotransmitters (Schumacher 1990, Chadwick & Widdows 1990, Majewska 1992, Baulieu & Robel 1995, this volume). Because of the widespread distribution of hormone-sensitive neurotransmitter receptors, such as the type A γ-aminobutyric acid (GABA$_A$) receptor, sex steroids influence neuronal activity within large parts of the nervous system, where they exert a variety of effects that are not necessarily related to reproduction. However, this does not imply that all the effects sex steroids exert outside brain regions involved in reproduction, such as the hypothalamus, are mediated by membrane sites. In fact, with the advent of sensitive biochemical and immunocytochemical techniques, it has become obvious that intracellular receptors for sex steroids are more widely distributed throughout the nervous system than has been thought. Thus, oestrogen receptors have been detected very recently in nerve growth factor (NGF)-sensitive neurons of the basal forebrain and dorsal root ganglia (Toran-Allerand et al 1992, Sohrabji et al 1994). In addition, work in our laboratory has shown that receptors of sex steroids, such as oestradiol and progesterone, are not only present in neurons, but also in glial cells, where they mediate steroid effects on important glial functions (Jung-Testas et al 1991, 1994).

Research performed over the past few years has shown that sex steroids continue to exert neurotrophic effects in the adult nervous system by influencing changes in the morphology and connections of nerve cells. Indeed, despite the apparent appearance of stability, we now know that there is a continuous remodelling of neuronal connections within adult nervous tissues. For example, androgens increase the number of synaptic contacts in a group of spinal motor neurons (Matsumoto et al 1988), and oestrogens increase the density of dendritic spines of hypothalamic and hippocampal neurons (McEwen et al 1995, this volume). These observations have very recently stimulated research to exploit the trophic effects of sex steroids as therapeutic agents in neuronal injury. In these studies, testosterone has been identified as a trophic factor for facial and sciatic motor neurons. Indeed, this 'male sex steroid' accelerates motor neuron axonal regeneration following crush injury of the hamster hypoglossal and the rat sciatic nerves (Jones 1993). The 'female sex steroid' progesterone has been shown to prevent death of facial motor neurons after nerve transection (Yu 1989). These observations suggest that sex steroids may be therapeutically useful in activating or accelerating the reparative response of neurons to injury.

Neurosteroids: concept and biosynthesis

In the light of the various effects of sex steroids on both the developing and the adult nervous systems, it was a significant finding that some steroids, called

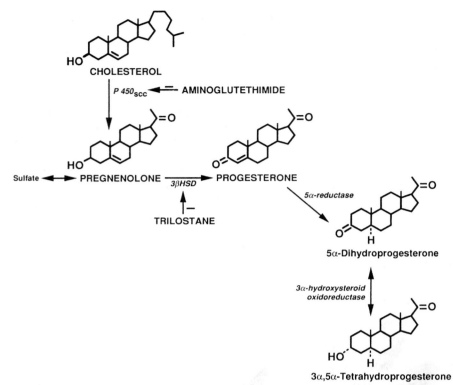

FIG. 1.   Biosynthesis of neurosteroids in glial cells. Enzymes are shown in italics.
P450$_{scc}$, cytochrome P450$_{scc}$; *3β-HSD*, Δ5-3β-hydroxysteroid dehydrogenase isomerase.
Aminoglutethimide and trilostane are inhibitors of P450$_{scc}$, and 3β-HSD, respectively.

'neurosteroids', are synthesized within the brain and peripheral nerves by glial
cells (Baulieu 1991, Robel & Baulieu 1994). The term neurosteroid (Baulieu
1981) does not designate a particular class of steroids, but only refers to their
site of synthesis—the nervous system. Progesterone, for example, which is a
hormone produced and secreted by the ovaries and adrenal glands, is
considered to be a neurosteroid if it is synthesized within the brain or
peripheral nerves.

Steroidogenesis begins with the conversion of cholesterol to pregnenolone by
the side-chain cleavage cytochrome P450$_{scc}$ (Fig. 1). This enzymic step, which
takes place in the inner mitochondrial membrane, is characteristic of the
steroidogenic cells of endocrine glands, including the testes, ovaries and
adrenal glands. The first indication that pregnenolone may also be synthesized
within the nervous system came from the unexpected observation that its levels
are much higher in the brain than in blood (Corpéchot et al 1983). This finding

could not be explained by the cerebral retention of circulating hormone, as pregnenolone persisted in the brain days after castration and adrenalectomy, and this in spite of a very rapid cerebral clearance (Robel & Baulieu 1985, Baulieu 1991). It was then shown by immunocytochemistry that the cytochrome $P450_{scc}$ is expressed in the white matter throughout the brain (Le Goascogne et al 1987). The biosynthesis of pregnenolone was demonstrated by incubating newborn rat glial cells in the presence of [^{3}H]mevalonate, a precursor of cholesterol which easily enters cells and mitochondria (Jung-Testas et al 1989). More recently, the $P450_{scc}$ mRNA has been detected by reverse transcriptase (RT)-PCR both in the rat brain and in cultured glial cells (Mellon & Deschepper 1993). In agreement with the predominant localization of $P450_{scc}$ immunoreactivity in white matter, the enzyme is present in cultured oligodendrocytes, which are the myelinating glial cells of the CNS (Jung-Testas et al 1989). However, $P450_{scc}$ immunoreactivity and mRNA have also been detected in type I astrocytes (Mellon 1994). So far, the synthesis of pregnenolone has not been demonstrated in neurons.

Pregnenolone can be converted by the $\Delta5$-3β-hydroxysteroid dehydrogenase isomerase (3β-HSD) to progesterone, which in turn is metabolized successively to 5α-dihydroprogesterone (5α-pregnane-3,20-dione) by the 5α-reductase and to $3\alpha,5\alpha$-tetrahydroprogesterone (3α-hydroxy-5α-pregnan-20-one = allopregnanolone) by the 3α-hydroxysteroid oxidoreductase (Fig. 1). These three enzymic reactions are present in mixed cultures of oligodendrocytes and astrocytes (Jung-Testas et al 1989) and in type I astrocytes prepared from embryonic or neonatal rat brain (Kabbadj et al 1993, Akwa et al 1993a). Whether or not oligodendrocytes have the capability to synthesize progesterone and its reduced metabolites remains to be established. By using testosterone as a substrate, Celotti et al (1992) showed that myelin-forming oligodendrocytes possess a high 5α-reductase activity and that this enzyme may be incorporated into central and peripheral myelin. *In vivo*, progesterone may be the physiological substrate for this enzyme.

From pregnenolone and progesterone, 7α- and 20α-reduced metabolites are also formed in the brain (for review, see Baulieu 1991, Robel & Baulieu 1994). What seems to be missing in the nervous system is the cytochrome $P450_{17\alpha}$, a 17α-hydroxylase which converts pregnenolone to dehydroepiandrosterone (DHEA) or progesterone to androstenedione, which are the obligatory precursors of androgens and oestrogens (Baulieu 1991, Mellon 1994). Thus, *a priori*, testosterone and oestradiol cannot be synthesized *de novo* from cholesterol or from pregnenolone within the brain, and these hormones, or their immediate precursors, have to be provided by the gonads. However, it is a puzzling observation that DHEA remains present in the brain after removal of the steroidogenic endocrine glands and its source remains unknown (Baulieu 1991). Measurements of corticosterone levels strongly suggest that glucocorticosteroids too are not synthesized within the nervous system (Baulieu 1991, Akwa et al 1993b, see also Fig. 4), although the activity of a 21-hydroxylase and the mRNA of the $P450_{11\beta}$, two enzymes involved in corticosterone biosynthesis, have been detected in the brain (Mellon & Deschepper 1993).

The physiological stimuli which regulate neurosteroid formation are still unknown. Whether second messengers or trophic factors, which stimulate steroid biosynthesis in classic steroidogenic organs (Oonk et al 1989, Magoffin et al 1990), also increase neurosteroid formation by glial cells remains to be explored. A recent study suggests a role for autocrine factors in the regulation of neurosteroid synthesis. Thus, astrocytes preferentially convert [^{3}H]pregnenolone to [^{3}H]progesterone at low cell density and to 7α-hydroxylated metabolites in dense cultures (Akwa et al 1993a). In addition, the synthesis of neurosteroids may be influenced by the endocrine milieu. So far, however, only the peripheral-type benzodiazepine receptor (PBR) has been shown to play a significant role in the neurosteroidogenic activation of glial cells. This receptor is found primarily on the outer mitochondrial membrane. By increasing the intramitochondrial cholesterol transport, PBR ligands stimulate pregnenolone formation in the C6-2B glioma cell line (Costa et al 1991, Papadopoulos 1993).

Neurosteroids: biological significance

As reported in the previous section, it is likely that pregnenolone, progesterone and their reduced metabolites are the only steroids which can be formed *de novo* from cholesterol within the brain. An important question is the biological significance of these neurosteroids. Probably the best guess is that they may be essential for important nervous functions and their local synthesis may protect the brain from drops in circulating steroid levels that occur, for example, during the oestrous cycle or during ageing. Also, blood levels of pregnenolone and progesterone are very low in males and their nervous system may rely on a local production of these steroids. In fact, as we shall see, progesterone synthesized by glial cells plays an important role during nerve regeneration in male rodents.

Over the past few years, particular attention has been paid to the actions of pregnenolone, progesterone and their reduced metabolites on membrane receptors for neurotransmitters. Progesterone inhibits the neuronal nicotinic acetylcholine receptor (Valera et al 1992) and activates hypothalamic oxytocin receptors (Schumacher et al 1990), whereas its 5α,3α-reduced metabolite 3α,5α-tetrahydroprogesterone activates the chloride channel of the $GABA_A$ receptor complex (Majewska et al 1986). Other neurotransmitter receptors have been identified as targets for neurosteroids and these results have been extensively reviewed (Chadwick & Widdows 1990, Robel & Baulieu 1994).

However, although it has been demonstrated that neurosteroids such as pregnenolone or progesterone influence the activity of neurotransmitter receptors, it has not been shown that their production by oligodendrocytes or astrocytes plays an important role in regulating receptor functions through paracrine actions. The same is true for the effects of neurosteroids on neuronal functions and behaviour. In rats, the local infusion of sulphated pregnenolone, an inhibitor of the $GABA_A$ receptor, into the nucleus basalis magnocellularis, enhances memory performance. Infusion of $3\alpha,5\alpha$-tetrahydroprogesterone, an activator of the $GABA_A$ receptor, has the opposite effect (Mayo et al 1993). Pregnenolone and steroids metabolically derived from it also have memory-enhancing effects in mice (Flood et al 1992). However, as these steroids are also produced by the gonads and adrenal glands, it is not clear whether their synthesis by glial cells is significant. This could be shown by selectively blocking or increasing the synthesis of neurosteroids within specific brain regions.

So far, little attention has been paid to the possible trophic effects of neurosteroids. As described above, steroids exert a variety of neurotrophic effects, particularly during early development, but also in the adult. Especially after the lesion of nervous tissues, locally produced neurosteroids may stimulate by paracrine or autocrine actions the reparative responses of neurons and glial cells. Such an important role for neurosteroids is suggested by cell culture experiments. When added to the culture medium, neurosteroids enhance neuronal survival (Bologa et al 1987) and increase the synthesis of myelin-specific proteins by oligodendrocytes, namely, MBP and CNPase (Jung-Testas et al 1994). Recently, we have begun to explore the role of neurosteroids in peripheral nerve repair (see below). Our results demonstrate for the first time that the synthesis of neurosteroids by glial cells plays an important role in the regeneration of the nervous system (Koenig et al 1995).

The rodent sciatic nerve: a system for studying the trophic actions of neurosteroids

One advantage in studying neurosteroid functions in peripheral nerves is their relatively simple structure. Sensory and motor nerve fibres are associated with a single glial cell type, the Schwann cells, which myelinate the large axons. In addition, it is easy to prepare pure cultures of Schwann cells from neonatal rat sciatic nerves or from embryonic rat dorsal root ganglia (DRG) (Brockes et al 1979, Kleitman et al 1991). From the latter, it is also possible to establish cultures of sensory neurons (Kleitman et al 1991). However, the most important feature of peripheral nerves is their remarkable regenerative capacity. Following injury, axons and their myelin sheaths distal to the lesion degenerate by a process known as Wallerian degeneration (Fawcett & Keynes 1990), leaving behind the dividing Schwann cells, which produce a

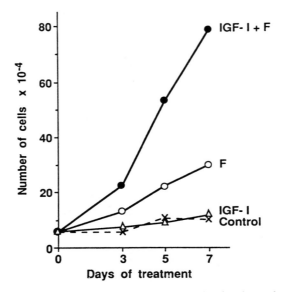

FIG. 2. Insulin-like growth factor 1 (IGF-1; 20 ng/ml) stimulates the proliferation of Schwann cells in the presence of forskolin (F, $5 \mu M$). Schwann cells were prepared from neonatal rat sciatic nerves and cultured in Dulbecco's modified Eagle's medium (DMEM) containing 10% heat-inactivated fetal calf serum. Media were changed and growth factors were added every two days. At the days indicated, cells were counted using a haemocytometer. The mean values of triplicate cultures are indicated, not differing one from another by $> 10\%$. One-way ANOVA showed a significant effect of treatment on days 3, 5 and 7 ($P \leqslant 0.001$) (control versus F: $P \leqslant 0.01$; IGF-1 + F versus F: $P \leqslant 0.01$ on all days by Tukey's tests). Modified from Jung-Testas et al 1994.

large number of neurotrophic peptides. Schwann cells play a crucial role during the regeneration of nerve fibres, which begins within a few hours of lesion of the nerve by axotomy, local crushing or freezing. In addition, their multiplication is critical for allowing axons to regenerate across gaps. Later, Schwann cells remyelinate the regenerating axons and, under favourable circumstances, the appropriate neuromuscular connections may be restored (Fawcett & Keynes 1990, Raivich & Kreutzberg 1993). There are thus several steps at the level of which the actions of 'sex steroids', and in particular of neurosteroids, could promote the repair of peripheral nerves: Schwann cell proliferation, the production of neurotrophic factors, axonal growth and remyelination.

Cultured Schwann cells, prepared from neonatal rat sciatic nerves, proliferate only in the presence of elevated levels of cAMP and specific peptide growth factors (Eccleston 1992). We have identified IGF-1 as a potent mitogen for Schwann cells in the presence of serum and dibutyryl cAMP or

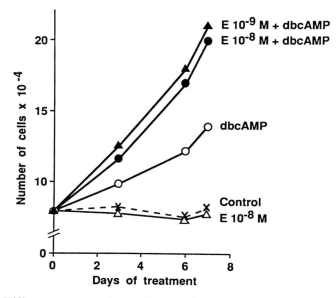

FIG. 3. Different concentrations of oestradiol (E) stimulate the proliferation of Schwann cells in culture in the presence of dibutyryl cyclic AMP (dbcAMP; 1 μM). Schwann cells were prepared from neonatal rat sciatic nerves and cultured in DMEM without phenol red and containing 10% heat-inactivated, charcoal-extracted calf serum. Media were changed and growth factors were added every two days. At the days indicated, cells were counted using a haemocytometer. The mean values of triplicate cultures are indicated, not differing one from another by >10%. One-way ANOVA showed a significant effect of treatment on days 3, 6 and 7 ($P \leqslant 0.001$) (control versus dbcAMP: $P \leqslant 0.01$; oestradiol [all concentrations] + dbcAMP versus dbcAMP: $P \leqslant 0.001$ on all days by Tukey's tests). Modified from Jung-Testas et al 1994.

forskolin, a reversible activator of adenylate cyclase (Fig. 2). Indeed, Schwann cells express the receptor for IGF-1, whose number increases if the cells are cultured for several days in the presence of forskolin (Schumacher et al 1993). By binding to IGF-1 receptors, insulin also stimulates Schwann cell proliferation at micromolar concentrations. Thus, in the presence of forskolin and a high concentration of insulin, it is possible to expand rat Schwann cell cultures and to rapidly obtain pure cells in sufficient number to study the metabolism and actions of steroids.

Schwann cells are a target for oestradiol

Cultures of rat Schwann cells, expanded in the presence of insulin and forskolin as described above, have been used to study the effects of oestradiol, a sex steroid produced by the ovaries. Several lines of evidence indicated that Schwann cell tumours may be sensitive to oestrogen: neurofibromas contain binding sites for oestradiol, they usually develop at puberty, and the anti-oestrogen tamoxifen has been shown to reduce their incidence in an animal model (Jay et al 1986).

Our results show that oestradiol, in the presence of dibutyryl cAMP, stimulates the proliferation of rat Schwann cells in culture (Jung-Testas et al 1993) (Fig. 3). Its mitogenic effect is already maximal at low nanomolar concentrations and can be blocked by the pure anti-oestrogen ICI 164 384. In agreement with this observation, Schwann cells contain oestrogen receptors and the binding of [^{3}H]oestradiol is increased by elevating intracellular levels of cAMP (Jung-Testas et al 1993). However, it has to be pointed out that in contrast to IGF-1, which potentiates Schwann cell proliferation even in the presence of large doses of forskolin, the mitogenic effect of oestradiol can only be observed in the presence of a moderate increase in cAMP. These findings suggest that clinical conditions associated with oestrogen deficiency could contribute to a reduced capacity for nerve regeneration. On the other hand, increased levels of oestrogen may favour the malignant proliferation of Schwann cells.

Schwann cells synthesize neurosteroids

Schwann cells are not only a target for circulating steroids such as oestradiol: similarly to the glial cells of the CNS, they can synthesize neurosteroids, including pregnenolone, progesterone and their reduced metabolites. The synthesis of pregnenolone within peripheral nerves was first suggested by the high levels of this steroid found in the human sciatic nerve (Morfin et al 1992). In sciatic nerves of adult male rats, the concentration of pregnenolone is about ten times higher than in plasma, and is not reduced five days after castration and adrenalectomy (Akwa et al 1993b). In contrast to pregnenolone, the concentration of corticosterone, a steroid of adrenal origin, is much higher in plasma than in nerves, and its levels drop almost to zero after removal of the steroidogenic glands. As steroids are rapidly cleared from nervous tissues (Raisinghani et al 1968, Robel & Baulieu 1985), these observations strongly suggest a local synthesis of pregnenolone that is independent of glandular sources. In fact, Schwann cells prepared from neonatal rat sciatic nerves convert [^{3}H]-25-OH cholesterol to pregnenolone (Akwa et al 1993b), but only when cultured in the presence of micromolar concentrations of forskolin and insulin, which are both mitogens for these cells. Thus, pregnenolone may play a role during Schwann cell proliferation. Interestingly, cAMP and IGF-1, whose effects can be mimicked by insulin, also both stimulate the expression of the cytochrome P450$_{scc}$ in classical steroidogenic cells (Oonk et al 1989, Magoffin

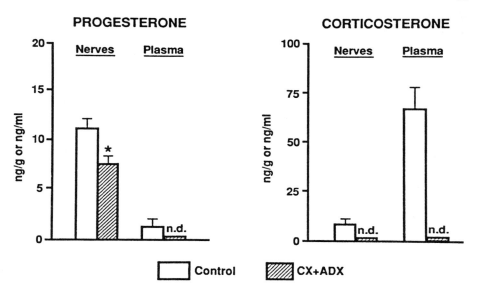

FIG. 4. Levels of progesterone are about 10-fold higher in sciatic nerves than in plasma of sham-operated adult male mice (control) and remain elevated five days after castration and adrenalectomy (CX + ADX) (n.d. = not detectable) (mean ± SEM, $n = 4$, *$P \leqslant 0.05$ when compared with controls by Student's t-test). In contrast, levels of corticosterone are much higher in plasma than in sciatic nerves and its levels became undetectable after CX + ADX. Modified from Koenig et al 1995.

et al 1990). However, even in the presence of insulin and forskolin, the rate of pregnenolone formation by Schwann cells is very low *in vitro*, not exceeding 25 fmol/μg DNA per 24 h. This suggests that other factors may be required for optimal P450$_{scc}$ activity in glial cells.

In sciatic nerves of adult male rats and mice, the concentration of progesterone is also ten times higher (about 10 ng/g) than in plasma, and remains high after removal of the steroidogenic endocrine glands (Koenig et al 1995) (Fig. 4). However, pure cultures of Schwann cells prepared from neonatal rat sciatic nerves do not convert [3H]pregnenolone to progesterone, even when cultured at different cell densities in the absence or presence of the mitogens insulin and forskolin (Akwa et al 1993b, Y. Akwa, M. Schumacher & E. E. Baulieu, unpublished work 1994). In contrast, Schwann cells isolated from embryonic (E18) rat DRG explants, which have been co-cultured for four weeks with sensory neurons, produce significant amounts [3H]progesterone and its reduced metabolites 5α-dihydroprogesterone and 3α,5α-tetrahydro-progesterone when incubated with [3H]pregnenolone (Koenig et al 1995) (Fig. 5). We also confirmed by immunocytochemistry that Schwann cells, which have been cultured for several weeks in the presence of DRG neurons, contain the 3β-HSD enzyme which converts pregnenolone to progesterone. Taken together, these results strongly suggest that neurons induce the biosynthetic pathway of progesterone in Schwann cells.

Progesterone may regulate Schwann cell functions through autocrine actions. Indeed, these cells not only synthesize progesterone, they also express the intracellular receptor for this neurosteroid, as shown by binding of the selective ligand [³H]ORG 2058 and by RT-PCR using primers complementary to the ligand-binding domain of the rat progesterone receptor (I. Jung-Testas, R. Fiddes, K. Shazand, M. Schumacher & E. E. Baulieu, unpublished work 1994).

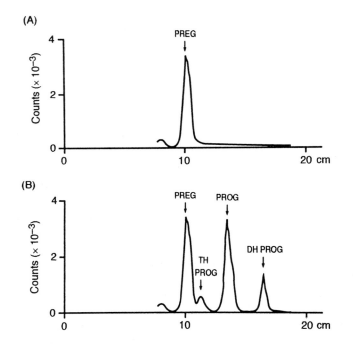

FIG. 5.   Metabolism of pregnenolone by rat Schwann cells in culture. The thin layer chromatograms show that (A) [³H]pregnenolone (PREG; 100 nM) is not metabolized by pure cultures of rat Schwann cells prepared from neonatal rat sciatic nerves, (B) but is converted by Schwann cells isolated from DRG explants and grown in the presence of sensory neurons for four weeks to [³H]progesterone (PROG; 14.3 ± 0.6 pmol/μg DNA per 24 h), [³H]5α-dihydroprogesterone (DH PROG; 3.9 ± 0.2 pmol/μg DNA per 24 h) and [³H]3α,5α-tetrahydroprogesterone (TH PROG; 2.6 ± 0.1 pmol/μg DNA per 24 h, mean ± SEM).

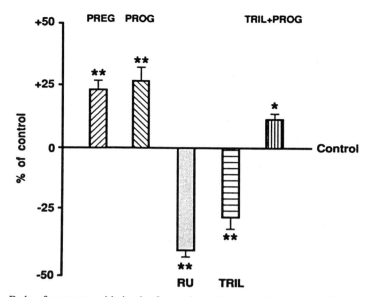

FIG. 6.   Role of neurosteroids in the formation of new myelin sheaths. The number of myelin lamellae formed 15 days after cryolesion of the right male mouse sciatic nerve was measured by electron microscopy. The number of lamellae was increased by pregnenolone (PREG) and progesterone (PROG) and decreased by the antiprogestin RU486 (RU) and by trilostane (TRIL), an inhibitor of the conversion of pregnenolone to progesterone. The inhibitory effect of trilostane was reversed by the concomitant administration of progesterone. Steroids and their inhibitors (100 μg in 50 μl sesame oil) were repeatedly applied to the lesioned site immediately after surgery (day 0) and on days 5, 10 and 14. Results are expressed as percentage of control (mean $\pm$ SEM, $n = 5$, **$P \leqslant 0.01$ and *$P \leqslant 0.05$ when compared to the corresponding control by Dunnett's multiple comparison tests after ANOVA). Modified from Koenig et al 1995.

Progesterone synthesized by Schwann cells promotes myelination

The production of neurosteroids is thus a feature of myelinating glial cells: oligodendrocytes in the CNS and Schwann cells in the peripheral nervous system (PNS). Therefore, we hypothesized that neurosteroids, and in particular progesterone, may play an important role during the process of myelination. To test this hypothesis, we examined the relationship between neurosteroids and myelin formation in the regenerating sciatic nerve of male mice after cryolesion (Koenig et al 1995). As with other types of lesion, axons and their myelin sheaths degenerate quickly after local freezing within the distal segment of the nerve. However, the advantage of this type of lesion is that the endoneurial tubes and the Schwann cell basal lamina remain intact and guide the regenerating nerve fibres directly back to their targets. In this system,

Schwann cells begin to remyelinate the regenerating axons after one week and the new myelin sheaths reach approximately one-third of their final size after two weeks.

First, we showed that levels of pregnenolone and progesterone remain elevated within the regenerating mouse nerve one or two weeks after cryolesion—that is, during the period of active remyelination (pregnenolone ~ 6 ng/g and progesterone ~ 10 ng/g tissue weight). We then demonstrated that the high levels of endogenous progesterone, most likely synthesized by Schwann cells, are necessary for an efficient remyelination of the regenerating axons. In fact, local application of 100 μg trilostane, an inhibitor of the conversion of pregnenolone to progesterone, or of RU486, a potent competitive antagonist of progesterone, dramatically decreased the thickness of the regenerating myelin sheaths when observed two weeks after lesion on electron microscopic cross-sections (Fig. 6). Both inhibitors were injected at the site of lesion four times during the two-week interval. The inhibitory effect of trilostane was not a toxic one since it could be reversed by the simultaneous administration of progesterone. On the other hand, repeated injections of a high dose of progesterone (100 μg), or its direct precursor pregnenolone, significantly enhanced the number of lamellae in each myelin sheath.

It is likely that progesterone directly stimulates myelin formation by acting on Schwann cells rather than by stimulating axonal growth. This was shown by using co-cultures of DRG neurons and Schwann cells. After four weeks in culture, elongation of the DRG neurites is maximal and Schwann cells myelinate the sensory axons in the presence of serum and ascorbic acid (Eldridge et al 1987). At this stage of culture, myelination, but not neurite extension, was dramatically increased when a physiological concentration of progesterone (20 nM) was daily added to the culture medium for two weeks (Koenig et al 1995).

Conclusions and perspectives

The designation of androgens, oestrogens and progestagens as 'sex steroids' or 'gonadal steroids' is too restrictive. First, steroids belonging to these three classes exert multiple effects on the nervous system that are not related to reproduction, including the potentiation of neuronal survival, axonal growth and myelin formation. Second, some of these steroids are not only produced by the testes or ovaries, but also are synthesized *de novo* from cholesterol in the brain and peripheral nerves by glial cells. These 'neurosteroids' influence neurotransmission by acting on neurotransmitter receptors and they activate important glial functions, such as myelination. Thus, progesterone synthesized by Schwann cells increases the formation of new myelin sheaths after lesion of the rodent sciatic nerve. As Schwann cells also express the intracellular receptor for progesterone, it is likely that this steroid activates the process of myelination through autocrine actions, probably by stimulating the synthesis of specific myelin proteins or lipids. From this it follows that neurosteroids, like many peptide growth factors, belong to the important group of autocrine/ paracrine factors, which regulate vital functions within the nervous system. This is a significant finding, because it is hoped that it will be possible to use these molecules to treat diseases and injuries of the nervous system, in particular those of peripheral nerves. In fact, in spite of their great capacity to regenerate, there are no efficient treatments for accelerating the process of nerve repair after injury. This is an urgent problem, since a delay in re-innervation often results in atrophy of the denervated muscles.

It is likely that neurosteroids exert their trophic effects by acting in concert, or even in synergy, with peptide growth factors such as IGF-1 or NGF. The study of these interactions will become an important field of neuroendocrine research. Another critical question is the regulation of neurosteroid biosynthesis by glial cells. There are two possible ways to use neurosteroids as therapeutic agents: either directly, by the administration of steroid compounds, or indirectly, by the use of agents which stimulate their synthesis. Finally, an important problem that remains to be explored is the possibility of interactions between circulating steroids, which are secreted into the bloodstream by the gonads, and neurosteroids, which are synthesized within the nervous system.

The observation that neurosteroids modulate the activity of neurons and increase the myelination of axons may have important clinical implications. High levels of steroids are indeed found in the brain and sciatic nerve of humans. Further investigations should assess the pathophysiological significance of the synthesis of steroids within the peripheral and central nervous systems.

Acknowledgements

This work was partially supported by the Association Française contre les Myopathies (AFM).

References

Akwa Y, Sananès N, Gouezou M, Robel P, Baulieu EE, Le Goascogne C 1993a Astrocytes and neurosteroids: metabolism of pregnenolone and dehydro-epiandrosterone. Regulation by cell density. J Cell Biol 121:135–143

Akwa Y, Schumacher M, Jung-Testas I, Baulieu EE 1993b Neurosteroids in rat sciatic nerves and Schwann cells. C R Acad Sci Ser III Sci Vie 316:410–414

Baulieu EE 1981 Steroid hormones in the brain: several mechanisms? In: Fuxe K, Gustafsson JA, Wetterberg L (eds) Steroid hormone regulation of the brain. Pergamon, Oxford, p 3–14

Baulieu EE 1991 Neurosteroids: a new function in the brain. Biol Cell 71:3–10

Baulieu EE, Robel P 1995 Non-genomic mechanisms of action of steroid hormones. In: Non-reproductive actions of sex steroids. Wiley, Chichester (Ciba Found Symp 191) p 24–42

Bologa L, Sharma J, Roberts E 1987 Dehydroepiandrosterone and its sulfated derivative reduce neuronal death and enhance astrocytic differentiation in brain cell cultures. J Neurosci Res 17:225–234

Brockes JP, Fields KL, Raff MC 1979 Studies on cultured rat Schwann cells. I. Establishment of purified populations from cultures of peripheral nerve. Brain Res 165:105–118

Celotti F, Melcangi RC, Martini L 1992 The 5α-reductase in the brain: molecular aspects and relation to brain function. Front Neuroendocrinol 13:163–215

Chadwick D, Widdows K (eds) 1990 Steroids and neuronal activity. Wiley, Chichester (Ciba Found Symp 153)

Corpéchot C, Synguelakis M, Talha S et al 1983 Pregnenolone and its sulfate ester in the rat brain. Brain Res 270:119–125

Costa E, Romeo E, Auta J, Papadopoulos V, Kozikowski A, Guidotti A 1991 Is there a pharmacology of brain steroidogenesis? In: Costa E, Paul SM (eds) Neurosteroids and brain function. Thieme Medical, New York, p 171–176

De Vries GJ, De Bruin JPC, Uylings HBM, Corner MA (eds) 1984 Sex differences in the brain (Prog Brain Res, vol 61). Elsevier, Amsterdam

Eccleston PA 1992 Regulation of Schwann cell proliferation: mechanisms involved in peripheral nerve development. Exp Cell Res 199:1–9

Eldridge CF, Bartlett M, Bunge RP, Wood PM 1987 Differentiation of axon-related Schwann cells in vitro. I. Ascorbic acid regulates basal lamina assembly and myelin formation. J Cell Biol 105:1023–1034

Fawcett JW, Keynes RJ 1990 Peripheral nerve regeneration. Annu Rev Neurosci 13: 43–60

Flood JF, Morley JE, Roberts E 1992 Memory-enhancing effects in male mice of pregnenolone and steroids metabolically derived from it. Proc Natl Acad Sci USA 89:1567–1571

Jay JR, MacLaughlin DT, Badger TM, Miller DC, Martuza RL 1986 Hormonal modulation of Schwann cell tumors. Ann N Y Acad Sci 486:371-382

Jones KJ 1993 Gonadal steroids and neuronal regeneration. A therapeutic role. Adv Neurol 59:227-240

Jung-Testas I, Hu ZY, Baulieu EE, Robel P 1989 Neurosteroids: biosynthesis of pregnenolone and progesterone in primary cultures of rat glial cells. Endocrinology 125:2083–2091

Jung-Testas I, Renoir JM, Gasc JM, Baulieu EE 1991 Estrogen-inducible progesterone receptor in primary cultures of rat glial cells. Exp Cell Res 193:12–19

Jung-Testas I, Schumacher M, Bugnard H, Baulieu EE 1993 Stimulation of rat Schwann cell proliferation by estradiol: synergism between the estrogen and cAMP. Dev Brain Res 72:282–290

Jung-Testas I, Schumacher M, Robel P, Baulieu EE 1994 Actions of steroid hormones and growth factors on glial cells of the central and peripheral nervous system. J Steroid Biochem Mol Biol 48:145–154

Kabbadj K, El-Etr M, Baulieu EE, Robel P 1993 Pregnenolone metabolism in rodent embryonic neurons and astrocytes. Glia 7:170–175

Kleitman N, Wood PM, Bunge RP 1991 Tissue culture methods for the study of myelination. In: Banker G, Goslin K (eds) Culturing nerve cells. Bradford Books, London, p 337–377

Koenig H, Schumacher M, Ferzaz B et al 1995 Progesterone synthesis and myelin formation by Schwann cells. Science, in press

Le Goascogne C, Robel P, Gouézou M, Sananès N, Baulieu EE, Waterman M 1987 Neurosteroids: cytochrome P450scc in rat brain. Science 237:1212–1215

Magoffin DA, Kurtz KM, Erickson GF 1990 Insulin-like growth factor-I selectively stimulates cholesterol side-chain cleavage expression in ovarian theca-interstitial cells. Mol Endocrinol 4:489–496

Majewska MD 1992 Neurosteroids: endogenous bimodal modulators of the GABA$_A$ receptor. Mechanism of action and physiological significance. Prog Neurobiol 38:379–395

Majewska MD, Harrison NL, Schwartz RD, Barker JL, Paul SM 1986 Steroid hormone metabolites are barbiturate-like modulators of the GABA receptor. Science 232:1004–1007

Matsumoto A, Micevych PA, Arnold AP 1988 Androgen regulates synaptic input to motoneurons of the adult rat spinal cord. J Neurosci 8:4168–4176

Mayo W, Dellu F, Robel P et al 1993 Infusion of neurosteroids into the nucleus basalis magnocellularis affects cognitive processes in the rat. Brain Res 607: 324–328

McEwen BS 1991 Steroid hormones are multifunctional messengers to the brain. Trends Endocrinol Metab 2:62–67

McEwen BS, Gould E, Orchinik M, Weiland NG, Woolley CS 1995 Oestrogens and the structural and functional plasticity of neurons: implications for memory, ageing and neurodegenerative processes. In: The non-reproductive actions of sex steroids. Wiley, Chichester (Ciba Found Symp 191) p 52–73

Mellon SH 1994 Neurosteroids: biochemistry, modes of action, and clinical relevance. J Clin Endocrinol & Metab 78:1003–1008

Mellon SH, Deschepper CF 1993 Neurosteroid biosynthesis: genes for adrenal steroidogenic enzymes are expressed in the brain. Brain Res 629:283–292

Morfin R, Young J, Corpéchot C, Egestad B, Sjövall J, Baulieu EE 1992 Neurosteroids: pregnenolone in human sciatic nerves. Proc Natl Acad Sci USA 89:6790–6793

Oonk RB, Krasnow JS, Beattie WG, Richards JS 1989 Cyclic AMP-dependent and -independent regulation of cholesterol side chain cleavage cytochrome P-450 (P-450scc) in rat ovarian granulosa cells and corpora lutea. J Biol Chem 264: 21934–21942

Papadopoulos V 1993 Peripheral-type benzodiazepine/diazepam binding inhibitor receptor: biological role in steroidogenic cell function. Endocr Rev 14:222–240

Priest CA, Pfaff DW 1995 Actions of sex steroids on behaviours beyond reproductive reflexes. In: The non-reproductive actions of sex steroids. Wiley, Chichester (Ciba Found Symp 191) p 74–89

Raisinghani KH, Dorfman RI, Forchielli E, Gyermek L, Genther G 1968 Uptake of intravenously administered progesterone, pregnanedione and pregnanolone by the rat brain. Acta Endocrinol 57:393–404

Raivich G, Kreutzberg GW 1993 Peripheral nerve regeneration: role of growth factors and their receptors. Int J Neurosci 11:311–324

Robel P, Baulieu EE 1985 Neuro-steroids: 3β-hydroxy-Δ5-derivatives in the rodent brain. Neurochem Int 7:953–958

Robel P, Baulieu EE 1994 Neurosteroids. Biosynthesis and function. Trends Endocrinol Metab 5:1–8

Schumacher M 1990 Rapid membrane effects of steroid hormones: an emerging concept in neuroendocrinology. Trends Neurosci 13:359–362

Schumacher M, Coirini H, Pfaff DW, McEwen BS 1990 Behavioral effects of progesterone associated with rapid modulation of oxytocin receptors. Science 250:691–694

Schumacher M, Jung-Testas I, Robel P, Baulieu EE 1993 Insulin-like growth factor I: a mitogen for rat Schwann cells in the presence of elevated levels of cyclic AMP. Glia 8:232–240

Sohrabji F, Miranda RC, Toran-Allerand DC 1994 Estrogen differentially regulates estrogen and nerve growth factor receptor mRNAs in adult sensory neurons. J Neurosci 14:459–471

Toran-Allerand CD, Miranda RC, Bentham WDL et al 1992 Estrogen receptors colocalize with low-affinity nerve growth factor receptors in cholinergic neurons of the basal forebrain. Proc Natl Acad Sci USA 89:4668–4672

Truss M, Candau R, Chávez S, Beato M 1995 Transcriptional control by steroid hormones: the role of chromatin. In: The non-reproductive actions of sex steroids. Wiley, Chichester (Ciba Found Symp 191) p 7–23

Valera S, Ballivet M, Bertrand D 1992 Progesterone modulates a neuronal nicotinic acetylcholine receptor. Proc Natl Acad Sci USA 89:9949–9953

Yu WH 1989 Survival of motoneurons following axotomy is enhanced by lactation or by progesterone treatment. Brain Res 491:379–382

DISCUSSION

Manolagas: Were your experiments which show the mitogenic effect of oestrogen on Schwann cells and the increase in myelin formation by progesterone in dorsal root ganglia cultures carried out in serum-free medium?

Schumacher: The mitogenic effect of oestradiol was studied in the presence of calf serum (10%), which had been treated with charcoal to remove endogenous steroids (Jung-Testas et al 1993). Untreated fetal calf serum is required for the myelination of sensory axons by Schwann cells in culture (Kleitman et al 1991).

Manolagas: I was not concerned that oestrogen might be present—I was just wondering whether the effect you saw might have been mediated through another factor.

Schumacher: It is very likely that many of the effects that steroids exert on cells of the nervous system could be mediated by peptide growth factors or cytokines. Oestrogens and progestagens may induce their synthesis or modulate their effects. Thus, oestrogen and growth factors have been shown to act in an interactive manner on other target tissues such as the uterus (Murphy & Ghahary 1990), mammary glands (Rochefort 1995, this volume) and bone cells (Ernst & Rodan 1991, Horowitz 1993, Manolagas et al 1995, this volume).

Beato: This seems to me to be a very nice model with which to study the mechanisms of interactions between neurons and Schwann cells. Do you have any information about the mechanisms by which the neurons activate the Schwann cells to convert pregnenolone to progesterone?

Schumacher: We still ignore the nature of the neuronal signal that induces steroid biosynthetic pathways in Schwann cells. This may be a diffusible factor, secreted by the sensory neurons. Alternatively, the induction of progesterone synthesis may require direct contact between neurons and Schwann cells. We are now doing experiments on this. We have already tested the effect of culture medium conditioned by sensory neurons—this had no effect. It is however possible that the neuronal factor is labile and/or produced in small amounts. We plan to explore this possibility in a co-culture system in which Schwann cells and neurons are separated by a microporous membrane.

It is interesting that Schwann cells also synthesize myelin only in response to a neuronal signal, which has recently been shown to be a diffusible molecule (Bolin & Shooter 1993). It is thus possible that the synthesis of both neurosteroids and myelin are induced by the same factor. This would strengthen the relationship between neurosteroids and myelination.

Beato: But in terms of mechanism in the Schwann cells, is the level of 3β-HSD altered by the co-culture paradigm? Is the enzyme induced?

Schumacher: Yes, it is. We not only measured the conversion of [³H]pregnenolone to [³H]progesterone by the Schwann cells; also, by using immunocytochemistry, *in situ* hybridization and RT-PCR, we showed that the expression of the 3β-HSD is increased in Schwann cells by neurons (our unpublished results). The mechanism of this induction is not yet understood.

Horwitz: You described that oestradiol stimulates Schwann cell proliferation in synergy with cAMP and that this second messenger increases the binding of [³H]oestradiol. The synergistic effect between cAMP and steroid receptors is a very interesting phenomenon. Have you tried adding cAMP to your cells with progesterone to see if there is either an acceleration of the rate of myelination, or an increase in the thickness of myelin deposition?

Schumacher: We have not done this, but it has been reported that cAMP increases the synthesis of myelin proteins (Morgan et al 1991, Leblanc et al 1992).

Horwitz: Is there an interaction between progesterone and cAMP in the activation of myelination?

Schumacher: It is possible that progesterone and cAMP may activate the process of myelination by acting in concert, but this has not been demonstrated as such.

Oelkers: Can Schwann cells also convert cholesterol to pregnenolone?

Schumacher: We have shown that Schwann cells in culture, prepared from neonatal rat sciatic nerves, convert [³H]-25-OH cholesterol to [³H]pregnenolone, but only in small amounts (Akwa et al 1993).

Oelkers: Do you think that most of the pregnenolone is being taken up from plasma?

Schumacher: I don't think so. Levels of pregnenolone remain high in the sciatic nerve of rats days after castration and adrenalectomy, in the absence of circulating hormone (Akwa et al 1993), suggesting that it is synthesized locally. The fact that we did not observe an important metabolic activity in our culture conditions does not necessarily argue against this. Thus, Schwann cells convert [³H]-25-OH cholesterol to [³H]pregnenolone only when cultured in the presence of forskolin and insulin (Akwa et al 1993), which are mitogens for these cells (Schumacher et al 1993). Other factors may be required for an optimal activity of the P450$_{scc}$ enzyme.

Oelkers: Did you look at whether adrenocorticotrophic hormone (ACTH) or luteinizing hormone have any effect on the formation of pregnenolone by Schwann cells?

Schumacher: We have not tried this. However, some experiments have been done on oligodendrocytes, the myelinating glial cells of the central nervous system.

Baulieu: There was a reproducible increase of pregnenolone synthesis by cAMP in brain glial cells. However, the experimental conditions were not good because mixed cultures of glial cells, consisting of oligodendrocytes, astrocytes and microglia were used. This has to be worked out properly by using pure cultures of glial cells.

Thijssen: I have a question on pregnenolone and its conversion to progesterone. You said that in the nerves, there are about five to 10 times higher concentrations of the endogenous hormones than in plasma. We have recently been measuring pregnenolone and progesterone in human adipose tissue, and the concentrations of pregnenolone were about 100 times higher than in plasma (Thijssen et al 1994). Does adipose tissue surround the nerves? Could the pregnenolone present in adipose tissue contaminate your assay?

Schumacher: The sciatic nerves were well cleaned from surrounding tissues prior to steroid assays. However, about 70% of myelin is made up of lipids, which may retain steroids such as pregnenolone. Nevertheless, I do not think that the myelin sheaths accumulate and retain steroids for days. That is, [³H]pregnenolone and [³H]progestagens do not accumulate after their *in vivo* administration within the brain, which is rich in myelin and lipids. They are cleared from nervous tissues within hours (Raisinghani et al 1968, Robel & Baulieu 1985).

Thijssen: I have been listening for several years to people from Utrecht who have been working with neuropeptides in a very similar system (De Koning & Gispen 1987). They have claimed in the past that one of the neuropeptides, ACTH$_{4-9}$ (Org 2766), was able to stimulate nerve repair. In clinical studies they showed that it did have some effect under experimental conditions (Gerritsen van der Hoop et al 1990). But at the moment, almost everybody agrees that the effects are not very large. Now I'm looking for an explanation for why you have similar findings.

Schumacher: The role of ACTH in peripheral nerve repair has been investigated by several groups for years (for review, see Strand et al 1989). However, their work has focused on the effects of neuropeptides on the regeneration of nerve fibres, whereas we are studying the role of neurosteroids during the process of myelin formation. Why were the clinical studies with peptides not successful?

Thijssen: They showed in diabetic rats that the long-term nerve damage caused by diabetes was reduced after treatment with the peptide (Bravenboer et al 1994). They also showed that it reduced nerve damage in people who had undergone chemotherapy (Gerritsen van der Hoop et al 1990). The reason this treatment hasn't reached the clinic was because the effects were marginal.

Muramatsu: Is the effect on myelination you have described specific for progesterone and related steroids? Have you tried oestrogen or glucocorticosteroids?

Schumacher: This needs to be done. The only steroids we have tried so far are progesterone and its direct precursor pregnenolone. The effect of progesterone is inhibited by RU486, which blocks its action at the receptor level (Koenig et al 1995). We intend to test a wide range of steroids in our *in vitro* and *in vivo* systems (in collaboration with H. Koenig, University Bordeaux I).

Muramatsu: Have you looked to see whether MBP mRNA is activated after treatment with progesterone?

Schumacher: Work is now in progress to study the effects of neurosteroids on the expression of different myelin proteins in the PNS. In oligodendrocytes, we have shown by immunocytochemistry that progesterone increases the number of cells expressing MBP (Jung Testas et al 1994). However, the levels of MBP mRNA appear not to be affected by the steroid (our unpublished results). Thus, progesterone could increase the synthesis of MBP by acting at a post-transcriptional level. Such a mechanism is also suggested by another study (Verdi & Campagnoni 1990).

Rochefort: At the Schwann cell level, do both males and females make the same amount of progesterone?

Schumacher: We still do not know this.

Rochefort: Our laboratory showed several years ago that progesterone, via its receptor, induces fatty acid synthetase in hormone-dependent breast cancer and the endometrium (Chalbos et al 1987, Joyeux et al 1990). Because myelin contains fatty acids, it might be interesting to look in your cells to see whether or not fatty acid synthetase is regulated by progesterone.

Pfaff: Rupprecht et al (1993) have suggested that certain neurosteroids could act through the classical nuclear progesterone receptor. How do your findings agree with theirs?

Schumacher: Two observations strongly suggest that progesterone increases myelination by binding to its intracellular receptor. First, RU486 blocks the effect of progesterone. Second, Schwann cells express the receptor for this neurosteroid, as we have shown by binding studies and by RT-PCR (our unpublished results). However, these observations do not rule out additional mechanisms of action of progesterone, for example at the level of the cell membrane.

Baulieu: The paper you refer to (Rupprecht et al 1993) shows that allopregnanolone, which acts on the $GABA_A$ receptor, can also activate the progesterone receptor and the transcription of progesterone-sensitive genes after its conversion to 5α-dihydroprogesterone. That is, 5α-dihydroprogesterone has some binding affinity for the progesterone receptor.

In Schwann cells, as Michael Schumacher showed, progesterone can be converted to dihydroprogesterone and allopregnanolone. We still have not studied whether these 5α-reduced metabolites of progesterone play a role in myelination.

Schumacher: We have some preliminary results which show that the conversion of progesterone to allopregnanolone is increased (about 10-fold) in cultures of dorsal root ganglia explants if Schwann cells are induced to myelinate the sensory axons by the addition of serum and ascorbic acid to the medium (in collaboration with N. A. Do-Thi, Collège de France). However, we still do not know whether allopregnanolone is an active metabolite of progesterone in myelination.

Pfaff: What about the relative affinities of progesterone and its reduced metabolites for the progesterone receptor? When we think of other data, does that mean that the 5α-reduced metabolites are not physiologically important? As I recall, these compounds don't have a very high affinity for the progesterone receptor?

Baulieu: Classically, 5α-dihydroprogesterone doesn't have a very high affinity for the progesterone receptor compared with progesterone itself. However, we have some preliminary evidence that the progesterone receptor that is expressed in glial cells may not have the same binding specificity as the classical one (R. Fiddes, K. Shazand, I. Jung-Testas, M. Schumacher, E. E. Baulieu, unpublished results). We are now cloning and characterizing this glial receptor.

McEwen: Michael, do you want to speculate about what's going on in the CNS? Are oligodendrocytes in possession of the same kind of system as the Schwann cells? Is there a similar system that might operate for myelination in the brain itself?

Schumacher: Glial cells in culture prepared from neonatal rat brain synthesize progesterone and express its cognate receptor (Jung-Testas et al 1989, 1991) and the addition of progesterone to these cultures increases the expression of myelin-specific proteins such as MBP and CNPase (Jung-Testas et al 1994).

McEwen: There is a related puzzle that's been around for some time. I think Jerry Meyer described that when you adrenalectomize rats, they develop a 'fatter' brain in the sense that they actually produce more myelin lipids and more myelin (Meyer & Fairman 1985).

Baulieu: Male or female rats?

McEwen: Both sexes as far as I know.

Oelkers: But do you have to substitute adrenalectomized rats with corticosterone or some related steroids?

McEwen: I think they were maintained on low levels of a long-lasting neural corticosteroid to keep them alive. The speculation was that gluco-corticosteroids regulate glycerol phosphate dehydrogenase, which is an oligodendrial enzyme that has the function of converting glycerol away from the formation of lipids (Meyer et al 1982, Kumar et al 1989). When you have higher levels of this enzyme, you get more dihydroxyacetone phosphate and you get more metabolism. But now with your observations that the adrenals are also producing progesterone, there may be an entirely different explanation.

References

Akwa Y, Schumacher M, Jung-Testas I, Baulieu EE 1993 Neurosteroids in rat sciatic nerves and Schwann cells. C R Acad Sci Ser III Sci Vie 316:410–414

Bolin LM, Shooter EM 1993 Neurons regulate Schwann cell genes by diffusible molecules. J Cell Biol 123:237–243

Bravenboer B, Hendriksen PH, Oey PL, van Huffelen AC, Gispen WH, Erkelens DW 1994 Randomized double-blind placebo-controlled trial to evaluate the effect of the ACTH(4–9) analogue Org2766 in type I diabetic patients with neuropathy. Diabetologia 37:408–413

Chalbos D, Chambon M, Ailhaud G, Rochefort H 1987 Fatty acid synthetase and its mRNA are induced by progestins in breast cancer cells. J Biol Chem 262:9923–9926

De Koning P, Gispen WH 1987 Org 2766 improves functional and electrophysiological aspects of regenerating sciatic nerves in the rat. Peptides 8:415–422

Ernst M, Rodan GA 1991 Estradiol regulation of insulin-like growth factor-1 expression in osteoblastic cells: evidence for transcriptional control. Mol Endocrinol 5:1081–1089

Gerritsen van der Hoop R, Vecht ChJ, Van den Burg MEL et al 1990 Prevention of cispatinin neurotoxicity with an ACTH(4–9) analogue in patients with ovarian cancer. N Engl J Med 322:89–94

Horowitz MC 1993 Cytokines and estrogen in bone: anti-osteoporotic effects. Science 260:626–627

Joyeux C, Chalbos D, Rochefort H 1990 Effects of progestins and menstrual cycle on fatty acid synthetase and progesterone receptor levels in human mammary glands. J Clin Endocrinol & Metab 70:1438–1444

Jung-Testas I, Hu ZY, Baulieu EE, Robel P 1989 Neurosteroids: biosynthesis of pregnenolone and progesterone in primary cultures of rat glial cells. Endocrinology 125:2083–2091

Jung-Testas I, Renoir JM, Gasc JM, Baulieu EE 1991 Estrogen-inducible progesterone receptor in primary cultures of rat glial cells. Exp Cell Res 193:12–19

Jung-Testas I, Schumacher M, Bugnard H, Baulieu EE 1993 Stimulation of rat Schwann cell proliferation by estradiol: synergism between the estrogen and cAMP. Dev Brain Res 72:282–290

Jung-Testas I, Schumacher M, Robel P, Baulieu EE 1994 Actions of steroid hormones and growth factors on glial cells of the central and peripheral nervous system. J Steroid Biochem Mol Biol 48:145–154

Kleitman N, Wood PM, Bunge RP 1991 Tissue culture methods for the study of myelination. In: Banker G, Goslin K (eds) Culturing nerve cells. Bradford Books, London, p 337–377

Koenig H, Schumacher M, Ferzaz B et al 1995 Progesterone synthesis and myelin formation by Schwann cells. Science, in press

Kumar S, Cole R, Chiappelli F, de Vellis J 1989 Differential regulation of oligodendrocyte markers by glucocorticoids: post-transcriptional regulation of both proteolipid protein and myelin basic protein and transcriptional regulation of glycerolphosphate dehydrogenase. Proc Natl Acad Sci USA 86:6807–6811

Leblanc AC, Windebank AJ, Podulso JF 1992 P_0 gene expression in Schwann cells is modulated by an increase of cAMP which is dependent on the presence of axons. Mol Brain Res 12:31–38

Manolagas SC, Bellido T, Jilka RL 1995 Sex steroids, cytokines and the bone marrow: new concepts on the pathogenesis of osteoporosis. In: Non-reproductive actions of sex steroids. Wiley, Chichester (Ciba Found Symp 191) p 187–202

Meyer JS, Fairman KR 1985 Early adrenalectomy increases myelin content of rat brain. Dev Brain Res 17:1–9

Meyer JS, Leveille PJ, de Vellis J, Gerlach JL, McEwen BS 1982 Evidence for glucocorticoid target cells in the rat optic nerve. Hormone binding and glycerophosphate dehydrogenase induction. J Neurochem 39:423–434

Morgan L, Jessen KR, Mirsky R 1991 The effects of cAMP on differentiation of cultured Schwann cells: progression from an early phenotype (04[+]) to a myelin phenotype (Po[+], GFAP[−], N-CAM[−], NGF-receptor[−]) depends on growth inhibition. J Cell Biol 112:457–467

Murphy LJ, Ghahary A 1990 Uterine insulin-like growth factor-1: regulation of expression and its role in estrogen-induced uterine proliferation. Endocr Rev 11:443–453

Raisinghani KH, Dorfman RI, Forchielli E, Gyermek L, Genther G 1968 Uptake of intravenously administered progesterone, pregnanedione and pregnanolone by the rat brain. Acta Endocrinol 57:393–404

Robel P, Baulieu EE 1985 Neuro-steroids: 3β-hydroxy-Δ5-derivatives in the rodent brain. Neurochem Int 7:953–958

Rochefort H 1995 Oestrogen- and anti-oestrogen-regulated genes in human breast cancer. In: Non-reproductive actions of sex steroids. Wiley, Chichester (Ciba Found Symp 191) p 254–268

Rupprecht R, Reul JMHM, Trapp T et al 1993 Progesterone receptor-mediated effects of neuroactive steroids. Neuron 11:523–530

Schumacher M, Jung-Testas I, Robel P, Baulieu EE 1993 Insulin-like growth factor I: a mitogen for rat Schwann cells in the presence of elevated levels of cyclic AMP. Glia 8:232–240

Strand FL, Rose KJ, King JA, Segarra AC, Zuccarelli LA 1989 ACTH modulation of nerve development and regeneration. Prog Neurobiol 33:45–85

Thijssen JHH, Maitimu-Smeele I, Symczack J, Blankenstein MA 1994 Lipoidal and free steroids in human adipose tissue of different origin. Eur J Endocrinol 130(suppl 2):95

Verdi JM, Campagnoni AT 1990 Translational regulation by steroids. J Biol Chem 265:20314–20320

General discussion II

Sex steroids and memory

Robel: In studies of memory processes, it is necessary to take into account not only sex steroid hormones, but also their precursors or metabolites. I have carried out two sorts of experiments in collaboration with Willy Mayo, Michel Le Moal, and Hervé Simon (INSERM U259, Domaine de Carreire, Bordeaux, France). We have attempted to relate one precursor (pregnenolone sulphate) and one metabolite (allopregnanolone) of progesterone to the regulation of memory processes in rats.

Considerable attention has focused on the nucleus basalis of Meynert (NBM) since the discovery of marked cell loss and various pathological alterations of this structure in patients suffering from senile dementia of Alzheimer's type. The NBM is regarded as the main source of hippocampal–cortical cholinergic innervation. The GABAergic neurons originating in the nucleus accumbens have been shown to make synaptic contacts on cholinergic cells of the NBM, and modulation of GABAergic inputs to the NBM may be involved in memory processes.

In the first set of experiments, we examined the effect on memory processes of local injection into the NBM of two neurosteroids that act in opposite ways on the type A γ-aminobutyric acid ($GABA_A$) receptor complex. Allopregnanolone is a sedative, anxyolytic steroid, whose benzodiazepine-like properties, linked to the potentiation of GABAergic neurotransmission, might be expected to produce proamnestic effects. Pregnenolone sulphate, on the contrary, has an antagonistic, excitatory, proconvulsant action on $GABA_A$ receptors, analogous to those of β-carbolines.

In our study, allopregnanolone or pregnenolone sulphate were stereotaxically infused into the NBM of the rat, and their effects were examined in a two-trial memory task (Mayo et al 1993). The results showed that allopregnanalone (2 ng in 0.5 μl) disrupted performance when injected before an acquisition trial. In contrast, pregnenolone sulphate (5 ng in 0.5 μl) enhanced memory performance when injected after an acquisition trial.

In a second set of experiments, we evaluated the cognitive performances of ageing rats. Ageing is associated with impairment of cognitive functions and particularly with a decline in memory. However, there are considerable inter-individual differences in the severity of age-related impairments in both humans and animals. Some older subjects are impaired, whereas others perform as well as young ones. Memory performances of 29 older rats (about 24 months old) were measured in a water maze and in a two-trial recognition task. At the completion of the behavioural study, the rats were killed and their brains were removed for analysis of pregnenolone sulphate concentrations in selected areas. The results showed that there were considerable inter-individual differences in memory performances of older rats and that scores were correlated in the two tasks, suggesting true memory evaluation. A striking observation was the significant positive correlation between the concentration of pregnenolone sulphate in the hippocampus and memory performance— namely the animals with the better performances had the greater levels of pregnenolone sulphate.

Although we have still to show a cause–effect relationship between memory performance and neurosteroid concentrations in the hippocampus, these results support the possible neuroprotective role of pregnenolone sulphate (and likewise dehydroepiandrosterone [DHEA] sulphate) against neurodegenerative processes.

We have to perform more experiments to prove that pregnenolone sulphate is able, by itself, to help maintain memory performance during ageing. We feel that this is a very promising approach, which might also be used to prevent development of Alzheimer's disease in humans.

McEwen: Owen Wolkowitz is conducting a trial, not with pregnenolone, but with DHEA (Wolkowitz et al 1994). He is administering it and looking at memory functions, over a relatively short term. As you may know, there are correlations in human ageing between lower levels of plasma DHEA and greater cognitive impairment and other medical problems.

Baulieu: We have also done experiments with DHEA. Naturally, the decrease in DHEA that occurs during ageing is that of plasma DHEA sulphate, which is not necessarily correlated to a local concentration of this steroid in the brain. J. M. Mienville showed that, like pregnenolone sulphate, DHEA sulphate acts as an antagonist of the $GABA_A$ receptor. However, several recent results (E. E. Baulieu, F. Monnet et al, unpublished results) indicate that pregnenolone sulphate also has effects opposite to DHEA sulphate. DHEA sulphate increases and pregnenolone sulphate decreases the N-methyl-D-aspartate (NMDA)- evoked release of noradrenaline in hippocampus slices. It is clear that these effects are not just due to the sulphate moiety because there is steroid specificity.

McEwen: Going back to pregnenolone sulphate, what is your speculation as to its site and mechanism of action? Is it a specific activator of excitatory amino acids or Ca^{2+} channels, for instance?

Robel: Pregnenolone sulphate is a prototypical naturally excitatory neuro- steroid (Majewska 1992). At low micromolar concentrations, it antagonizes $GABA_A$ receptor-mediated $^{36}Cl^-$ uptake into synaptoneurosomes and Cl^- conductance in cultured neurons. Pregnenolone sulphate bimodally modulates [^3H]muscimol binding to synaptosomal membranes, slightly potentiates benzodiazepine binding and inhibits the binding of the convulsant [^{35}S]TBPS to the $GABA_A$ receptor chloride channel. Pregnenolone sulphate also augments glutamate-induced depolarization mediated by the NMDA subtype of glutamate receptor in chick spinal cord neurons (Wu et al 1991). Thus, the excitatory properties of pregnenolone sulphate may be due to both a reduction in GABA-mediated inhibitory events and an enhancement of glutamate-mediated excitation. There are other targets yet to be defined. There is no evidence for a direct effect on Ca^{2+} uptake but, of course, pregnenolone sulphate augments NMDA-receptor-mediated elevations in the intracellular free Ca^{2+} concentrations.

McEwen: Spence et al (1991) reported an effect of pregnenolone sulphate on a Ca^{2+} channel.

Oelkers: Morales et al (1994) have recently shown that treatment of elderly people with 50 mg DHEA per day leads to a slight increase in insulin-like growth factor 1 (IGF-1) and a decrease in IGF-1-binding protein, so that more IGF-1 might be available. The subjective effects, and somehow also the objective effects, on the relationship between lean body mass and body fat were similar to those of low-dose growth hormone treatment.

Baulieu: 50 mg is a very small dose compared with previous animal studies in which huge doses of DHEA were administered.

Oelkers: This dose raised the low levels of DHEA in elderly people to those of younger people.

Baulieu: In fact, Dr Yen and I thought of that dose in order to compensate for defects brought about by non-reproductive activities of steroids as a consequence of ageing. The increase in IGF-1 and change of the correlated binding protein seen by Morales et al (1994) is, to my knowledge, the first observed effect of DHEA in humans when given at a 'reasonable' dose. We have obtained similar results in Paris. We found also that, while the concentration of DHEA sulphate decreases with age in all men and women, it is an individual marker in the sense that when categorized as high or low, it remains in the same group for the period of observation (Thomas et al 1994).

The DHEA problem is difficult since there is no DHEA or DHEA sulphate in the plasma of rodents, and extrapolating human physiology from animal experiments is hazardous. Let me add that even though we found DHEA sulphate in the brain of rodents and there is evidence that it has neuromodulatory activity, it has not been proved definitively that it is synthesized in the nervous system.

Thijssen: So far this meeting we have been focusing almost exclusively on oestrogens. However, there seem to be large differences between the different oestrogens. I'm referring to, for instance, the synthetic ethinyloestradiol, oestradiol, oestriol or even the DHEA-related 5-androstenediol. Is there any evidence of different mechanisms of action for different steroids?

Jensen: Liao has found that 5-androstene-3β,17β-diol binds to a different protein in the vaginal epithelium than oestrodiol does (Shao et al 1975).

Thijssen: We have done experiments on the uterus in which we administered androstenediol and oestradiol in different ratios (Coosen 1986). By analysing the data we got the impression that they were both competing for the same oestradiol receptor. We did not look at the androgen receptor, which is also activated by androstenediol.

Jensen: Clark has shown that oestriol is different from oestradiol, not in its binding to the receptor, but in the length of time the hormone–receptor complex remains in the nucleus (Clark et al 1977). The retention of oestriol after a single administration is of short duration, but one can obtain good oestrogenic response by giving the same total dose of oestriol in multiple sequential injections. This is a pharmacological difference rather than a different basic mechanism.

Thijssen: Data were presented in previous papers showing the difference between the binding of tamoxifen, nafoxidine and oestradiol in the nucleus. Is binding comparable or are there differences between oestrogens?

Baulieu: There are a number of aspects to this. First, there are the pharmacological aspects, with essentially metabolic differences, and thus different availability of the compound to the cellular machinery. It's clear that the constant presence of a hormone, as compared with its transient presence, may generate a different type of response. The difference between oestriol and oestradiol is not trivial. In the classical experiments performed by the Huggins group, they showed that you do not get protein synthesis after a single shot of oestriol, contrary to the result obtained with oestradiol, but you stimulate protein synthesis if you prolong the administration of oestriol. At the target level, since the ligand is different sterically, we expect that the subsequent formation of the ligand–receptor complex is different. There are not many known differences between the natural steroids as far as changes in the intracellular receptor conformation. It is a little clearer with synthetic steroid hormone analogues, including antihormones. Different anti-oestrogens apparently elicit different types of response at the same genome locus.

Horwitz: One can create mutations in the hormone binding domains of steroid receptors that will knock out the binding of one ligand, but permit the binding of another ligand. This suggests that the ligand binding sites for different steroids are not identical—they may overlap, but they're not exactly the same. You could envision the possibility that one ligand binds at a slightly different site from another, which in turn produces structural changes in the receptors that lead to entirely different responses. Similarly, just as with these receptor mutants, when receptors are occupied by different ligands, one sees subltle differences in their DNA binding characteristics. It is believed that receptors assume different allosteric conformations depending on the ligand bound at the hormone binding domain. Most of this work has been done with synthetic steroids and we do not know whether some of the natural ligands also cause these subtle changes, but it wouldn't be surprising if they did.

Manolagas: Is this true for synthetic anti-oestrogens?

Horwitz: Yes.

Manolagas: In the bone field, there is a lot of excitement over the bone- and lipid-specific anti-oestrogen, raloxifene. In a recent abstract, workers from Eli Lilly claim that raloxifene acts via a different type of receptor to that of oestradiol.

Baulieu: Besides its anti-oestrogen activity, tamoxifen may be used to modify membrane stability, because it is very sticky. There are even people who envisage using it to favour the entry of chemotherapeutic drugs into cells.

Jensen: One should not forget the phenomenon discovered by Rob Sutherland, who showed that there is a separate protein with a high affinity for type I anti-oestrogens, which does not bind oestradiol (Sutherland et al 1980). Do type II anti-oestrogens also react with this anti-oestrogen binding protein?

Sutherland: No.

Jensen: Recently, we found that the guinea-pig uterus contains a substance that has a high affinity for hydroxytamoxifen but practically none for oestradiol (unpublished results). Because of its sedimentation at about 4.5S, this does not appear to be the classical anti-oestrogen-binding protein, although it may be related. It seems that there are a number of proteins that like to bind to anti-oestrogens.

Manolagas: What is the latest on the TAF-1 and TAF-2 with respect to anti-oestrogens?

Jensen: The current explanation, based on molecular biological studies, of how a compound can be both an agonist and an antagonist is that there are two transactivation domains in the receptor (Berry et al 1990). One is ligand dependent and is activated by oestradiol but not by tamoxifen, whereas the other is constitutively active and is responsible for the agonistic action of anti-oestrogens. However, this nice model does not explain the dose dependency of agonism versus antagonism. It has been found in several systems that low doses of hydroxytamoxifen and related type I anti-oestrogens act as agonists but at higher levels they become antagonists (Black et al 1983, Katzenellenbogen et al 1987, Poulin et al 1989). We feel that this can be explained by the presence of two binding loci in the receptor, a primary site that reacts with either oestradiol or anti-oestrogen, and a secondary, weaker binding site that selectively recognizes anti-oestrogen and which becomes occupied only at higher doses.

Thijssen: Tamoxifen has a higher affinity for the anti-oestrogen binding protein than for the oestrogen receptor—it's 1–2 nM.

Jensen: I'm talking about two binding sites in the receptor molecule itself. Tamoxifen can be an agonist at lower concentrations because only the primary site is occupied. Then, whether oestrogen or anti-oestrogen is present at the primary site, when a secondary site becomes occupied at higher levels of anti-oestrogen, one sees antagonism. As mentioned earlier, our recent findings that the oestrogen receptor of MCF-7 cells binds more anti-oestrogen than it does oestradiol provides direct evidence for a secondary, anti-oestrogen-specific binding locus.

Parker: I don't agree. I don't dispute the second binding site, but I take a slightly different view to the one presented so far. While I accept that antagonist can induce a different conformation in the receptor and block certain activities, I am not as convinced that there are certain conformational changes that are induced by any agonist. There are several reasons for saying this. The first is that there is a fantastic variety of different chemicals that can induce the activity of oestrogen receptors—from phenol right through to oestradiol, including lots of synthetic oestrogens, herbicides and alkyl phenols. If there's a subtle allosteric change that is induced by the ligand, I can't imagine how such a variety of different chemicals would induce such a change. Then there are some functional data that we've generated, when we've looked at the affects of different mutations and their ability to respond to each of these different types of ligands. The mutations respond identically. In other words, a mutation that abolishes the response to oestradiol, abolishes the affect of all those ligands that are able to stimulate the activity. Those mutants that have no affect in modulating activity of the receptor respond in a similar manner. The way I'm beginning to think about this stems from the idea that there are a number of activation domains in the receptors, one of which is in the hormone binding domain, and one of which is in a region that can generate an amphipathic α-helix. Until very recently I used to think this helix was probably induced following the binding of the hormone. I don't think that's so any longer—I think it is a pre-formed helix that's masked in the inactive complex by heat shock proteins, and following the binding of the hormone the complex associates, dimerization occurs, and this amphipathic α-helix is exposed. It can then interact with whatever target proteins in the basal transcription machinery it has to. This is based on a couple of observations. The amphipathic α-helix I'm talking about is probably present in every nuclear receptor we could name today, and in a number of them this short region has been shown to confer transcriptional activity in transient transfection experiments. So it has the ability to stimulate transcription in isolation. I suspect that it's present as an isolated structure in the receptors but it's not able to function until the ligand is bound. All the different chemicals that can induce the activity of the receptor do so because, providing you can get them to bind to the receptor by administering enough of them (providing they dissociate heat shock proteins), the rest will follow. There will be dimerization of the receptor, the receptor will bind to DNA, and this pre-formed α-helix will be able to stimulate transcription.

Jensen: I agree with you completely. The key role of the oestrogen is to dissociate the steroid-binding subunit from these other components. We didn't know what they were until people such as Etienne Baulieu, David Toft and Bill Pratt showed that they are primarily heat shock proteins. To activate (or transform) the receptor is the primary function of the hormone.

Parker: But I'm not suggesting that this can be done by hydroxytamoxifen.

Jensen: But it must be done by hydroxytamoxifen if there is to be agonist activity.

Parker: There are other ways in which you can derive agonist activity from the receptors, not from the hormone binding domain but from the N-terminal activation domain.

Jensen: Don't you think that hydroxytomoxifen will get rid of the heat shock proteins?

Parker: I do, yes. But I don't think it will induce activation through the hormone-binding domain; I think that this could arise in different ways. Transcriptional effects may be mediated by the N-terminal TAF-1 domain of they may be indirect, for example, by protein–protein interactions with other transcription factors, such as AP-1.

Manolagas: How can the same ligand bind on the same receptor with the same conformation and still have different effects in these tissues?

Parker: The receptor has to interact in some way with the basic transcription machinery, and in different types of cells it will be talking to different types of proteins. People are looking for those sorts of proteins.

Manolagas: But coming back to your original statement, how can the same ligand (or the same family of ligands), inducing the same conformation, make the requirement for a transcription factor, present in one tissue and not in another, relevant for agonistic or antagonist activity?

Parker: It may depend on whether the receptor is binding to sites that are close to other sites, such as AP-1 sites. Receptors can interfere with the activity of AP-1, depending on the presence or absence of different family members of the AP-1 complex.

Manolagas: How can a ligand behave as agonist or antagonist in the two different tissues, even though it may bind different sites?

Parker: Because it may depend on how the receptor interacts with other proteins.

Jensen: I would interpret this by the relative affinities of the substance for the primary versus the secondary site. In tissues where the receptor has a very weak secondary site, or perhaps lacks it altogether, type I anti-oestrogens should act only as agonists, as they do in the mouse uterus. In tissues where the two sites have comparable affinities, both become occupied concurrently, so the agonist configuration (i.e. only the primary site occupied) is never formed. In such tissues, tamoxifen should act as a pure antagonist, as is seen in chick oviduct. This is all hypothetical so far, but it should be subject to experimental test.

Rochefort: I agree with Malcolm Parker that any ligand—including a low-affinity ligand like phenol red or adrenal androgens like 5-androstenediol—is able to trigger several oestrogenic responses. But androstenediol also has the ability to bind to other receptors, such as androgen receptor and progesterone receptor (for review, see Rochefort & Garcia 1984). Some of the ligands can bind to several receptors, some of them can bind only to one receptor. That might explain differences in their physiological effects.

References

Berry M, Metzger D, Chambon P 1990 Role of the two activating domains of the oestrogen receptor in the cell-type and promoter-context dependent agonistic activity of the antioestrogen 4-hydroxytamoxifen. EMBO J 9:2811–2818

Black LJ, Jones CD, Falcone JF 1983 Antagonism of estrogen action with a new benzothiophene-derived anti-estrogen. Life Sci 32:1031–1036

Clark JH, Paszko Z, Peck EJ Jr 1977 Nuclear binding and retention of the receptor–estrogen complex: relation to the agonistic and antagonistic properties of estriol. Endocrinology 100:91–96

Coosen R 1986 Estrogen and androgen action on human breast cancer *in vitro*. PhD Thesis, Utrecht University, Utrecht, The Netherlands

Katzenellenbogen BS, Kendra KL, Norman MJ, Berthois Y 1987 Proliferation, hormonal responsiveness, and estrogen receptor content of MCF-7 human breast cancer cells grown in short-term and long-term absence of estrogens. Cancer Res 47:4355–4360

Majewska MD 1992 Neurosteroids: endogenous bimodal modulators of the $GABA_A$ receptor. Mechanism of action and physiological significance. Prog Neurobiol 38:379–395

Mayo W, Dellu F, Robel P et al 1993 Infusion of neurosteroids into the nucleus basalis magnocellularis affects cognition processes in the rat. Brain Res 607:324–328

Morales AJ, Nolan JJ, Nelson JC, Yen SSC 1994 Effects of replacement dose of dehydroepiandrosterone in men and women of advancing age. J Clin Endocrinol & Metab 78:1360–1367

Poulin R, Merand Y, Poirier D, Levesque C, Dufour J-M, Labrie F 1989 Antiestrogenic properties of keoxifene, *trans*-4-hydroxytamoxifen, and ICI 164384, a new steroidal antiestrogen, in ZR-75-1 human breast cancer cells. Breast Cancer Res Treat 14: 65–76

Rochefort H, Garcia M 1984 The estrogenic and antiestrogenic activities of androgens in female target tissues. Pharmacol Ther 23:193–216

Shao T-C, Castañeda E, Rosenfeld RL, Liao S 1975 Selective retention and formation of a Δ^5-andtrostenediol–receptor complex in cell nuclei of the rat vagina. J Biol Chem 250:3095–3100

Spence KT, Plata-Salaman CR, Ffrench-Mullen JMH 1991 The neurosteroids pregnanolone and pregnenolone sulphate, but not progesterone, block Ca^{2+} currents in acutely isolated hippocampal CA1 neurons. Life Sci 49:PL235–PL239

Sutherland RL, Murphy LC, Foo MS, Green MD, Whybourne AM 1980 High-affinity anti-oestrogen binding site distinct from the oestrogen receptor. Nature 288: 273–275

Thomas G, Frenoy N, Legrain S, Sebag-Lanöe R, Baulieu EE, Debuire B 1994 Serum dehydroepiandrosterone sulfate levels as an individual marker. J Clin Endocrinol & Metab 79:1273–1276

Wolkowitz O, Reuss V, Roberts E et al 1994 Pharmacologic normalization of DHEA and DHEA-S in middle aged and elderly depressives. Abstracts, Int Soc Psycho-neuroendocrinology, Seattle, WA

Wu FS, Gibbs, TT, Farb DH 1991 Pregnenolone sulfate: a positive allosteric modulator at the N-methyl-D-aspartate receptor. Mol Pharmacol 40:333–336

Oestrogen synthesis, oestrogen metabolism and functional oestrogen receptors in bovine aortic endothelial cells

Francis Bayard, Simone Clamens, Georges Delsol*, Nelly Blaes, Arlette Maret and Jean-Charles Faye

INSERM U397, Institut Louis Bugnard, CHU Rangeuil, F-31054 Toulouse Cedex and *Groupe d'Etudes des Lymphomes Malins (CIGH, UPR-8291), CHU Purpan, F-31059 Toulouse Cedex, France

Abstract. In order to investigate the mechanisms by which oestrogenic hormones influence the vascular system, we have studied their metabolism and the functioning of oestrogen receptors in bovine aortic endothelial cells from primo-secondary cultures, a widely studied model of vascular pathophysiology. We have demonstrated the enzymic activity of oestradiol-17β-hydroxysteroid dehydrogenase, 17-ketoreductase and aromatase in these cells. Immunocytochemical analyses, using two different monoclonal antibodies that recognize epitopes in the A/B domain of the oestrogen receptor, showed that this molecule has a predominantly cytoplasmic localization even after the addition of oestrogen to the culture medium. We showed that the hormone–receptor complexes were functional by demonstrating their transactivating ability in transfection experiments using the luciferase gene reporter and an oestrogen-responsive element transcriptional enhancer, although the amplitude of the response was in the range of only 140–150%: this was not a consequence of the presence of a specific limiting factor, but instead might be related to the peculiar subcellular localization of the oestrogen receptor.

1995 Non-reproductive actions of sex steroids. Wiley, Chichester (Ciba Foundation Symposium 191) p 122–138

The incidence of cardiovascular disease, the leading cause of mortality in western societies (Ross 1993), is higher in men than in premenopausal women, but increases in postmenopausal women. An abundance of epidemiological data supports a role for oestrogens in this atheroprotective effect, prompting recommendations for their widespread use as a postmenopausal replacement therapy (Lobo & Speroff 1994). However, the mechanism by which this protection is mediated has remained obscure. It has traditionally been thought to be a result of potentially favourable changes in blood lipids and lipoproteins

(Lobo & Speroff 1994), but a number of animal studies strongly suggest a direct effect on the vascular system (Adams et al 1990, Hayashi et al 1992, Hough & Zilversmit 1986, Wagner et al 1991). Two components play a role in the molecular mechanism of action of steroid hormones at the tissue level: their metabolism and the presence of receptors. Oestrogens may derive from circulating hormones and be metabolized by oestradiol-17β-hydroxysteroid dehydrogenase, the enzyme responsible for the interconversion of oestrone and oestradiol. Oestrogens can also be derived from *in situ* conversion of androstenedione, a reaction mediated by the aromatase enzyme complex. To our knowledge, these reactions, which have been characterized in adipose tissue and probably also exist in other tissues including bone (Purohit et al 1992), have not been studied in the vascular parietal wall. Binding studies using radiolabelled oestradiol have suggested the presence of oestrogen receptors in vascular cells of different species (Colburn & Buonassissi 1978, Horwitz & Horwitz 1982), but their characterization, in terms of protein and mRNA, has been reported only recently by Orimo et al (1993) in rat cells and by our group in human and bovine cells (Bayard et al 1994). Further reports have since confirmed these observations *in vitro* (Karas et al 1994) as well as *in vivo* (Losordo et al 1994). Because bovine aortic endothelial cells are a commonly used model of vascular pathophysiology, we present results of studies on the metabolism of oestrogens and the functional ability of oestrogen receptors in these cells.

Materials and methods

Cell cultures

Primo-secondary cultures of bovine vascular endothelial cells were established as previously described (Gospodarowicz et al 1976). Cells at passages four to 12 were seeded at a density of 5×10^4 cells/cm^2 in 60 mm diameter dishes (Nunc, Roskilde, Denmark). Cells were grown in Dulbecco's modified Eagle's medium (DMEM; Gibco BRL) supplemented with 10% calf serum, 2 mM glutamine, 50 ng/ml gentamycin, 2.5 μg/ml amphotericin and 1 ng/ml recombinant basic fibroblast growth factor (bFGF). They were maintained at 37 °C in a humidified atmosphere with 5% CO_2. Cells were subcultured every week. One week before the experiments, cells were switched to phenol red-free DMEM containing glutamine, gentamycin, amphotericin, bFGF and 10% charcoal-treated fetal calf serum (CTFCS-DMEM).

Assay of oestradiol-17β-hydroxysteroid dehydrogenase and aromatase activities

Confluent cells, grown in 60 mm diameter petri dishes in CTFCS-DMEM, were washed twice with phosphate-buffered saline (PBS). The enzymic conversion of steroids from the precursor to the product was studied using intact whole-cell monolayers. 1β-[^{3}H]androstenedione (45 pmol, 27.5 Ci/mmol), (1,2,6,7)[^{3}H]-oestrone (5 pmol, 90 Ci/mmol) or (1,2,6,7)[^{3}H]oestradiol (4 pmol, 102 Ci/mmol), obtained from Amersham, were added to the cultures in 2.5 ml serum-free DMEM. After a 20 h incubation, aliquots of medium were removed and the aromatase, oestradiol-17β-hydroxysteroid dehydrogenase (reductive, oestrone→oestradiol) or oestradiol-17β-hydroxysteroid dehydrogenase (oxidative, oestradiol→oestrone) activities were determined. After extraction with ethyl acetate, steroids were separated by high-pressure liquid chromatography (HPLC; Hewlett Packard 1090M) with a 5 μm spherical C18 resolve TM column (Waters, Millipore) and the radioactivity was measured with a radioactivity monitor (LB 506D, Berthold) as described by Parinaud et al (1988). As a control, reagents were incubated in culture dishes in the absence of cells. More than 98.5% of the radioactive steroids eluted with the expected retention time; this was taken into account in the measurement of enzyme activities. Aromatase activity was assayed by measuring the production of ^{3}H$_2$O from 1β-[^{3}H]androstenedione (Ackerman et al 1981) in the presence or absence of the specific aromatase inhibitor 4-hydroxyandrostenedione (1 μM). The number of cells in each dish was measured with trypsinized suspensions using a Coulter counter (Coultronics, model ZM). Results are expressed as fmol or pmol of product per 10^6 cells per 20 h (mean $\pm$ SD, $n = 3$).

Immunostaining of cytospin preparations

Cells grown in CTFCS-DMEM were washed three times in cold Ca^{2+}- and Mg^{2+}-free PBS, trypsinized, resuspended in cold CTFCS-DMEM and kept at 4 °C until centrifugation (Cytospin 3, Shandon). Cytospin preparations of the cultured cells were immunostained using the alkaline phosphatase and anti-alkaline phosphatase technique (Dakopatts, Lab-Trappes, France) as described by Cordell et al (1984). Supernatants of hybridoma cultures producing the anti-oestrogen receptor monoclonal antibodies 1D5 and 1C5 (IgG1 isotype) were used undiluted (Al Saati et al 1993). As negative controls, we either used an unrelated anti-T-lymphocyte antibody, Dako-T3, at 1 : 100 dilution (IgG1 class; Dakopatts) or carried out immunostaining after omitting the primary antibody. All cytospin preparations were immunostained using both the conventional method and the antigen retrieval method described by Shi et al (1991), with some modifications. In the latter case, cytospin preparations were placed in plastic jars containing 10^{-3} M sodium citrate buffer, pH 6. The jars were heated in a microwave oven (Toshiba, ER-7720 W) at the highest power setting (650 W) for two 5 min cycles with an interval of 1 min between cycles when the jars were refilled with distilled water. They were then removed from the oven and allowed to cool for 20 min at room temperature. Immunostaining with 1D5 and 1C5 anti-oestrogen receptor antibodies has been reported to be stronger after microwave heating of tissue sections (Al Saati et al 1993).

Transfection experiments

We transfected cells, which had been grown in CTFCS-DMEM, with 10 μg of the circular pERE-tkFLuc plasmid per 2×10^6 cells, using Lipofectin (in accordance with the manufacturer's instructions; Gibco-BRL). This reporter plasmid was obtained by modification of the ptkFLuc plasmid (Gouilleux et al 1991). The fragment -331 to -87 of the vitellogenin A2 gene, containing the palindromic core sequence 5'-GGTCACAGTGACC-3' (Klein-Hitpaß et al 1986), was inserted as an oestrogen-responsive element (ERE) in the 5' position of the partial promoter sequence of the herpes simplex virus thymidine kinase (tk) driving the luciferase marker (FLuc). Hormones, diluted in ethanol, were added for 24 h. Cells were harvested and aliquots of the cell extracts were assayed for luciferase activity (Nguyen et al 1988). pERE-tkFLuc reporter activity was expressed as a percentage of the activity of control cells treated with ethanol alone (1 : 1000, volume : volume). Only 2–4% of the cells displayed a typical nuclear staining after transfection with a constitutively expressing β-galactosidase gene, but this percentage was sufficient for luciferase activity measurements (data not shown).

To determine the oestrogen receptor-independent luciferase activation, we used the ptkFLuc plasmid (10 μg) instead of pERE-tkFLuc. As a positive control for ethinyloestradiol responsiveness, we overexpressed the oestrogen receptor in cells co-transfected with 5 μg of the pSG5-HEG0 plasmid containing the human oestrogen receptor cDNA (Green et al 1986, 1988) and 5μg pERE-tkFLuc.

Results

Aromatase and oestradiol-17β-hydroxysteroid dehydrogenase activities

The activity of the oestradiol-17β-hydroxysteroid dehydrogenase enzyme complex was found to be similar in both the reductive and oxidative directions (0.31 ± 0.03 and 0.26 ± 0.08 pmol/10^6 cells per 20 h, respectively; mean $\pm$ SD, $n = 3$). As shown in Fig. 1, between a quarter and a third of the radioactive oestrogens recovered from the culture medium could not be extracted with ethyl acetate. These hydrophilic metabolites were not characterized further. HPLC analysis of the conversion products of tritiated androstenedione also showed a 17-ketoreductase enzyme activity, with the generation of testosterone. A significant aromatase activity was also detected (10 ± 4 fmol/10^6 cells per 20 h; mean $\pm$ SD, $n = 3$); this activity was completely inhibited by 4-hydroxyandrostenedione.

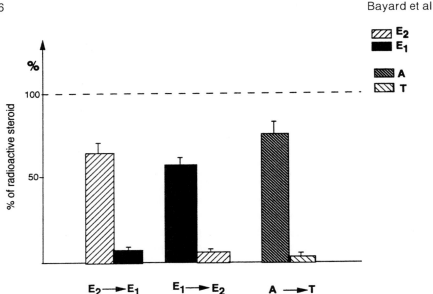

FIG. 1.   Assay of enzyme activities. ABAE (aortic bovine arch endothelial) cells were grown in CTFCS-DMEM until confluent and the conversion of the steroid precursors to the product steroids was carried out using $(1,2,6,7)[^3H]$oestrone $(E_1, 5\,pmol, 90\,Ci/mmol)$, $(1,2,6,7)[^3H]$oestradiol $(E_2, 4\,pmol, 102\,Ci/mmol)$, or $1\beta\text{-}[^3H]$androstene-dione $(A, 45\,pmol, 90\,Ci/mmol)$. After incubating the cells for 20 h, we used aliquots of medium to determine oestradiol-17β-hydroxysteroid dehydrogenase (reductive, $E1{\rightarrow}E2$), oestradiol-17β-hydroxysteroid dehydrogenase (oxidative, $E2{\rightarrow}E1$) and 17-ketoreductase activities after separation by HPLC. Results are expressed as percentages of the radioactivity recovered in the ethyl acetate extracts (means $\pm$ SD of three experiments with measurements made in duplicates). T, testosterone.

Immunocytochemical studies

To test the specificity of the antibodies, we first carried out immuno-cytochemical analyses on sections of bovine uterus. Specific staining was detected clearly in the nuclei of endometrial cells (data not shown). As shown in Plate I, we discovered a weak-to-moderate granular cytoplasmic staining when we used either 1C5 or 1D5 antibodies in vascular endothelial cells, both of which recognize epitopes in the A/B domain of the oestrogen receptor (amino acids 118–140). Occasional nuclei displayed some tiny red granules but this nuclear staining could not be confirmed with certainty. Microwave heating of the cytospin preparations did not improve the staining. Addition of 10^{-7} M ethinyloestradiol to the culture medium one hour before trypsinization had no significant effect on either the intensity or the compartmentalization of the staining pattern. Cells were found to be

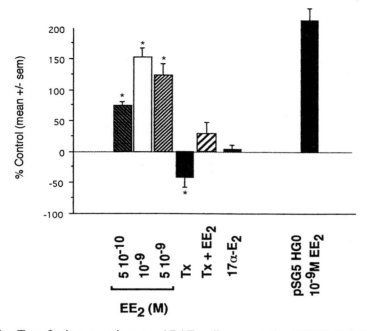

FIG. 2. Transfection experiments. ABAE cells, grown in CTFCS-DMEM, were transfected using Lipofectin with 10 μg circular DNA of the pERE-tkFLuc plasmid for 2×10^6 cells. Eighteen hours later the indicated concentrations of hormones were added. Ethinyloestradiol (EE$_2$) was used in this series of experiments to prevent interference resulting from the metabolism of the ligand. Cells were harvested 30 h later and aliquots of the cell extracts were assayed for luciferase activity. pERE-tkFLuc reporter activity is expressed as a percentage of the activity of control treated cells (1 : 1000 ethanol, volume : volume). Tamoxifen (Tx) was used at a concentration of 10^{-6} M in the absence or in the presence of ethinyloestradiol (10^{-9} M). Specificity was tested using 17α-oestradiol (17α-E$_2$, 10^{-7} M). Positive controls for ethinyloestradiol responsiveness were obtained by oestrogen receptor overexpression in cells co-transfected with 5 μg pSG5-HEG0 and 5 μg pERE-tdFLuc. Data represent means $\pm$ SEM from two different experiments with observations made in quadruplicate in each experiment. * indicates significantly different ($P < 0.05$) from control.

unreactive when the Dako-T3 antibody was used as negative control or when the primary antibody was omitted (Plate I [c]).

Oestrogen-responsive element luciferase expression

To determine whether the oestrogen receptor protein was capable of specific, oestrogen-induced transcriptional activation, we carried out transfections using the reporter plasmid pERE-tkFLuc. Ethinyloestradiol was used in this series of experiments to prevent interference from metabolism of the ligand. After the cells had been incubated for 24 h with concentrations of ethinyl-oestradiol ranging from 0.5 to 5×10^{-9} M, the induction of reporter activity increased in a dose-dependent fashion (Fig. 2). A saturable response was observed with maximum stimulation at 10^{-9} M. Treatment with the anti-oestrogen tamoxifen inhibited the stimulation induced by 10^{-9} M ethinyl-oestradiol, providing further evidence for the specificity of the ethinyl-oestradiol-mediated transactivation of the ERE reporter. Tamoxifen alone resulted in a 10% reduction in luciferase activity, suggesting the presence of some residual oestrogenic activity in the culture medium and indicating that tamoxifen acted as a pure anti-oestrogen in these conditions. Luciferase activation was specifically induced by ethinyloestradiol. The isomer 17α-oestradiol had no effect at a concentration of 10^{-7} M.

We co-transfected the cells with oestrogen receptor cDNA and the pERE-tkFLuc plasmid to determine whether the oestrogen-dependent induction of reporter activation might be limited by the abundance of endogenous receptors. In cells which overexpressed oestrogen receptor, ethinyloestradiol stimulation resulted reproducibly in a higher induction of reporter activity than in cells transfected with pERE-tkFLuc alone (Fig. 2). Finally, use of the basic reporter plasmid ptkFLuc instead of pERE-tkFLuc abolished the ethinyloestradiol-dependent luciferase induction. This confirmed ptkFLuc as unresponsive to oestrogenic stimulation and indicated that for oestrogenic stimulation to occur, the presence of ERE, as a *cis*-acting enhancer element, is required.

Taken together, these results indicate that the ethinyloestradiol-dependent induction of luciferase activity reflects the presence of functional oestrogen receptors in vascular endothelial cells.

Discussion

In this series of experiments we have demonstrated for the first time that vascular endothelial cells synthesize oestrogens and exhibit oestradiol-17β-hydroxysteroid dehydrogenase and 17-ketoreductase activities, although we have not determined whether the latter two activities are mediated by the same or different enzymes. The production of polar metabolites also suggests sulphatase and/or glucuronidase activities. These data confirm and amplify previous observations made with human and rat aortic smooth muscle cells (Bayard et al 1994, 1995). Although adipose tissue is generally considered to be the primary site for the peripheral formation of oestrogens, our results indicate that the vascular wall is also a candidate for extra-ovarian oestrogen production, in agreement with the suggestion made by Ackerman et al (1981). These observations suggest that the vascular generation and metabolism of oestrogens may play a role in the molecular mechanisms of action of these hormones. The regulation of these enzymic activities will have to be studied. Growth factors, such as insulin-like growth factor 1 and bFGF, and the cytokines interleukins 1 and 6, have been found to regulate oestrogen synthesis in breast cancer cells (Schmidt & Loffler 1994) and are produced in the vascular wall.

We have also shown that functional oestrogen receptors are present in vascular endothelial cells. We achieved this by demonstrating the trans-activating capability of these receptors after transient transfection of cultured endothelial cells with a plasmid containing both the luciferase reporter gene and an ERE as a transcriptional enhancer. As expected for steroid hormone-regulated gene expression, the ethinyloestradiol-induced luciferase activity was specific to the ligand structure. Basal luciferase activity was similar in transfection experiments using both ptkFLuc and pERE-tkFLuc plasmids (data not shown). ptkFLuc-transfected cells were unresponsive to oestrogenic stimulation, demonstrating the specific role of the *cis*-acting DNA sequence (ERE) despite the presence of a cryptic AP-1 binding site located in the 5′ position of this ERE in the reporter plasmid (Kushner et al 1994, Lopez et al 1993). Maximal induced luciferase activity appeared reproducibly in the range of 140–150% of the basal activity. These results again confirm and amplify our previous observations (Bayard et al 1994, 1995). They are also in agreement with the observations of Karas et al (1994), in human vascular smooth muscle cells, but clearly differ from those from similar studies conducted in osteoblastic cells, where a seven- to eightfold increase in the activity of a similar gene receptor was found (Ernst et al 1991). Overexpression of the transacting factor (oestrogen receptor) enhanced ethinyloestradiol-dependent expression of the reporter plasmid, thus showing that there was no specific limiting factor in the endothelial cells. The low intensity of the response may reflect the low concentration of the oestrogen receptor protein or may be related to its peculiar cell compartmentalization. In binding studies and ELISA of the soluble fraction of cell homogenates, we detected a concentration of oestrogen receptor equivalent to about 300 binding sites per cell (our unpublished data), which is similar to the concentration measured in osteoblasts (Eriksen et al 1988, Komm et al 1988) and thus does not provide an explanation for the phenomenon.

Our immunocytochemical analyses, in which we used two different monoclonal antibodies recognizing epitopes in the A/B domain of the oestrogen receptor molecule, showed a predominant cytoplasmic localization in vascular endothelial cells, even after the addition of ethinyloestradiol to the culture medium. This peculiar compartmentalization has previously been observed in the brain of different species (Lehman et al 1993). It has also been observed in vascular smooth muscle cells by Orimo et al (1993) and by ourselves (Bayard et al 1994). In rat aortic smooth muscle cells we have excluded the deletion of proto-signals for nuclear localization, located in domains C, D and E of the oestrogen molecule and corresponding to exon 4 of the human gene (Ylikomi et al 1992), as an explanation of these observations (Bayard et al 1995). Such a deletion has been observed in the brains of lizards and rats, as well as in mammary carcinoma cells (Pfeffer et al 1993, Skipper et al 1993) and should result in the cytoplasmic localization of a functionally silent receptor (Koehorst et al 1994). The transactivating ability of oestrogen receptor suggests that such a deletion can also be excluded in vascular endothelial cells, but also indicates that oestrogen receptors are present in the nucleus in too low a concentration to be detectable. Further work is necessary for us to understand the mechanisms and the biological implications of this peculiar compartmentalization of oestrogen receptors.

Acknowledgements

We thank Dr P. Chambon for providing the expression vector pSG5-HEG0, Drs J. Parinaud and C. Clamagirand for their help in analysing steroid metabolism, Dr P. Augereau for providing the fragment -331 to -87 of the vitellogenin A2 gene, and G. Tatinian, K. Kovacs and M. Larribe for technical and secretarial assistance. This work was supported in part by INSERM, the Ministère de la Recherche et de la Technologie and the Conseil Régional Midi-Pyrénées.

References

Ackerman GE, Smith ME, Mendelson CR, MacDonald PC, Simpson ER 1981 Aromatization of androstenedione by human adipose tissue stromal cells in monolayer culture. J Clin Endocrinol & Metab 53:412–417

Adams MR, Kaplan JR, Manuck SB et al 1990 Inhibition of coronary artery atherosclerosis by 17-β estradiol in ovariectomized monkeys. Lack of an effect of added progesterone. Arteriosclerosis 10:1051–1057

Al Saati T, Clamens S, Cohen-Knafo E et al 1993 Production of monoclonal antibodies to human estrogen-receptor protein (ER) using recombinant ER (RER). Int J Cancer 55:651–654

Bayard F, Clamens S, Delsol G, Faye J-C 1994 Estrogen receptor in vascular endothelial and smooth muscle cells. J Cell Biochem Suppl 18A:282(abstr)

Bayard F, Clamens S, Meggetto F, Blaes N, Delsol G, Faye J-C 1995 Estrogen synthesis, estrogen metabolism, and functional estrogen receptors in rat arterial smooth muscle cells in culture. Endocrinology 136:1523–1529

Colburn P, Buonassissi V 1978 Estrogen-binding sites in endothelial cell cultures. Science 201:817–819

Cordell JL, Falini B, Erber WN et al 1984 Immunoenzymatic labeling of monoclonal antibodies using immune complexes of alkaline phosphatase and anti-alkaline phosphatase (APAAP complexes). J Histochem Cytochem 32:219–229

Eriksen EF, Colvard DS, Berg NJ et al 1988 Evidence of estrogen receptors in normal human osteoblast-like cells. Science 241:84–86

Ernst M, Parker M, Rodan GA 1991 Functional estrogen receptors in osteoblastic cells demonstrated by transfection with a reporter gene containing an estrogen response element. Mol Endocrinol 5:1597–1606

Gospodarowicz D, Moran J, Braun D, Birdwell CR 1976 Clonal growth of bovine endothelial cells in tissue culture: fibroblast growth factor as a survival agent. Proc Natl Acad Sci USA 73:4120–4124

Gouilleux F, Sola B, Couette B, Richard-Foy H 1991 Cooperation between structural elements in hormono-regulated transcription from the mouse mammary tumor virus promoter. Nucleic Acids Res 19:1563–1569

Green S, Walter P, Kumar V, Krust A, Bornert J-M, Argos P, Chambon P 1986 Human oestrogen receptor cDNA: sequence, expression and homology to v-erb-A. Nature 320:134–139

Green S, Issemann I, Sheer E 1988 A versatile *in vivo* and *in vitro* eukaryotic expression vector for protein engineering. Nucleic Acids Res 16:369

Hayashi T, Fukuto JM, Ignarro LJ, Chaudhuri G 1992 Basal release of nitric oxide from aortic rings is greater in female rabbits than in male rabbits: implications for atherosclerosis. Proc Natl Acad Sci USA 89:11259–11263

Horwitz KB, Horwitz LD 1982 Canine vascular tissues are targets for androgens, estrogens, progestins, and glucocorticoids. J Clin Invest 69:750–758

Hough IL, Zilversmit DB 1986 Effect of 17β-estradiol on aortic cholesterol content and metabolism in cholesterol-fed rabbits. Arteriosclerosis 6:57–61

Karas RH, Patterson BL, Mendelsohn ME 1994 Human vascular smooth muscle cells contain functional estrogen receptor. Circulation 89:1943–1950

Klein-Hitpaß L, Schorpp M, Wagner U, Ryffel GU 1986 An estrogen-responsive element derived from the 5′ flanking region of the *Xenopus* vitellogenin A2 gene functions in transfected human cells. Cell 46:1053–1061

Koehorst SGA, Cox JJ, Donker GH et al 1994 Functional analysis of an alternatively spliced estrogen receptor lacking exon 4 isolated from MCF-7 breast cancer cells and meningioma tissue. Mol Cell Endocrinol 101:237–245

Komm BS, Terpening CM, Benz DJ et al 1988 Estrogen binding, receptor mRNA, and biologic response in osteoblast-like osteosarcoma cells. Science 241:82–84

Kushner PJ, Baxter JD, Duncan KG et al 1994 Eukaryotic regulatory elements lurking in plasmid DNA—the activator protein-1 site in *Puc*. Mol Endocrinol 8:405–407

Lehman MN, Ebling FJP, Moenter SM, Karsh FJ 1993 Distribution of estrogen receptor-immunoreactive cells in the sheep brain. Endocrinology 133:876–886

Lobo RA, Speroff L 1994 International consensus conference on postmenopausal therapy and the cardiovascular system. Fertil Steril 61:592–595

Lopez G, Schaufele F, Webb P, Holloway M, Baxter JD, Kushner PJ 1993 Positive and negative modulation of Jun action by thyroid hormone receptor at a unique AP1 site. Mol Cell Biol 13:3042–3049

Losordo DW, Kearney M, Kim EA, Jekanowski J, Isner JM 1994 Variable expression of the estrogen receptor in normal and atherosclerotic coronary arteries of premenopausal women. Circulation 89:1501–1510

Nguyen VT, Morange M, Bensaude O 1988 Firefly luciferase luminescence assays using scintillation counters for quantitation in transfected mammalian cells. Anal Biochem 171:404–408

Orimo A, Inoue S, Ikegami A et al 1993 Vascular smooth muscle cells as target for estrogen. Biochem Biophys Res Commun 195:730–736

Parinaud J, Beaur A, Bourreau E, Vieitez G, Pontonnier G 1988 Effect of a luteinizing hormone-releasing hormone agonist (Buserelin) on steroidogenesis of cultured human preovulatory granulosa cells. Fertil Steril 50:597–602

Pfeffer U, Fecrotta E, Castagnetta L, Vidali G 1993 Estrogen receptor variant messenger RNA lacking exon 4 in estrogen-responsive human breast cancer cell lines. Cancer Res 53:741–743

Purohit A, Flanagan AM, Reed MJ 1992 Estrogen synthesis by osteoblast cell lines. Endocrinology 131:2027–2029

Ross R 1993 The pathogenesis of atherosclerosis: a perspective for the 1990s. Nature 362:801–809

Schmidt M, Löffler G 1994 Induction of aromatase in stromal vascular cells from human breast adipose tissue depends on cortisol and growth factors. FEBS Lett 341:177–181

Shi S-R, Key ME, Kalra KL 1991 Antigen retrieval in formalin-fixed, paraffin-embedded tissues: an enhancement method for immunohistochemical staining based on microwave oven heating of tissue sections. J Histochem Cytochem 39:741–748

Skipper JK, Young LJ, Bergeron JM, Tetzlaff MT, Osborn CT, Crews D 1993 Identification of an isoform of the estrogen receptor messenger RNA lacking exon 4 and present in the brain. Proc Natl Acad Sci USA 90:7172–7175

Wagner JD, Clarkson TB, St Clair RW, Schwenke DC, Shively CA, Adams MR 1991 Estrogen and progesterone replacement therapy reduces LDL accumulation in the coronary arteries of surgically postmenopausal cynomolgus monkeys. J Clin Invest 88:1995–2002

Ylikomi T, Bocquel MT, Berry M, Gronemeyer H, Chambon P 1992 Cooperation of proto-signals for nuclear accumulation of estrogen and progesterone receptors. EMBO J 11:3681–3694

DISCUSSION

McEwen: In relation to the aromatization that you referred to at the beginning, what's the situation in males? The story that I've heard is that the females are protected because they have oestrogens. If there is a significant amount of aromatization in the male, there should be some production of oestrogens; the question is, is there some programming of the endothelial cell because of sexual differentiation that makes it respond differently to those oestrogens in males versus females?

Bayard: To start with, I didn't know whether oestradiol was coming directly from aromatization of testosterone or from aromatization of androstenedione and conversion of oestrone. The aromatase activity could then generate higher concentrations of these hormones in the vascular compartment than may be present in the circulation. Indeed, we were surprised to find it. This is why we need to study the regulation of this enzyme's activity because it may be that in some conditions men can produce enough of these hormones to be protected. We know that there is a genetic factor that influences whether or not individuals develop atherosclerosis.

Thijssen: How did you measure the aromatase synthesis? Good aromatase assays are very difficult to do.

Bayard: We have measured this activity by the conversion of substrate to product as analysed by HPLC and also by the generation of ^3H_2O. Now I would like to confirm these observations by molecular biological techniques and look at the mRNAs for the different enzymes.

James: What may be more important is what happens to the oestrogen after its produced. You demonstrated that there is a 17-hydroxysteroid dehydrogenase present, and you implied that the main direction of metabolism was towards oestrone. It may be that the role of the oestradiol dehydrogenase is to protect the tissues by converting the oestradiol to a less active oestrogen, oestrone. What regulates that activity?

Bayard: All we can say is that the enzyme activity is present: we have yet to study its regulation.

Horwitz: It is interesting that you only find oestrogen receptors in the cytoplasm: if you add oestradiol, do the receptors translocate to the nucleus?

Bayard: Adding oestradiol doesn't change anything.

Horwitz: That is important, because it suggests that you really are dealing with an unusual variant of the oestrogen receptor. Despite the fact that you see apparently normal mRNAs, they may not encode normal proteins, so these may not be *bona fide* oestrogen receptor-positive cells. You may be looking at membrane effects, like the ones we were discussing yesterday, or some other unusual oestrogen receptor. In that case, a search for the usual transcriptional effects may be uninformative.

Do you see any difference between arteries from male and female animals?

Bayard: All these studies have been done with bovine and rat endothelial and smooth muscle cells from both male and female animals, and we didn't see any difference. That's why I feel that there is no sex difference. As far as the cytoplasmic localization is concerned, we have used an inducible promoter to obtain different levels of receptor in these cells. At low levels of expression we see cytoplasmic as well as a nuclear staining. At high levels of expression the staining is clearly nuclear. Our interpretation is that even if we cannot exclude different receptor types, there might also be a factor which limits the translocation process. Obviously, in the vascular cells the receptors can be transcriptionally active and must go to the nucleus. Some of them are cycling to the nucleus, but most of them stay outside the nucleus.

Foegh: I have a couple of comments before you begin to speculate whether or not the oestrogen receptor on smooth muscle cells in your culture system differs from the classical receptor. We have been studying the oestradiol receptor on rat coronary smooth muscle cells. We've been using rat-specific oestrogen receptor oligonucleotide primers and a cDNA fragment that was amplified by PCR. Nucleotide sequence analysis confirmed that this was a portion of the complete rat oestrogen receptor cDNA sequence. We also find that it has the same specificity and affinity as the oestradiol receptor you usually find in breast cancer cells. We also agree with you that the density of the receptor is very low.

Jensen: The conclusion that the native oestrogen receptor is exclusively nuclear, because one sees very little extranuclear staining with most immunohistochemical techniques, is a fallacy, which I have discussed at greater length in a chapter in Malcolm Parker's book (Jensen 1991). Most claims of exclusive nuclear localization ignore the fact that the extranuclear volume is eight to 10 times as large as the nuclear volume, so, with a procedure that measures concentration rather than quantity, a significant amount of extranuclear receptor could go undetected. But I am surprised that you don't see anything in the nuclei of your vascular cells. Could it be that when a receptor is bound in the nucleus the epitope for your particular antibody, which recognizes the A/B region of the receptor, is somehow masked? Have you tried any other antibodies, such as the one in the Abbott kit, which usually shows 'exclusive' nuclear staining?

Bayard: We have used the H222 antibody and we got similar results, with no staining in the nucleus.

Jensen: Your receptor must be different.

Rochefort: Your oestrogen receptor looks like as if it may be lysosomal. Have you tried to identify its exact localization?

Bayard: No.

Baulieu: When you transfect cells with *bona fide* receptor, which becomes localized to the nucleus, if you then look at the dose–response to oestrogen, do you observe an increased response to the same concentration of hormone?

Bayard: We have not yet done that type of study.

Foegh: We did a simpler experiment in which we incubated pig coronary artery explants *in vitro* in the presence of thymidine-labelled oestradiol. We found by autoradiography that oestradiol localized in the nucleus, not in the cytoplasm (Vargas et al 1993).

Manolagas: It would be worth looking to see if the production of interleukin 6 (IL-6) by endothelial cells is regulated by oestrogen. There is evidence that FGF is a potent stimulant of IL-6 production, and as you know there is evidence that IL-6 is present in atherosclerotic lesions.

Baulieu: In which cells?

Manolagas: I believe the endothelial cells, the macrophages and some stromal cells of the vessel produce IL-6 in large quantities.

Have you had the opportunity to investigate what happens when you remove oestrogens? Can you see any changes in the amount of the receptor or any of the factors that you're looking at upon ovariectomy?

Bayard: We haven't looked at the *in vivo* situation.

Pfaff: Recently, Thomas Lüscher reported an increase in oestrogen-induced endothelial nitric oxide synthase (NOS) RNA. Are you aware of this?

Bayard: I am aware that there are conflicting results on this subject. We have measured the production of NOS mRNA and protein, without finding any variation. This contrasted with an increased production of NO by the

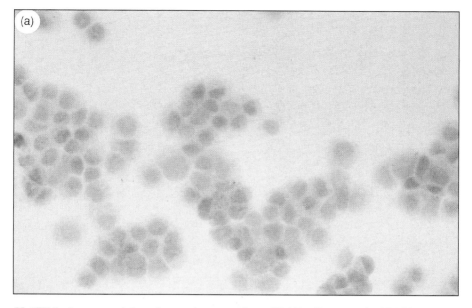

PLATE 1 Immunostaining of cytospin preparations. ABAE cells, grown in CTFCS-DMEM, were trypsinized, resuspended and centrifuged at 4 °C. We immunostained the cytospin preparations by the alkaline phosphatase and anti-alkaline phosphatase technique using the 1D5 monoclonal antibody which recognizes epitopes in the A/B domain of the oestrogen receptor (a). Note the weak to moderate cytoplasmic staining and the absence of nuclear staining. The absence of nuclear staining was further demonstrated in cytospin preparations without nuclear counterstaining (b, overleaf). Anti-T-lymphocyte antibody (Dako-T3) of IgG1 class was used as a negative control (c, overleaf).

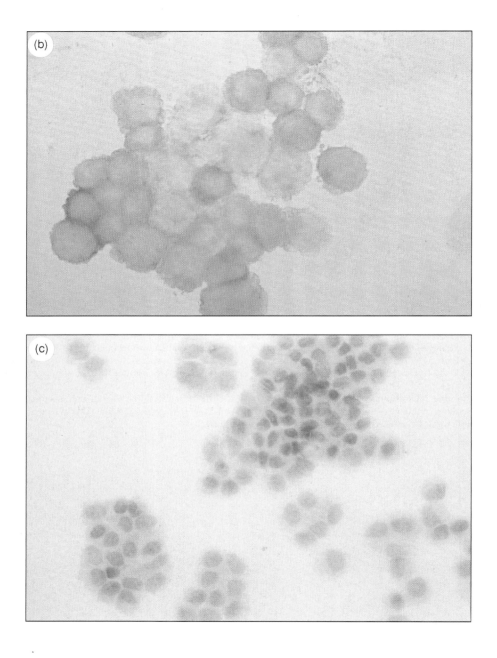

endothelial cells, and appeared to be related to a decrease of superoxide anions in these cells, then probably decreasing the formation of peroxinitrites.

Pfaff: There's a third gene for NOS which was first characterized in the nervous system. Have you ever looked at it?

Bayard: We just looked at the gene which has been described as the 'endothelial' constitutive NOS. In endothelial cells, it has also been reported that there may be an inducible NOS that we didn't see in our experiment.

Bonewald: It is interesting that you claim that oestrogen has an effect on oxygen radical production. Some groups are studying bone-resorbing cells, the osteoclasts, which are also terrific generators of oxygen radicals (Garrett et al 1990, Key et al 1990). Oestrogen, again, seems to inhibit osteoclastic activity (Oursler et al 1993). Is this another general mechanism of oestrogen action for any cells which generate these ions? Has anybody looked to see how oxygen radical production is down-regulated? Is oestrogen increasing superoxide dismutase or some other inhibitor? Is this a rapid effect or do the cells have to go through transcription?

Bayard: From literature that I have read and from our own experience I haven't found any solution to this problem. Peroxide anions are produced by many oxidoreductase activities and this makes them difficult to study, because we have to look at many enzymes.

Inoue: Recently, we found that endothelin 1 is down-regulated by oestrogen in smooth muscle cells (M. Akishita et al, unpublished results). We therefore think that some oestrogen-responsive genes exist in smooth muscle cells.

Castagnetta: Keeping with the effects observed in breast cancer patients, anti-oestrogen treatment exerts a protective effect against cardiovascular disease. We also have convincing evidence for functional high-affinity binding sites of oestrogen in smooth muscle cells of the human aorta (Campisi et al 1993); in addition, use of the D5 monoclonal antibody revealed a concurrent, intensive staining for the oestrogen receptor-associated 27 kDa heat shock protein (Hsp27), which is believed to be a marker of oestrogen sensitivity (King et al 1987). The critical point remains how oestrogens exert their action. It might not just be through the control of proliferative activity of cells.

Bayard: I agree with your observation, and reports from different groups that tamoxifen could be protective against atherosclerosis. But the activity of tamoxifen in this situation may be more complicated than just being mediated by an antagonism or partial agonism of oestrogens. Tamoxifen by itself can be an being antioxidant, so this may be another way it could act in these cells. On the other hand, the fact that tamoxifen behaved like an anti-oestrogen on the reporter gene that we used doesn't mean that on another promoter it couldn't be active as an agonist.

Horwitz: Larry Horwitz and I showed quite a long time ago that oestradiol, acting through the oestrogen receptors, could induce progesterone receptors in vascular smooth muscle cells (Horwitz & Horwitz 1982). Perhaps you're looking at the wrong hormone: maybe you should be treating with progesterone, not with oestradiol!

Oelkers: You started off by emphasizing the difference in the incidence of atherosclerosis in women and men. Are there any direct effects of androgens on your system? In hypertension-prone strains of rats, blood pressure is increased by testosterone and not by oestrogens (Chen & Meng 1991). In spontaneously hypertensive rats, the males always have a blood pressure about 40 mmHg or so higher than the females. If you castrate them, blood pressure develops as in females, but ovariectomy of the female has no effect. Testosterone substitution leads to an increase in blood pressure. Are there androgen receptors in the endothelial cells or in smooth muscle cells?

Bayard: We have tried to look at this point, but so far we have been unable to characterize this type of receptor in our vascular cells.

Manolagas: In response to the speculation that the anti-atherosclerotic effect might be mediated by progesterone, it is important to note that in the big nurses trial (Stampfer et al 1991) where they followed up almost 50 000 women on hormone replacement therapy (HRT), the majority of the subjects were only taking oestrogen. Under those circumstances they found a 50% reduction in the incidence of cardiovascular deaths.

Horwitz: Was their progesterone suppressed?

Manolagas: I don't think they looked at this issue.

Horwitz: So if they were being given oestrogens, the oestrogens could induce progesterone receptors, leading to enhanced sensitivity to endogenous progesterone. I'm just saying that this clinical trial is, by itself, not entirely conclusive.

Bayard: I think Stavros Manolagas is right, because what worries the people working in the field is to have to give progesterone in association with oestradiol to prevent endometrial cancer. At the present time, progesterone is considered to be a negative factor in the prevention of atherosclerosis in postmenopausal women.

Manolagas: There is some evidence that combination of progesterone with oestradiol in HRT cancels out the beneficial effect of oestrogens on lipoprotein production.

Thijssen: That's not completely true; it depends on the progestin you are looking at. They have very big differences in their metabolic profiles.

Jensen: Can you maintain the favourable effects of oestrogen on the cardiovascular system using combined HRT?

Thijssen: Yes.

Jensen: The usual progestin used is provera (medroxyprogesterone acetate)—is that good or bad?

Manolagas: I think the evidence that progesterone negates the beneficial effects of oestrogen comes from studies where subjects were using provera.

Thijssen: Norgestrel and desogestrel have less effect on the oestrogen effects.

Vickers: There are some recent studies which show that norgestrel doesn't have the large adverse effects on cholesterol that it was originally thought to have (Nabulsi et al 1993, Seed 1994) and some of the less androgenic progestogens such as medroxyprogesterone acetate have no effect on high-density lipoprotein. Further, our own studies show that adding a progestogen seems to attenuate the adverse effect of oestrogen on serum triglyceride levels. So overall, the combined oestrogen and progestogen preparations may be at least as effective as oestrogen-only HRT in protecting against cardiovascular disease.

Bäckström: New progestogenic compounds, like desorgestrel and gestodene actually have very small androgenic effects. Micronized progesterone is also available; this is a natural substance that may be of interest. I don't think that the last word has been said on this issue.

Jensen: It would appear that in premenopausal women, physiological amounts of progesterone do not negate the beneficial effects of physiological amounts of oestrogens, or else they would have an incidence of cardiovascular problems comparable to that of postmenopausal women.

References

Campisi D, Cutolo M, Carruba G et al 1993 Evidence for soluble and nuclear site I binding of estrogens in human aorta. Atherosclerosis 103:267–277

Chen Y, Meng QC 1991 Sexual dimorphism of blood pressure in spontaneously hypertensive rats is androgen dependent. Life Sci 48:85–96

Garrett IR, Boyce BF, Oreffo ROC, Bonewald L, Poser J, Mundy GR 1990 Oxygen-derived free radicals stimulate osteoclastic bone resorption in rodent bone *in vitro* and *in vivo*. J Clin Invest 85:632–639

Horwitz KB, Horwitz LD 1982 Canine vascular tissues are targets for androgens, estrogens, progestins, and glucocorticoids. J Clin Invest 69:750–758

Jensen EV 1991 Overview of the nuclear receptor family. In: Parker MG (ed) Nuclear hormone receptors. Academic Press, London, p 1–13

Key LL, Ries WL, Taylor RG, Pitzer BL 1990 Oxygen-derived free radicals in osteoporosis: the specificity and location of the nitroblue tetrazolium reaction. Bone 11:115–119

King RBJ, Finlay JR, Coffer AE, Nillis RR, Rubens RD 1987 Characterization and biological relevance of a 29-kDa oestrogen-related protein. J Steroid Biochem 27:471–475

Nabulsi AA, Folsom AR, White A et al 1993 Association of hormone-replacement therapy with cardiovascular risk factors in postmenopausal women. N Engl J Med 328:1069–1075

Oursler MJ, Pederson L, Pyfferoen J, Osdoby PH, Fitzpatrick L, Spelsberg TC 1993 Estrogen modulation of avian osteoclast lysosomal gene expression. Endocrinology 132:1373–1380

Seed M 1994 Postmenopausal hormone replacement therapy, coronary heart disease and plasma lipoproteins. Drugs 47(suppl 2):25–34

Stampfer MJ, Colditz GA, Willett WC et al 1991 Postmenopausal estrogen therapy and cardiovascular disease—ten-year follow-up from the Nurses' Health Study. N Engl J Med 325:756–762

Vargas R, Wroblewska B, Rego A, Hatch J, Ramwell PW 1993 Oestradiol inhibits smooth muscle cell proliferation of pig coronary artery. Br J Pharmacol 109:612–617

Oestradiol inhibition of vascular myointimal proliferation following immune, chemical and mechanical injury

Marie L. Foegh, Yejun Zhao, Michel Farhat* and Peter W. Ramwell*

*Departments of Surgery and of *Physiology and Biophysics, Georgetown University Medical Center, 4000 Reservoir Road NW, Washington DC 20007, USA*

Abstract. The lower incidence of coronary artery disease in premenopausal women and in postmenopausal women treated with oestrogen supports the hypothesis that oestrogen protects the vasculature from injuries or from responding to injuries with arteriosclerosis. The mechanism remains unknown, although currently the most frequent suggestion is that oestrogen induces beneficial quantitative and qualitative changes in serum lipoprotein concentrations. We studied other mechanisms and in particular the direct effects of 17β-oestradiol on vascular smooth muscle and the endothelium. Our focus has been on the vascular response to injury by myointimal and medial thickening, leading to narrowing or occlusion of the vessel. This is frequently seen in coronary arteries within months following angioplasty and a few years following cardiac transplantation. We find that oestradiol treatment protects and reduces the vascular response to injury in three *in vivo* and one *in vitro* models: (1) in the rabbit cardiac allograft where oestrogen inhibits accelerated graft atherosclerosis; (2) in monocrotaline- or hypoxia-induced pulmonary hypotension in the rat, where oestrogen attenuates pulmonary artery pressure, right ventricular hypertrophy and medial thickening of the pulmonary artery; (3) oestrogen protects against balloon injury in rabbit aorta; and (4) it inhibits smooth muscle cell proliferation in the porcine left anterior descending coronary artery.

1995 Non-reproductive actions of sex steroids. Wiley, Chichester (Ciba Foundation Symposium 191) p 139–149

Sexually dimorphic syndromes of the vascular system are of considerable interest in that they provide insight into disease mechanisms. There are a number of such cardiovascular syndromes; these include migraine, cluster headaches, Raynaud's disease, Bürger-Grütz syndrome, systemic hypertension, primary pulmonary hypertension, hot flushes and atherosclerosis.

Of particular interest to the clinician is the putative protection against atherosclerosis by the secretion of oestrogen in young women, since coronary artery disease is the major cause of death in postmenopausal women, and hormone replacement is a very simple and inexpensive therapy. This effect of oestrogen in young women relates to their superior lipid metabolism, especially during pregnancy, when this is due to the metabolic demands of the fetus for glucose. However, in addition to the impact of oestrogen on lipid and carbohydrate metabolism, there are grounds for believing that oestrogen has a direct protective effect on the arterial wall.

This protection against injury may be mediated through the endothelial secretion of antiproliferative factors that prevent the change from the contractile to the secretory smooth muscle cell phenotype. Oestrogen may also have a direct antiproliferative effect on the smooth muscle itself.

A more complex situation which will be considered first here is the potential role of oestrogen therapy in women recipients of whole organ transplants.

Immune injury

Women exhibit enhanced humoral- and cellular-mediated immune responses compared with men. Women are known to have a higher incidence of autoimmune disease (Grossman 1984) and oestrogen appears to promote these diseases (Screpanti et al 1991), with the exception of rheumatoid arthritis where oestrogen therapy suppresses symptoms (Holmdahl et al 1989). Consequently, the effect of sex and oestrogen replacement therapy is a matter of concern in clinical transplantation.

After organ transplantation, women recipients are reported to endure more frequent rejection episodes than men (Esmore et al 1991). This observation is supported by experimental data in rat cardiac transplantation where grafts survive longer in male recipients than females (Enosawa & Hirasawa 1989). It is important to know whether this increased rejection in female cardiac graft recipients relates to increased risk of developing accelerated arteriosclerosis, the major cause of death or regrafting. A recent report on cardiac transplant patients (Mehra et al 1995) showed that the highest risk of accelerated transplant arteriosclerosis was in male recipients of grafts from women, followed by men receiving a graft from men. The lowest risk was in female recipients who received grafts from men. Unfortunately, the investigators did not have enough female recipients of grafts from women to interpret whether the sex differences are due to the sex of the donor or whether the largest determinant is the sex of the recipient. It is also important to know whether the women in this study were pre- or postmenopausal: the mean age of the women was 47, which suggests they were mainly premenopausal.

Thus, it is difficult to predict the exact effect of oestradiol replacement therapy in female transplant recipients and whether it might increase the risk of rejection or obtain the desired effect of inhibiting transplant arteriosclerosis. We have explored the effect of oestradiol in a rabbit heterotopic cardiac allograft model in which substantial coronary arteriosclerosis develops within five to six weeks.

In this model, the allograft is placed in the neck of the rabbit with an end-to-side anastomosis between the aorta and the carotid artery, and between the pulmonary artery and the jugular vein. Both the donor and recipient rabbits are fed a 0.5% cholesterol diet from one week prior to transplantation until five to six weeks after transplantation, when the recipients are killed. This diet increases the normal rabbit serum cholesterol level from 30 mg/dl to 400–600 mg/dl. The rabbits are treated daily with cyclosporin A to prevent rejection in this strong histo-incompatible model.

We evaluated the coronary arteries 3, 7, 21 and 40 days after transplantation by light and electron microscopy as well as immunocytochemical staining in order to assess the evolution of the myointimal hyperplasia (Kuwahara et al 1991).

The first interesting finding was that the myointimal hyperplasia occurred under an ultrastructurally normal endothelium. Further, during the first three weeks this hyperplasia consisted mainly of smooth muscle cells, but after the third week an increasing number of macrophages appeared in the intima. The effect of daily treatment with a dose of 100 μg/kg per day 17β-oestradiol was evaluated (Foegh et al 1987); we found that oestradiol significantly inhibited myointimal hyperplasia six weeks after transplantation.

These studies were repeated and confirmed in a different transplant model in which the rabbit abdominal aorta was transplanted to the neck of a recipient rabbit (Cheng et al 1991). The myointimal hyperplasia was evaluated by morphometry, histology and electron microscopy three weeks after transplantation. Recipient rabbits received cyclosporin A and one of four different doses of oestradiol cypionate (1, 10, 100 or 1000 μg/kg per day) or placebo. All four doses of oestradiol inhibited myointimal hyperplasia even though the lowest dose maintained serum oestradiol levels only slightly above the normal male rabbit levels (29.6 ± 2.7 versus 21.1 ± 0.7 pg/ml, respectively). However, at the lowest dose of oestradiol, macrophages appeared occasionally in the myointima, whereas in the three groups receiving the higher doses of oestradiol no macrophages were seen. It is interesting that the endothelium of the transplanted aorta of the oestradiol-treated rabbits appeared ultra-structurally normal compared with the control group in which degenerative changes were observed in the endothelial cells. The myointima of the control rabbits contained lipid-laden macrophages as well as lipid-laden smooth muscle cells, whereas no macrophages and only a few smooth muscle cells were found in the intima of aortae from the oestradiol-treated rabbits.

There are many potential mechanisms for the inhibitory effect of oestrogen on transplant arteriosclerosis. One mechanism may be the promotion of endothelial prostacyclin and nitric oxide. Schray-Utz et al (1993) report that oestradiol induces the constitutive nitric oxide synthase, and other reports indicate that oestrogen increases vascular prostacyclin (Fogelberg et al 1990, Steinleitner et al 1989). Oestrogen may inhibit free radicals that induce smooth muscle proliferation by a direct effect on O_2 or by increasing enzymes such as superoxide dismutase, catalase and peroxidases.

Finally, it is noteworthy that inhibition of graft arteriosclerosis is associated with decreased vascular insulin-like growth factor (IGF)-1 and other vascular growth factors (Hayry et al 1993).

Chemical injury

Primary pulmonary hypertension is a syndrome over-represented in women. The female predominance appears at puberty. In young women the histopathology is characterized by pulmonary vascular medial hypertrophy associated with concentric laminar intimal fibrosis and plexiform lesions. The pathogenetic denominator is thought to be an undefined injury to the pulmonary endothelium. Thus, a model of this disease has been developed in rats by injuring the pulmonary vascular endothelium by a single subcutaneous injection of monocrotaline (Ghodsi & Will 1981). This causes pulmonary oedema, inflammatory invasion of lung parenchyma, pulmonary hypertension and right ventricular hypertrophy. The development of monocrotaline-induced pulmonary hypertension is sexually differentiated in that female rats develop a lower degree of pulmonary hypertension than their male counterparts do. This sex difference is partially explained by sex differences in the ability of the liver to convert monocrotaline to monocrotaline pyrrole. The latter is the substance that causes the injury to the endothelial cells of the pulmonary vasculature (Kiyatake et al 1994).

In this model we have found oestradiol treatment to attenuate medial thickening in the pulmonary vessels and decrease the right ventricular hypertrophy 30 days after the monocrotaline injury. Oestradiol also prevented the oedema induced by monocrotaline (Farhat et al 1993). Recently, we have shown oestradiol also to protect in hypoxia-induced pulmonary hypertension in rats.

We have evaluated the early ultrastructural changes in the pulmonary vascular endothelium in rats injected with monocrotaline. Oestradiol treatment did indeed preserve the endothelium and prevent the probable consequences of the endothelial injury. Light and electron micrographs showed leukocytic infiltration in the lung parenchyma and pulmonary vessels of the rats given only monocrotaline. Oestradiol treatment also prevented the endothelial cell ultrastructural changes characterized by blebbing, change in height, detachment of adjacent cells, and the appearance of perinuclear vacuoles and cytoplasmic filaments. These changes, demonstrating endothelial dysfunction, were not seen in the group treated with oestradiol (T. Bhatti et al, unpublished results).

Mechanical injury with balloon catheter

In order to study the effect of oestradiol on smooth muscle cell proliferation in native vessels that are not immunologically compromised, we have used the rabbit balloon-injury model. In this model smooth muscle cell proliferation in the media begins 24–48 h following the injury. Thereafter, the smooth muscle cells migrate to the intima and continue proliferating. We subjected rabbits to ballon injury of the aorta and iliac artery. Rabbits pretreated with 100 μg/kg per day oestradiol exhibited a significant decrease in myointimal hyperplasia. In the aorta, the myointimal thickness decreased from 18.6% in the control group to 5.7% in the oestradiol-treated group. The myointimal hyperplasia is expressed as a percentage of area of myointimal hyperplasia over total vessel area. The mechanisms of action of the oestradiol inhibition of the myointimal hyperplasia were further studied in the same model, where thymidine incorporation was used as an index of cell proliferation. We subjected a further series of rabbits to balloon injury of the aorta and treated them with the same doses of oestradiol as were used previously. Forty-eight hours after balloon injury, radioactive thymidine was injected and the rabbits were killed 72 h later. We found that oestradiol significantly inhibited thymidine incorporation in the aorta (Foegh et al 1994).

This inhibitory effect of oestradiol on thymidine uptake *in vivo* was studied further *in vitro* in explants of pig coronary arteries. The object of these experiments was to determine whether or not oestrogen had a direct effect on vascular smooth muscle.

Hearts were obtained from the abattoir. The left anterior descending branch of the coronary tree was isolated and segments without the endothelium were placed in tissue culture media (for details, see Vargas et al 1993). Cell proliferation was measured as thymidine uptake per mg protein. 17β-oestradiol (180–360 nM) inhibited thymidine uptake by pig left anterior descending segments ($P < 0.05$). The inhibition was observed in the absence of phenol red, which is a weak oestrogen receptor agonist. The anti-oestrogens tamoxifen and its more potent metabolite 4-hydroxytamoxifen, both of which are partial oestrogen receptor agonists, also significantly inhibited thymidine uptake. However, neither tamoxifen nor 4-hydroxytamoxifen significantly blocked 17β-oestradiol-induced inhibition of thymidine uptake.

The data suggest that 17β-oestradiol inhibits smooth muscle cell proliferation in porcine left anterior descending segments, possibly through an oestrogen receptor mechanism. This supports the idea that 17β-oestradiol directly protects coronary arteries against myointimal proliferation in premenopausal women.

Conclusion

We have shown in four models that treatment with oestradiol protects the blood vessels against injuries. The first is a rabbit cardiac allograft model in which 17β-oestradiol inhibited accelerated graft atherosclerosis. The second model is that of pulmonary hypertension in the rat induced by either hypoxia or monocrotaline, in which 17β-oestradiol treatment attenuates pulmonary artery pressure, right ventricular hypertrophy and medial thickening of the pulmonary artery. The third model involves balloon injury of rabbit aorta where both morphometry and thymidine incorporation measurements indicate a protective effect of 17β-oestradiol. The fourth model utilizes explants of the porcine left anterior descending coronary artery wherein 17β-oestradiol inhibits thymidine incorporation. The precise mechanism of the antimitogenic effect of oestradiol in vascular tissue is obscure. Recent experiments in our laboratory indicate that the proliferative effect of the growth factor IGF-1, present in vascular tissue, is blocked by either an anti-IGF-1 monoclonal antibody or by oestrogen (M. L. Foegh, Y. Zhao & H. Lou, unpublished results).

Thus, in contrast to the common perception in gynaecology and oncology that oestrogen is mitogenic, our findings in different species and different *in vivo* and *in vitro* models indicate that oestradiol inhibits vascular smooth muscle proliferation. The notion that sex may be an independent determinant as well as oestrogen has some substance. We have found in a syngeneic (same strain) rat aorta transplant model that female vascular grafts respond with lesser myointimal proliferation than the male graft and this response is independent of the sex of the recipient (Foegh et al 1995). Thus, it is possible that the vessels themselves display a degree of sexual dimorphism.

Acknowledgement

This work was supported by NIHLB Program Project grant HL40069.

References

Cheng LP, Kuwahara M, Jacobsson J, Foegh ML 1991 Inhibition of myointimal hyperplasia and macrophage infiltration by estradiol in aorta allografts. Transplantation 52:967–972

Enosawa S, Hirasawa K 1989 Examination of sex-associated differences and organ specificities in the survival of cardiac and skin allografts in rats treated with cyclosporine. Transplantation 48:714–717

Esmore D, Keogh A, Spratt P, Jones B, Chang V 1991 Heart transplantation in females. J Heart Lung Transplant 10:335–341

Farhat M, Chen M, Bhatti T, Iqbal A, Cathapermal S, Ramwell P 1993 Protection by estradiol against the development of cardiovascular changes associated with monocrotaline pulmonary hypertension in rats. Br J Pharmacol 110:719–723

Foegh ML, Khirabadi BS, Nakanishi T, Vargas R, Ramwell PW 1987 Estradiol protects against experimental cardiac transplant atherosclerosis. Transplant Proc 19: 90–95

Foegh ML, Asotra S, Howell MH, Ramwell PW 1994 Estradiol inhibition of arterial neointimal hyperplasia after balloon injury. J Vasc Surg 19:722–726

Foegh ML, Lou H, Rego A, Katz N, Ramwell P 1995 Gender effects on graft myointimal hyperplasia. Transplant Proc, in press

Fogelberg M, Vesterqvist O, Diszsalusy U, Henriksson R 1990 Experimental atherosclerosis: effects of oestrogen and atherosclerosis on thromboxane and prostacyclin formation. Eur J Clin Invest 20:105–110

Ghodsi F, Will JA 1981 Changes in pulmonary structure and function induced by monocrotaline intoxication. Am J Physiol 240:H149–H155

Grossman CJ 1984 Regulation of the immune system by sex steroids. Endocr Rev 5:435–455

Hayry P, Raisanen A, Ustinov J, Menannder A, Paavonen T 1993 Somatostatin analog, lanreotide, inhibits myocyte replication and several growth factors in allograft arteriosclerosis. FASEB J 7:1055–1060

Holmdahl R, Carlsten H, Jansson L, Larsson P 1989 Oestrogen is a potent immunomodulator of murine experimental rheumatoid disease. Br J Rheumatol 28:54–58

Kiyatake K, Kakusaka I, Kasahara Y et al 1994 Relationship between the converting ability of liver microsomes and monocrotaline-induced pulmonary hypertension in male, female and castrated male rats. Jpn J Thoracic Dis 32:125–129

Kuwahara M, Jacobsson J, Kuwahara M, Kagan E, Ramwell PW, Foegh ML 1991 Coronary artery ultrastructural changes in cardiac transplant atherosclerosis in the rabbit. Transplantation 52:759–765

Mehra MR, Ventura HO, Escobar A, Cassidy CA, Smart FW, Stapleton DD 1995 Does donor and recipient sex influence the development of cardiac allograft vasculopathy? Transplant Proc, in press

Schray-Utz B, Zeiher AM, Busse R 1993 Expression of constitutive NO synthase in cultured endothelial cells is enhanced by 17-β estradiol. Circulation 88:80(abstr)

Screpanti I, Meco D, Morrone S, Gulino A, Mathieson BJ, Frati L 1991 *In vivo* modulation of the distribution of thymocyte subsets: effects of estrogen on the expression of different T-cell receptor V beta gene families in CD4$^-$, CD8$^-$ thymocytes. Cell Immunol 134:414–426

Steinleitner A, Stanczyk FZ, Levine JH et al 1989 Decreased *in vitro* production of 6-keto-prostaglandin F1 alpha by uterine arteries from post menopausal women. Am J Obstet Gynecol 161:1677–1688

Vargas R, Wroblewska B, Rego A, Hatch J, Ramwell PW 1993 Oestradiol inhibits smooth muscle cell proliferation of pig coronary artery. Br J Pharmacol 109: 612–617

DISCUSSION

Bonewald: It's known that oestradiol stimulates the production of transforming growth factor (TGF)-β (Komm et al 1988, Oursler et al 1991) and that TGF-β inhibits macrophage infiltration of tissues (Shull et al 1992, Kulkarni et al 1993). Have you looked at TGF-β?

Foegh: No, but I agree that TGF-β could easily be another factor involved in prevention of myointimal proliferation. In addition to the effect on macrophages you mentioned, it also inhibits smooth muscle cell proliferation under certain circumstances. However, we have been focusing on IGF-1 (Mii et al 1993, Majack et al 1990).

Rochefort: You have shown that oestradiol inhibits the mitogenic effect of IGF-1 (M. L. Foegh, unpublished results). Can you exclude an indirect effect on smooth muscle cells? Have you looked at single cell proliferation?

Foegh: Yes, we've looked at the effect of IGF-1 on smooth muscle cell proliferation in culture. In smooth muscle cell cultures we find that the anti-IGF-1 antibody abolishes the mitogenic effect.

Rochefort: This reminds me of our results on the anti-growth factor activity of anti-oestrogens, which is mediated by the oestrogen receptor (Vignon et al 1987). This may be explained by a negative cross-talk interaction between the oestrogen receptor 'activated' by anti-oestrogens and the Fos/Jun complex (Philips et al 1993).

Foegh: I envisage a direct effect on the smooth muscle cell, but also some cross-talk between cells. When we complete a more detailed analysis of transplant histology specimens by immunocytochemistry we expect to have more insight into what's actually going on.

Rochefort: I was talking about cross-talk between different transcription factors in the same cell.

Bayard: Have you tried to block the effect of interleukin 1 (IL-1), IGF-1 or IL-6 in your transplantation system?

Foegh: No, we have not done that yet.

Bayard: Have you tried blocking the action of oestradiol with anti-oestrogens?

Foegh: We have not done this in the *in vivo* model but we have done it in our smooth muscle cell cultures; there the specific anti-oestrogens will block the inhibitory effects of oestrogen.

Manolagas: Your first model is clearly a model where the effects of oestradiol are occurring against a background of major immunosuppression induced by cyclosporin A. Of course, this is not what happens in menopausal women. Is there a way you can avoid using immunosuppression, or find some other form of immunosuppression, to get away from this situation where you compromise your results dramatically by suppressing cytokine production and macrophage activation?

Foegh: In the rabbit model we can't get away from using immuno-suppression, because the rabbits are not inbred. However, we plan to begin using newer immunosuppressants like FK506 and rapamycin. The old-fashioned immunosuppressants, such as azathiopine, do not work sufficiently well in rabbits.

I admit that it may be difficult to relate findings in our transplant model to normal females; this is why we also have a non-transplant model. In this model, oestradiol has a direct effect on the vessel wall, preventing the myointimal hyperplasia.

Manolagas: This sounds a very important mechanism. The evidence that the action of exogenous cytokine can be blocked by oestradiol makes perfect sense.

Oelkers: If I understood correctly, in the transplantation experiments of the aorta you found that the intensity of proliferation was determined by the sex of the donor and not by the sex of the recipient. How do you explain this? I think it's paradoxical in terms of your other findings.

Foegh: It's not paradoxical, but it's not straightforward. First, we do think that rejection correlates with transplant atherosclerosis, and women have more rejection episodes so they should have more transplant atherosclerosis. There's not much information yet on whether there's a sex difference in transplant atherosclerosis, but there is a very recent study in which Mehra et al (1995) evaluated the effect of sex, with the exception of the combination of women to women, because there are very few of these. For the remaining three combinations, the most transplant arteriosclerosis is seen in the male recipient of a female graft; this agrees with the observation that women are more immune responsive. I've been talking to some of my colleagues who have more cases of female recipients of female grafts. They state that it looks as if these women may have less transplant atherosclerosis. In my non-transplant models the myointimal hyperplasia is not influenced by the immune status. In this model you see the vascular wall response to a mechanical injury, surgery, ischaemia and reperfusion. I showed the female vessel responds with a lesser degree of myointimal hyperplasia than the male vessel, independently of whether the environment is male or female.

Manolagas: The incidence of cardiovascular disease rises dramatically within a matter of months following the menopause. This suggests that it can't just be genetic—circulating oestrogen has to be involved.

Baulieu: Are there any practical applications of these results? What would you recommend to transplant patients?

Foegh: I would recommend oestrogen replacement therapy, since experimentally the oestrogen did not make the rejection worse, and it decreased the myointimal hyperplasia.

Baulieu: In both male and female patients?

Foegh: I don't think my male patients would like it! For men, I would like a non-feminizing oestrogen.

Manolagas: How about the risk of thrombophlebitis: would you worry about that?

Foegh: I would not worry about this risk with the doses of oestrogen we needed to inhibit myointimal hyperplasia. Oestrogen replacement therapy in postmenopausal women is not shown to cause thrombosis.

Manolagas: It would be nice to know whether the incidence of re-occlusion in women receiving transplants before or after the menopause is different.

Foegh: The trouble is that very few women receive cardiac transplants.

Baulieu: In the past few years, we have found molecular interactions between steroid receptors and a novel immunophilin. This protein, originally called p59, binds the immunosuppressants FK506 and rapamycin, and is thus a member of the FKBP (FK-binding protein) series. We first observed this so-called p59 protein when it was associated with heat shock protein 90 (Hsp90) in the 8S hetero-oligomeric form of steroid receptors, including in the nucleus of target cells (Renoir et al 1990). We then cloned this protein (Lebeau et al 1992), which we call FKBP59-HBI (heat shock protein-binding immunophilin). These observations have been confirmed in all the animal cells we have studied so far. The interaction between FKBP59-HBI and Hsp90 takes place at the level of a tetratricopeptide repeat domain (Radanyi et al 1994). Preliminary results indicate that another immunophilin, which is about 40 kDa and binds cyclosporin A (Kieffer et al 1993, Ratajczak et al 1993), is also an HBI and is present in steroid receptor oligomeric forms. We are currently studying the functional connections between immunosuppressant drugs and steroid function: we have found that FK506, rapamycin, cyclosporine and analogues do modify steroid activities such as transcriptional effects and growth (Jung-Testas et al 1993, Renoir et al 1994). There is correlation between the molecular data on immunophilins and receptors, and the functional results with immunosuppressant and steroid function, but as yet no demonstration of a cause–effect relationship.

References

Jung-Testas I, Delespierre B, Baulieu EE 1993 Inhibition par l'immunosuppresseur FK506 de l'effet antiglucocorticostérode du RU486 sur la croissance de fibroblastes de souris en culture. C R Acad Sci Ser III Sci Vie 316:1495–1499

Kieffer LJ, Seng TW, Li W, Osterman DG, Handschumacher RE, Bayney RM 1993 Cyclophilin-40, a protein with homology to the p59 component of the steroid receptor complex. J Biol Chem 268:12303–12310

Komm BS, Terpening CM, Benz DJ et al 1988 Estrogen binding, receptor mRNA, and biologic response in osteoblast-like osteosarcoma cells. Science 241:81–84

Kulkarni AB, Huh C-G, Becker D et al 1993 Transforming growth factor $\beta 1$ null mutation in mice causes excessive inflammatory response and early death. Proc Natl Acad Sci USA 90:770–774

Lebeau MC, Massol N, Herrick J et al 1992 p59, an hsp90 binding protein: cloning and sequencing of its cDNA. Preparation of a peptide-directed polyclonal antibody. J Biol Chem 267:4281–4284

Majack RA, Majesky MW, Goodman LV 1990 Role of PDGF-A expression in the contol of vascular smooth muscle cell growth by transforming growth factor β. J Cell Biol 111:239–247

Mehra MR, Ventura HO, Escobar A, Cassidy CA, Smart FW, Stapleton DD 1995 Does donor and recipient sex influence the development of cardiac allograft vasculopathy? Transplant Proc, in press

Mii S, Warre JA, Kent KC 1993 Transforming growth factor β inhibits human vascular smooth muscle cell growth and migration. Surgery 114:464–470

Oursler MJ, Cortese C, Keeting P et al 1991 Modulation of transforming growth factor beta production in normal human osteoblast-like cells by 17-[β]-estradiol and parathyroid hormone. Endocrinology 129:3313–3320

Philips A, Chalbos D, Rochefort H 1993 Estradiol increases and antiestrogens antagonize the growth factor-induced activator protein-1 activity in MCF7 breast cancer cells, without affecting c-*fos* and c-*jun* synthesis. J Biol Chem 268:14103–14108

Radanyi C, Chambraud B, Baulieu EE 1994 The ability of the immunophilin FKP59-HBI to interact with the 90-kDa heat shock protein is encoded by its tetratricopeptide repeat domain. Proc Natl Acad Sci USA 91:11197–11201

Ratajczack T, Carrello A, Mark PJ et al 1993 The cyclophilin component of the unactivated estrogen receptor contains a tetratricopeptide repeat domain and shares identity with p59 (FKBP59). J Biol Chem 2688:13187–13192

Renoir JM, Radanyi C, Faber LE, Baulieu EE 1990 The non-DNA binding heterooligomeric form of mammalian steroid hormone receptors contains a hsp90-bound 59 kDa protein. J Biol Chem 265:10740–10745

Renoir JM, Mercier-Bodard C, Le Bihan S et al 1994 Cyclosporin A, as FK506, potentiates the dexamethasone-induced MMTV-CAT activity in LMCAT cells. A possible role for different heat shock protein binding immunophilins (BBIs). Biochem Biophys Res Commun, in press

Shull GM, Ormsby I, Bier AB et al 1992 Targeted disruption of the mouse transforming growth factor-β1 gene results in multifocal inflammatory disease. Nature 359:693–699

Vignon F, Bouton MM, Rochefort H 1987 Antiestrogens inhibit the mitogenic effect of growth factors on breast cancer cells in total absence of estrogens. Biochem Biophys Res Commun 146:1502–1508

Hormone replacement therapy and cardiovascular disease: the case for a randomized controlled trial

M. R. Vickers, T. W. Meade and H. C. Wilkes

MRC Epidemiology and Medical Care Unit, The Wolfson Institute of Preventive Medicine, St Bartholomew's Hospital Medical College, Charterhouse Square, London EC1M 6BQ, UK

Abstract. The menopause is associated with an increased risk of developing cardiovascular disease. Oestrogen may influence various metabolic pathways which contribute to the pathogenesis of cardiovascular disease, and observational studies suggest that in postmenopausal women oral oestrogen replacement therapy confers some protection against coronary heart disease and to a lesser extent against stroke. What is not clear is the magnitude of the cardioprotective effect and the overall balance of long-term benefits and hazards. Research is also required to establish the relative effects of oestrogen replacement therapy and combined or opposed hormone replacement therapy (HRT) where progestogen is added to counter the proliferative action of oestrogen on the endometrium. A large randomized controlled trial is the only way to provide accurate estimates of the cardioprotective effect of HRT and of other long-term benefits and hazards. Feasibility studies undertaken through the UK Medical Research Council (MRC) General Practice Research Framework show that such a trial is acceptable to patients and their doctors. Recruitment and withdrawal rates indicate that a trial of sufficient size to show a 25% reduction in cardiovascular disease with 90% power at the 1% level would be feasible. The full trial is costly and it is proposed that the UK collaborates with other countries in a major international trial to complement the Women's Health Initiative trial in the USA. Feasibility studies in Europe are underway, the design and scientific rationale for the trial have been approved by the UK MRC and it is hoped that recruitment to the full-scale trial can begin soon.

1995 Non-reproductive actions of sex steroids. Wiley, Chichester (Ciba Foundation Symposium 191) p 150–164

Hormone replacement therapy (HRT) was initially marketed for the relief of short-term menopausal symptoms and its effectiveness for these indications is well established. The scientific rationale for its growing long-term use is not clear and the consequences of long-term use are unknown. The loss of ovarian function at the menopause with the resultant fall in endogenous oestrogen appears to contribute to the onset of cardiovascular disease, the combination of coronary heart disease and stroke, which is the leading cause of death and serious illness in postmenopausal women. Oestrogen may influence various metabolic pathways which contribute to the pathogenesis of cardiovascular disease, including lipid metabolism, coagulation and fibrinolysis, and carbohydrate metabolism; it may also affect blood flow and arterial vessel tone. Its overall effect on cardiovascular disease will depend on a complex balance of actions, some individually beneficial and others potentially adverse. Indeed, the number and variety of these actions and the impossibility of relying on them to predict confidently their net influence on the risk of clinical events is in itself a strong reason for the kind of trial advocated in this paper. If endogenous oestrogen in the premenopausal woman is cardioprotective, then it is a reasonable hypothesis that treatment of postmenopausal women with exogenous oestrogen will reduce the risk of developing cardiovascular disease. Observational studies based on clinical end points do suggest that HRT, in the form of oral oestrogen, reduces cardiovascular disease, but the magnitude of the benefit has varied considerably in different studies. The frequently quoted figure of a 50% reduction in cardiovascular disease, or even more (Stevenson & Baum 1994), is likely to be an overestimate because of selection bias in those receiving HRT and selective reporting of study results. It is generally accepted that an accurate estimate of the cardioprotective effect of oestrogen replacement therapy (ORT) can only be obtained from randomized controlled trials. We have been conducting feasibility studies for such trials in the UK and have been coordinating European studies. The design and scientific rationale for a major international trial, which would complement the Women's Health Initiative trial in the USA, has now been approved by the UK Medical Research Council (MRC). In addition to investigating the long-term effects of ORT, this trial would also establish the long-term effects of opposed HRT, where progestogen is added to counter the proliferative effects of oestrogen on the endometrium. This combination of progestogen plus ORT (PORT) is recommended for those women who have not had a hysterectomy and who comprise more than two-thirds of all HRT users. Few long-term observational studies have included women taking PORT because this formulation was introduced much later than ORT. By analogy with the actions of oral contraceptives (Wynn et al 1979, Wallace et al 1979), it was initially suggested that PORT would be less cardioprotective than ORT because of adverse effects on lipids. Early observational metabolic studies largely based on the actions of the more androgenic progestogens (Hirvonen et al 1981) appeared to confirm this, but more recent work discussed below, including randomized comparisons of ORT and PORT, indicates that this may not be so.

Observational studies

Observational studies, mostly using ORT, do suggest that HRT is cardioprotective, but there is considerable uncertainty about the magnitude of the effect. The recommendation that long-term preventive HRT should be considered for all postmenopausal women (American College of Physicians 1992) is not justified on present evidence. There are two main problems affecting the usefulness of the data from observational studies: first, the inability to allow fully for confounding factors and selection bias; and second, the lack of information on PORT.

Oestrogen-only preparations

In a recent overview of 31 studies, including hospital- and population-based case-control studies, prospective cohort studies, one very small randomized trial and surveys of women undergoing coronary angiography, Stampfer & Colditz (1991) reported a weighted summary relative risk of 0.56 in HRT users compared with non-users. A reduced risk of coronary heart disease in oestrogen users was found in 25 of the 31 studies but the results reached statistical significance in only 12 studies.

The main problem in the observational studies reviewed by Stampfer is selection bias. Initial interest in HRT use occurred at a time when, by analogy with the thromboembolic effects of the oral contraceptives then in use, HRT was not recommended for those at high risk of cardiovascular disease. Some of the early case-control and cohort studies accepted that those taking HRT were often at lower risk of cardiovascular disease than non-users, and they also tended to be of higher social class and more health conscious (Egeland et al 1988, Coope 1989). Other studies have used external controls, i.e. compared observations in a cohort of HRT users with expected data derived from rates for the whole population. For example, in the study by Hunt et al (1990), where observed mortality in women attending special menopause clinics was compared with mortality rates for the whole population, the selected nature of the HRT users could lead to biased results.

More recent cohort studies have attempted to avoid selection bias by taking the sample groups from a fairly homogeneous population and have adjusted for known confounding risk factors. The 10 year follow-up from the largest study—the Nurses' Health Study—reported, after adjustment for age and other risk factors, an overall relative risk of major coronary disease in women currently taking oestrogen of 0.56 (95% confidence interval 0.4–0.8) and in former users of 0.83 (95% confidence interval 0.65–1.05), but found no evidence for a protective effect on the risk of stroke (Stampfer et al 1991). The gradient of declining risk from non-users through former users to current HRT users does suggest a benefit in terms of coronary heart disease but does not justify the exaggerated claims of a reduction in risk of more than 50% (Stevenson & Baum 1994). However well designed, observational studies are unlikely to allow fully for confounding characteristics and selection bias and may easily give rise to twofold artefactual differences (Doll & Peto 1980). Thus, a clear example of how observational studies may overestimate benefit is provided by a comparison of the results of observational studies and randomized controlled comparisons of the value of oral anticoagulants after myocardial infarction (Doll & Peto 1980, Armitage 1980). Even though most of the observational studies attempted to adjust for confounding variables, the 50% benefit claimed was greater than the 20% benefit eventually established by randomized trials. It is therefore reasonable to assume that the cardioprotective benefit of HRT may be lower than the 50% claimed on the basis of observational studies. However, even if we accept a moderate decrease in risk of cardiovascular disease with ORT, this cannot necessarily be extrapolated to the opposed preparations used by the majority of women with a uterus.

Oestrogen–progestogen preparations

In the late 1970s/early 1980s it became clear that ORT increased the risk of developing endometrial cancer. Progestogen was added to counter the proliferative effect of oestrogen on the endometrium and PORT is standard treatment for women with a uterus. The majority of hysterectomized women continue to use ORT.

Early metabolic studies suggested, by analogy with the actions of oral contraceptives, that progestogen might have adverse effects on blood lipid metabolism and might thus counter any beneficial effects of ORT on this pathway (Hirvonen et al 1981). More recent data, reviewed below, show the picture is now far from clear.

Two prospective studies from Sweden actually suggest greater benefit from PORT than ORT on the risk of acute myocardial infarction (Falkeborn et al 1992) and all stroke (Falkeborn et al 1993). The studies compared cardiovascular disease rates in HRT users with regional rates, but if it is reasonable to assume that any selection biases have affected results for both preparations in a similar way, then the lower relative risk with PORT is interesting, particularly because the progestogen used was the relatively androgenic steroid norgestrel. However, another factor which should be considered is that the women taking ORT are more likely to be hysterectomized and are thus at higher risk of cardiovascular disease than women with a uterus for whom PORT would be prescribed.

Recent studies on cardiovascular risk factors, outlined below, also lend support to the concept that PORT may not be less cardioprotective that ORT. The effects of ORT and PORT on serum lipid levels and lipoprotein concentrations have been given greatest prominence, although other factors are being increasingly studied.

The menopause is associated with increased serum concentrations of total cholesterol, low-density lipoprotein (LDL)-cholesterol and high-density lipoprotein (HDL) subfraction 3, and lower concentrations of total HDL-cholesterol, particularly HDL subfraction 2 (Stevenson et al 1993). The effect of oral oestrogen on lipids has been studied extensively and, although there are some inconsistencies, the consensus in recent reviews (Rijpkema et al 1990, Lobo 1991, Walsh et al 1991) is that oral ORT raises HDL, particularly HDL subfraction 2, and lowers LDL, and that these changes indicate a cardioprotective effect. Progestogens do not appear to oppose the ability of oestrogen to lower LDL. The attenuation of the increase in HDL is not a consistent finding and may depend on the androgenicity of the progestogen, the dose and the regimen (Rijpkema et al 1990, Lobo 1991). Three recent studies, for example, report that in current users PORT is at least as effective as ORT in raising HDL and lowering LDL (Nabulsi et al 1993, Gambrell & Teran 1991, Barrett-Connor et al 1989).

The menopause is also associated with an increase in serum triglyceride levels, and an independent relationship of high triglyceride levels with ischaemic heart disease in women has been shown (Carlson & Böttiger 1985, Stensvold et al 1993). Our own randomized comparisons, described below, show that triglycerides rise in those treated with oral ORT but in women treated with PORT containing norgestrel there is little change in triglyceride level, suggesting a possible protective action of norgestrel.

The association of the menopause with particular effects on the haemostatic system is still the subject of much research. Oestrogen has procoagulatory actions but these are highly dose dependent and may be exerted only at doses higher than those used in HRT (Winkler 1992). Recent studies (Nabulsi et al 1993, Kroon et al 1994), however, show increases in factor VII_c and F1.2, which are markers of thrombin production and indicate that oral ORT may have potentially adverse effects on coagulation, and a small (though not significant) reduction in fibrinogen, which could be beneficial. If the net effect of ORT were procoagulatory, this might be countered by an increase in fibrinolytic activity demonstrated by a reduction in plasminogen activator inhibitor (PAI-1). The addition of a progestogen to ORT seems to attenuate the potentially adverse effect on factor VII (Kroon et al 1994) and the beneficial decrease in fibrinogen, but so far the effect on other markers of coagulation and on fibrinolytic activity is not known. The relative overall effects of ORT and PORT on haemostasis are not clear and more research is required, particularly involving randomized comparisons to avoid the problems of bias already discussed in relation to clinical endpoint studies.

Oestrogen may also influence cardiovascular disease by other routes, including beneficial effects on carbohydrate metabolism (Godsland et al

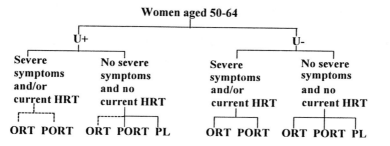

FIG. 1. MRC hormone replacement therapy trial design. - - -, exclude; ORT, oestrogen-only HRT; PORT, progestogen plus oestrogen HRT; PL, placebo.

1993) and blood flow (Ganger et al 1991, Collins et al 1993, Jiang et al 1992), but these factors are less well studied and there are few data on the relative effects of ORT and PORT. Our own randomized comparisons agree with the consensus that neither ORT nor PORT affects blood pressure or weight.

More detailed studies on individual risk factors will provide interesting and valuable information, and comparative studies of different formulations will be particularly useful, but a clear answer to the effects of ORT and PORT on cardiovascular disease will only be obtained by experimental evidence from a large randomized controlled trial.

Full-scale trial of the long-term effects of hormone replacement therapy

Feasibility studies

The MRC Epidemiology and Medical Care Unit began feasibility studies for a randomized controlled trial of the long-term effects of HRT in 1990. Patients have been identified through general practice using the MRC General Practice Research Framework (GPRF), a network of around 800 general practices throughout the UK. Two separate designs were piloted initially, one a double-blind comparison of ORT with PORT in women who had undergone hysterectomy, and the second a comparison of HRT (PORT for women with a uterus, ORT for hysterectomized women) with no HRT in women not requiring HRT for menopausal symptoms. These two approaches were then combined into a single design (Fig. 1), which is now undergoing feasibility testing and would be used in a full-scale trial.

The main objectives of the feasibility studies included establishing the response rate to invitations to participate, the proportion of patients eligible and willing to enter, and rates of withdrawal from randomized treatment.

Women were identified from age–sex registers of individual general practices and preliminary eligibility was assessed by record of hysterectomy status and/or major reasons for exclusion. The response rate to an invitation to potentially eligible women to attend the surgery for screening varied from practice to practice, but overall was around 50% with a higher response from hysterectomized women. Recruitment to a comparison of ORT and PORT in hysterectomized women was high, with 61% of those eligible entering the trial. The recruitment rate to an HRT versus no HRT comparison was lower; overall 32% of all eligible women screened entered, with a higher recruitment rate (45%) among hysterectomized women. As a proportion of the relevant age group, the number of women entering the trial is fairly low, but this is not unexpected considering the small proportion of long-term users in ordinary clinical practice in the UK. Recruitment rates are more than sufficient to make a large trial in the UK through the MRC GPRF feasible. The reasons given for not entering the trials include not wanting to undertake a long-term commitment, not wanting to take pills, concern over the breast cancer risk and (the main reason in women with a uterus) not wanting to restart periods. Two-year withdrawal rates in both pilot studies were of the order of 25%, with the highest rates in the first six months. These rates are comparable to those found in other major long-term preventive trials and are encouraging, especially since the women are not at high risk. We have taken account of withdrawal rates in calculating the sample size for the main trial.

Design of full-scale trial

We propose to recruit around 43 000 women aged 50–64 and to randomize them to treatment with ORT, PORT or placebo according to the design in Fig. 1. Treatment would be for 10 years using conjugated equine oestrogen as ORT and combining this with continuous medroxyprogesterone acetate as PORT. Follow-up would continue for a further 10 years. The cardiovascular endpoint evaluable at 10 years would be the sum of fatal and non-fatal coronary heart disease and stroke. Other main endpoints would be major fractures (vertebral, wrist and forearm at 10 years, hip at 20 years), all cancers (with particular emphasis on breast cancer) and death from all causes. Quality of life and psychological well-being would also be assessed. The health economic implications, costs of care and use of health services, would be estimated. In sub-studies the effects of ORT and PORT on various metabolic pathways— lipid and carbohydrate metabolism, coagulation and fibrinolysis and, possibly, blood flow—would be examined.

Cardiovascular endpoints

A sample size of 43 000 would allow the detection of a 25% reduction in cardio-vascular disease, with 90% power at the 1% significance level in those treated with HRT (either ORT or PORT) compared with those not receiving HRT. A comparison of PORT versus no HRT could be made with similar power since most HRT would be PORT. A direct comparison of ORT versus no HRT would be possible only if the trial were expanded to include randomization of women with a uterus to ORT, and this is unlikely to be acceptable in the UK because of the increased risk of endometrial cancer. In the Women's Health Initiative trial women with a uterus are randomized to ORT, PORT or placebo. The combination of data from the MRC trial in hysterectomized women with the USA data from the same age group would permit a direct comparison of ORT versus no HRT. The proposed trial would have some power to make a comparison of ORT versus PORT in hysterectomized women. For example, if PORT were more cardioprotective than ORT, a difference between the two treatments could be detected with 80% power at the 5% level if PORT gave a 36% reduction in cardiovascular disease and ORT a 12% reduction.

Other major endpoints

Whatever the magnitude of the cardioprotective effect, policy recommenda-tions on the long-term use of HRT should reflect the overall balance of long-term risks and benefits. The proposed trial could detect a 30% increase in breast cancer at 20 years, in those taking HRT compared with those not taking HRT, with 87% power at the 5% level. If we combined data from this trial with Women's Health Initiative data, the overall sample would be sufficient for us to detect a smaller increase. The proposed trial would also measure long-term effects of HRT on major fractures.

International collaboration

The full trial described above is costly: on the basis of recruitment through general practice in the UK it would require funding, excluding drug costs, in excess of £40 million over 25 years. An international collaborative trial is therefore proposed, possibly including researchers from Ireland, The Netherlands, Scandinavia, Australia, New Zealand and Canada. Coordinated feasibility studies using the proposed main trial design are already being conducted in the UK, Ireland and The Netherlands. It is anticipated that the UK would contribute 23 000 women to an international trial and this sample alone would be sufficient to detect a 25% reduction in cardiovascular disease in those taking HRT compared with those not taking HRT, with 80% power at the 5% level.

Timeliness of a randomized controlled trial

In the UK our studies show that about 20% of postmenopausal women are now using HRT at any one time, with the proportion who have ever used HRT being much higher. Use is increasing and the figure for current usage will probably reach or exceed 30% by the year 2000. The next few years provide the ideal 'window of opportunity' for recruitment to a randomized controlled trial. A cost-effectiveness analysis (J. Townsend, unpublished work 1994) indicates that despite the high costs, funding the trial would be a good use of public money. If the trial showed a favourable long-term balance of hazards and benefits with HRT, then money could be saved by extending a cost-effective treatment, whereas if the hazards outweighed the benefits, then money could be saved by curtailing a widespread but ineffective treatment. Without a randomized controlled trial the unevaluated, and possibly inappropriate, use of HRT will continue and will increase substantially. With a trial, the medical, social and economic consequences of HRT use can be determined accurately as a basis for informed and appropriate usage. An international trial with a major UK component should therefore be initiated soon.

References

American College of Physicians 1992 Guidelines for counselling postmenopausal women about preventive hormone therapy. Ann Intern Med 117:1038–1041

Armitage P 1980 Clinical trials in the secondary prevention of myocardial infarction and stroke. Thromb Haemostasis 43:90–94

Barret-Connor E, Wingard DL, Criqui MH 1989 Postmenopausal estrogen use and heart disease risk factors in the 1980s: Rancho Bernardo, Calif, revisited. JAMA 261:2095–2100

Carlson LA, Böttiger LE 1985 Risk factors for ischaemic heart disease in men and women. Results of the 19-year follow-up of the Stockholm prospective study. Acta Med Scand 218:207–211

Collins P, Rosano GMC, Jiang C, Lindsay D, Sarrel PM, Poole-Wilson PA 1993 Cardiovascular protection by oestrogen—a calcium antagonist effect? Lancet 341:1264–1265

Coope J 1989 Hormone replacement therapy. The Royal College of General Practitioners, London

Doll R, Peto R 1980 Randomised controlled trials and retrospective controls. BMJ 280:44

Egeland GM, Matthews KA, Kuller LH, Kelsey SF 1988 Characteristics of noncontraceptive hormone users. Prev Med 17:403–411

Falkeborn M, Persson I, Adami HO et al 1992 The risk of acute myocardial infarction after oestrogen and oestrogen–progestogen replacement. Br J Obstet Gynaecol 99:821–828

Falkeborn M, Persson I, Terent A et al 1993 Hormone replacement therapy and the risk of stroke. Arch Intern Med 153:1201–1209

Gambrell RD Jr, Teran AZ 1991 Changes in lipids and lipoproteins with long-term estrogen deficiency and hormone replacement therapy. Am J Obstet Gynecol 165:307–317

Ganger KF, Vyas S, Whitehead M, Crook D, Meire H, Campbell S 1991 Pulsatility index in internal carotid artery in relation to transdermal oestradiol and time since menopause. Lancet 338:839–842

Godsland IF, Ganger K, Walton C et al 1993 Insulin resistance, secretion, and elimination in postmenopausal women receiving oral or transdermal hormone replacement therapy. Metabolism 42:846–853

Hirvonen E, Malkonen M, Manninen V 1981 Effects of different progestogens on lipoproteins during postmenopausal replacement therapy. N Engl J Med 304: 560–563

Hunt K, Vessey M, McPherson K 1990 Mortality in a cohort of long-term users of hormone replacement therapy: an updated analysis. Br J Obstet Gynaecol 97: 1080–1086

Jiang C, Poole-Wilson PA, Sarrel PM, Mochizuki S, Collins P, Macleod KT 1992 Effect of 17β-oestradiol on contraction, Ca^{2+} current and intracellular free Ca^{2+} in guinea-pig isolated cardiac myocytes. Br J Pharmacol 106:739–745

Kroon U-B, Silfverstolpe G, Tengborn L 1994 The effects of transdermal estradiol and oral conjugated estrogens on haemostasis variables. Thromb Haemostasis 17:420–423

Lobo RA 1991 Clinical review 27: effects of hormonal replacement on lipids and lipoproteins in postmenopausal women. J Clin Endocrinol & Metab 73: 925–930

Nabulsi AA, Folsom AR, White A et al 1993 Association of hormone-replacement therapy with various cardiovascular risk factors in postmenopausal women. N Engl J Med 328:1069–1075

Rijpkema AHM, van der Sanden AA, Ruijs AHC 1990 Effects of postmenopausal oestrogen–progestogen replacement therapy on serum lipids and lipoproteins: a review. Maturitas 12:259–285

Stampfer MJ, Colditz GA 1991 Estrogen replacement therapy and coronary heart disease: a quantitative assessment of the epidemiologic evidence. Prev Med 20:47–63

Stampfer MJ, Colditz GA, Willett WC et al 1991 Postmenopausal estrogen therapy and cardiovascular disease—ten-year follow-up from the Nurses' Health Study. N Engl J Med 325:756–762

Stensvold I, Tverdal A, Urdal P, Graft-Iversen S 1993 Non-fasting serum triglyceride concentration and mortality from coronary heart disease and any cause in middle aged Norwegian women. BMJ 307:1318–1322

Stevenson JC, Baum M 1994 Hormone replacement therapy: should be used selectively. BMJ 309:191

Stevenson JC, Crook D, Godsland IF 1993 Influence of age and menopause on serum lipids and lipoproteins in healthy women. Atherosclerosis 98:83–90

Wallace RB, Hoover J, Barrett-Connor E et al 1979 Altered plasma lipid and lipoprotein levels associated with oral contraceptives and oestrogen use. Lancet I:111–114

Walsh BW, Schiff I, Rosner B, Greenberg L, Ravnikar V, Sacks FM 1991 Effects of postmenopausal estrogen replacement on the concentrations and metabolism of plasma lipoproteins. N Engl J Med 325:1196–1204

Winkler UH 1992 Menopause, hormone replacement therapy and cardiovascular disease: a review of haemostaseological findings. Fibrinolysis (suppl 6): 5–10

Wynn V, Adams PW, Godsland I et al 1979 Comparison of effects of different combined
   oral contraceptive formulations on carbohydrate and lipid metabolism. Lancet I:
   1045–1049

DISCUSSION

Sutherland: What sort of exclusion criteria do you have? What will you do
with women who develop breast cancer?

Vickers: In the feasibility studies we were very cautious and we excluded
more women than we needed to. For example, we excluded women with any
sort of breast abnormality, whereas in the main trial we would only exclude
women with breast cancer or with a first degree relative with breast cancer. It
could be argued that this limits our ability to detect an increase in breast
cancer, but we have to balance what is ethical with what we would like to do on
scientific grounds. There are proposals for separate trials looking at HRT and
breast cancer in the UK. There aren't that many other exclusion criteria. We
exclude people with severe liver or renal disease, active gall bladder disease and
those with otosclerosis. We are also excluding people who have had a previous
myocardial infarction or stroke. Again, you could argue the merits of including
this latter group of patients. There are a large number of contraindications to
HRT listed in the drug company product literature, but these are largely based
on contraindications for oral contraceptives and there are moves to modify the
lists.

Bäckström: What about risk for thromboembolic conditions?

Vickers: We're not excluding people who seem to be at high risk.

Manolagas: Concerning your bone studies: will you have subgroups with
Ca^{2+} supplements and no Ca^{2+}? Will you monitor vitamin D?

Vickers: We have no plans to factor in Ca^{2+} supplementation, although
we have considered both this and vitamin D supplementation. The World
Health Organization trial does have a dietary intervention arm and has been
criticized for being too complex. However, the design of our trial hasn't yet
been finalized. If we get funding for the main trial, we will consider adding
other sub-studies.

Baulieu: Would it cost much to include neuropsychological studies?

Vickers: Neuropsychological studies would certainly be worth doing. If the
studies involved questionnaires or fairly simple tests which the research nurse
could administer or oversee at the patients' regular four-monthly checks, then
the costs would be relatively low. For a sample of say 5000, and 15 min of
nurse time per patient per year, the costs would be in the region of £16 000 each
year.

Thijssen: Concerning the HRT itself: you mentioned that there is
discussion going on about whether you should use conjugated oestrogens or

pure oestrogens. Conjugated oestrogens are not natural human oestrogens. What is the reason for using them?

Vickers: We use conjugated oestrogen for the very practical reason that this is what most women use, at least in the UK. Some European countries are in favour of using natural oestrogens in the trial because the bulk of their market is oestradiol. All the evidence points to oestrogen and conjugated oestrogen having similar effects in HRT, so we see no scientific reason why we shouldn't use conjugated oestrogen in the UK and in other countries use oestradiol, and merge the data.

Thijssen: Other European countries chose oestradiol for obvious reasons—they wanted to get rid of these artificial oestrogens produced by the horse. Many people prefer the natural human oestrogen.

Vickers: As far as I know there is no definitive evidence that natural oestrogens and conjugated oestrogens act differently, for example, in their effects on lipids.

Oelkers: Would you give the women a choice between lower and higher doses? Some women get undesirable symptoms at the higher doses.

Vickers: We start everybody on the lower dose, which is 0.625 mg premarin a day. We allow the dose to be increased to 1.25 mg if necessary for control of short-term symptoms.

Oelkers: A serious problem in this study will be that women randomized within a placebo-controlled group will continuously worry about the possibility of being on placebo, because there's so much discussion going on about HRT.

Vickers: It is true that there will be discussion about HRT that the women will be aware of. One of the reasons this trial would be quite costly is that a long time would be spent making sure that the women are given understandable and accurate information. The women who go into the trial accept that there is uncertainty about the overall balance of long-term risks and benefits of HRT. From what we can tell from the feasibility studies, women are not constantly worrying that they might be on placebo. They really do accept that we don't know whether it's better to take HRT or have no treatment. We would of course be monitoring the quality of life and psychological well-being in the main trial.

Jensen: There are a number of women who do not tolerate premarin very well but who do tolerate the synthetic compound, oestrone sulphate, which is similar.

Manolagas: How are you going to handle the problem with hot flushes in your placebo group? If they do tolerate the hot flushes, they may represent an unusual group of women.

Vickers: You have to remember that women going into this comparison are not suffering from severe menopausal symptoms or at least from symptoms which they consider require HRT. A lot of menopausal women choose not to take HRT: only 17% of women in the UK with a uterus are taking HRT at any one time. Furthermore, in the trial we recruit women aged 50–64, so many women will not be experiencing short-term menopausal symptoms. Those that are and have chosen to come into the trial are all aware that they could be randomized to placebo.

Manolagas: Related to this is the issue of phyto-oestrogens. A number of women receive oestrogens from their diet, especially Asians.

Vickers: In a randomized comparison this will not make any difference: there should be equal numbers of such people in each group. You might have a problem with bias in an observational study, but not in a randomized trial.

Bäckström: Another possible error in this trial is that you are recruiting women who are not having hot flushes, because it's known that hot flushes only occur after a decrease in the endogenous production of oestrogen, and that oestrogen production is related to the body mass index (BMI). It could be that you are in fact recruiting those who have a high endogenous production of oestrogen. It's also known that smokers have a lower oestrogen concentration than non-smokers.

Vickers: There are two points: first, we're not excluding people who have hot flushes, we leave the choice to the women themselves. If they think their symptoms are sufficiently severe that they are definitely going to take HRT, they're not going to come into a placebo-controlled comparison anyway. We have to be practical. Secondly, in the hysterectomized women we do have a comparison of oestrogen only and combined oestrogen and progestogen for those who definitely wish to take HRT, and there is no evidence that BMI is different in this group compared with the BMI of those entering a placebo-controlled comparison.

Bäckström: Are you stratifying for the BMI?

Vickers: BMI is noted but we haven't stratified on this basis. In the feasibility studies so far we've shown no effect of either preparation on weight.

Thijssen: Aren't you introducing a bias by leaving the choice to the women? Those with more hot flushes will tend to be in favour of HRT.

Vickers: That's not true. This is the trouble with anecdotal evidence: when you actually look at women's responses in a trial there is no tendency for women with hot flushes to be more in favour of long-term HRT. There are lots of women who, when they are sufficiently informed, will choose not to take HRT, or will decide that they wish to participate in a trial.

Thijssen: There's not much evidence about heavier women having fewer hot flushes. This is a highly debated subject. Nobody has yet introduced the cultural aspect of hot flushes: different populations have completely different incidences of hot flushes (Martin et al 1993). Chinese women have far fewer hot flushes (Tang 1994).

Manolagas: That's not true. At a meeting a couple of years ago (National Institutes of Health Workshop on Menopause: Current Knowledge and Recommendations for Research, Bethesda, MD, March 22–24, 1993) a group of anthropologists reported studies in South America. From their findings they concluded that there was no incidence of the menopausal symptoms there whatsoever. But it was argued by others that their questionnaire could be wrong, because their interpreter was not using the right terms: these women were not calling their symptoms hot flushes. There are also anecdotal reports that Japanese women don't experience menopausal symptoms. One possibility is that they consume a lot of phyto-oestrogens. None the less, when I discussed this issue with Japanese endocrinologists, they very much doubted the anecdotal evidence and assured me that Japanese women do complain of hot flushes.

Thijssen: But there are quite a few arguments that the incidence of women complaining of flushes and who have, objectively, hot flushes, is lower in some cultures (Martin et al 1993, Tang 1994). It just means we have no real information on what's causing and influencing the hot flush. All we know is that it has something to do with oestrogen.

Bäckström: I wouldn't accept that completely. There is some validity to your argument, but it has been shown that women with high BMI have fewer hot flushes than those with low BMI.

Thijssen: It is debatable.

Vickers: We'll be able to look at this point when we analyse all the results from our screening and recruitment.

James: It looked as though you're going to pool all your data to achieve those powers of significance for interpretation. Over such a long time, isn't there the danger that fashions will change, new preparations will come along? For instance, transdermal preparations may become more fashionable. Are you going to have to break up your data into lots of subgroups? Isn't that going to damage your statistical interpretation?

Vickers: We hope to be able (as far as is possible) to stick to the preparations we decided on at the beginning. A lot of thought has been given to what will be the HRT of relevance in 15 years time. Most people think that a continuous combined preparation is going to be the preparation of choice and we would probably use this formulation in the main trial.

Manolagas: In the USA, new oestrogens are being developed with specific actions on bone and cholesterol and no affect on the breast and uterus. I believe that both Eli Lilly and Pfizer are coming up with these.
If an oestrogen comes on the market that gives protection from osteoporosis and heart disease and does not act on the breast and the uterus, how can you justify keeping your patients on existing preparations for another 20 years?

Vickers: The trouble with doing this sort of trial is that you can wait forever until the perfect preparation comes along. In the main trial there will be an independent data-monitoring and ethics committee, and if something comes along that is the answer to everyone's dreams and makes currently available preparations redundant, then the trial will have to change or stop. But to use that sort of argument to delay beginning the trial is invalid.

Manolagas: You're right.

References

Martin MC, Block JE, Sanchez SD, Arnaud CD, Beyene Y 1993 Menopause without symptoms: the endocrinology of the menopause among rural Mayan indians. Am J Obstet Gynecol 168:1839–1845
Tang G 1994 Menopause: the situation in Hong Kong chinese women. In: Berg G, Hammar M (eds) The modern management of the menopause. Parthenon, New York, p 47–55

General discussion III

Sutherland: I'm not aware of the tissue-specific oestrogens to which Stavros Manolagas referred in the last discussion. Do you have more information you could share with us?

Manolagas: As far as I know, these compounds bind to and activate the oestrogen receptor. I don't think anyone fully understands why they are tissue specific. From what I know from published information, the Eli Lilly compound—raloxifene—is thought to act via a different receptor to that for classical oestrogens. Personally, I am not convinced.

Baulieu: Have the raloxifene results been published?

Manolagas: The results were published by Black et al (1994). These experiments were carried out on rats and they showed a very nice protective effect on bone loss and a large effect on cholesterol. There was no effect on the proliferation of the endometrium. Preliminary results on the effects of raloxifene on women were reported at the Fourth International Symposium on Osteoporosis and Consensus Development Conference (Hong Kong, March 27–April 2, 1993). They were very convincing.

Baulieu: Have the structures of these oestrogens been published?

Bonewald: Yes. Raloxifene is a non-steroidal benzothiophene, so its structure is quite different from tamoxifen (Black et al 1994). Tamoxifen is a triphenyl ethylene.

Sutherland: Then it is related to the benzothiophene series of anti-oestrogens developed by Lilly in the 1980s (Black et al 1983). These compounds demonstrate both agonist and antagonist activity as do the other non-steroidal anti-oestrogens.

Bonewald: Raloxifene has mainly antagonistic activity on the uterus (Black et al 1994, Evans et al 1994).

A major question in the bone field concerns how these compounds that have anti-oestrogenic effects on the uterus have the same effect as oestrogen on bone cells. Are they using the same receptor-mediated pathway or are they using a different one? These are the questions that we're asking now.

Horwitz: We think that different tissues have distinct signalling systems, so that signalling pathways in bone probably differ from those in the uterus. We have evidence that steroid receptors don't act in isolation, but that their activity is modulated by other transcriptional signals. For example, in some tissues, the same gene might be activated both by cAMP acting through factor 'X', and independently by oestrogen receptors. The direction of that gene's transcription in response to anti-oestrogens is controlled by interactions between factor 'X' and the oestrogen receptors. If factor 'X' is present in bone but not in uterus, one might observe an agonist-like effect in the former but not in the latter. Thus, it is neither the unusual properties of the oestrogen receptors, nor of the ligands, that control tissue specificity, but the presence of other factors. Such a model can explain tissue-specific differences without invoking differences in oestrogen receptors, or in ligand actions among different tissues. The molecular tools are now in place to understand the mechanisms underlying tissue specificity.

Bonewald: Yang et al (1993) found that both oestrogen and raloxifene stimulate the production of the transforming growth factor (TGF)-β3 isoform, both *in vivo* and *in vitro*. It is proposed that TGF-β3 is the final target for both of these compounds and that TGF-β3 is important for maintaining bone mass.

Manolagas: In our hands, antibodies to TGF-β inhibit osteoclast development. From this evidence it is difficult to accept the view that the protective effect of oestrogens on bone is due to stimulation of TGF-β.

Bonewald: That's the effect on resorption: what about bone formation?

Manolagas: The concept that systemic hormones are a gross network of signalling information which needs to be fine-tuned at local level through cytokines is getting to be more and more evident. Specifically, I have presented evidence (Manolagas et al 1995, this volume) to suggest that one single molecular entity can integrate systemic (hormonal) and local (cytokine) signals. This is the signal transducer gp130.

Bayard: We looked at the oestrogen receptor in the uterus, and we found that there were two different species of the same receptor protein (Faye et al 1986). In the untreated animal we found that the oestrogen receptor protein has the A/B domain (67 kDa). A few days after treatment with a huge dose of oestradiol (25 μg/day) we found just the smaller form of the oestrogen receptor deleted from the A/B domain (45 kDa). We are still working on this because we suspected it to be a consequence of proteolysis of the oestrogen receptor, but we couldn't detect any proteolytic activity under our experimental conditions. We believe that a different mechanism might be operating that would explain why when the uterus is growing we find the complete oestrogen receptor, but when it is not growing any more we just find this 45 kDa protein. We got similar results with MCF-7 cells. In the log phase of growth we found the complete receptor, and in high-density growth phase cells we got this 45 kDa protein. So the same receptor may be produced in different molecular forms.

Muramatsu: I agree that different receptors and different signal transduction pathways are involved in the action of oestrogen. Oestrogen derivatives such as raloxifene may bind to the ErbB2 receptor (a tyrosine kinase-type receptor) and transmit signals. These signal transduction pathways are not well understood at the present time, so we cannot estimate exactly what those derivatives would do. We need to clarify what those pathways are.

Manolagas: I want to play the devil's advocate: are we trying to medicalize the menopause? A lot of people say that the menopause is a natural state of women in certain cultures. They gain decency, they gain respect, they gain wisdom. Are we trying to treat women with oestrogen replacement for a legitimate reason or for a false reason that's medically generated? This is a hot political issue in the USA.

Horwitz: It is only a political issue among men!

Jensen: One is not trying to eliminate the menopause, only its undesirable side effects.

Manolagas: One of the arguments in favour of your opinion is that if a woman loses oestrogen because of ovariectomy—say, at the age of 35—and is likely to break her bones a few years later, most physicians will consider it appropriate to administer oestrogens. Why do you not accept the same reasoning for a women who undergoes natural menopause? These are issues that many countries are going to have to tackle sooner or later.

Foegh: In the debate as to whether women should be given oestrogen replacement therapy after menopause or whether the menopause should be regarded as a natural process in which one should not interfere, it may be appropriate to bring up a completely different view. Before this century, many women died in childbirth from infections and thus did not live to the age of menopause. They did not experience long periods with menstrual cycles, because of their many pregnancies, which resulted in years of exposure to high oestrogen levels. Thus a continuous oestrogen exposure, as in postmenopausal oestrogen replacement therapy, may be a more natural state than the lack of oestrogen. So we are not medicalizing the menopause; we are just trying to eliminate a deficiency which is a consequence of the sudden increase in longevity of women, following the discovery by Simwelwis of the link between poor hygeine and infection.

Baulieu: So far, the issue of pheromones hasn't been discussed. Don Pfaff, weren't you interested in this issue a number of years ago?

Pfaff: In the animal literature there are examples of oestrogen-dependent pheromones to which males are innately attracted. Under the influence of oestrogen, female hamsters produce vaginal secretions. Even if you separate the male hamster at the time of weaning from all females, upon the very first introduction of this oestrogen-stimulated vaginal secretion the male will show a tremendous preference for it, providing he has normal testosterone levels.

Secondly, it has been reported that there are sexual dimorphisms in the ability of humans to detect steroids that have a low but non-zero vapour pressure. I don't know if that kind of study has been replicated.

More generally, the connection between olfaction and sex has been with us for a long time. If we go all the way back to aquatic animals, not only the ability of animals to detect pheromones, but also physical circumstances (such as hydrostatic pressure and salt content of the water and temperature of the water) which can affect the nasal epithelium could help govern reproductive cycles. We use this kind of teleological argument to rationalize gonadotropin-releasing hormone neurons borne on the olfactory placode in human beings as well as in mice, monkeys, guinea pigs, rats, fish and reptiles.

Uvnäs-Möberg: It's well known that if you put female rats together, they will start to cycle together. This is also the case with women.

Pfaff: Female undergraduates as well, under some circumstances (McClintock 1971).

Uvnäs-Moberg: I recently gave a lecture and told people about this synchronization of the menstrual cycle in female rats. One of the women listening came up and told me that at her workplace the women were cycling together and then asked me whether I could tell who was 'leading' the cycles. I told her that in rats it's always the largest female who leads. Since the largest rat is the most dominant individual in the group, I predicted that the most dominant woman in her workplace must be leading the cycles. She agreed and said it was not the boss, but the most dominant person in her group. How is this coordination brought about? Is it caused by pheromones?

Pfaff: I think authors like Martha McClintock might suspect that. More broadly, reflecting upon the subject of our meeting—the non-reproductive aspects of sex steroids—it seems to me that the potential practical importance of this might be clear if we think of the limbic system as the forebrain of the olfactory system, and in turn if we consider the amount of psychopathology that seems to depend primarily on dysfunctions of the limbic system.

Robel: The aggression described in Swiss mice, when females or castrated males are aggressive towards lactating female intruders, is definately due to an odorant present in the urine of these females. Intact males are not aggressive. Castrated males treated with oesradiol, testosterone, dehydroepiandrosterone or its analogue 3β-methyl-androst-5-en-17-one that cannot be converted to either androgens or oestrogens, cease to become aggressive (reviewed in Robel & Baulieu 1994). This model illustrates the relationships between sex steroids and the response to odorants.

Baulieu: There are effects of sex hormones on the production of phenomenal compounds. Female monkeys are attractive to males during oestrus, partly because of the red colour of the external female genitalia. During oestrus, females also produce relatively volatile fatty acids which when transferred to neutered female monkeys make them attractive to males. In one experiment, vaginal secretions were obtained from mid-cycle women and transferred to castrated female monkeys; this made them attractive to the male monkeys. This implies that some of these fatty acids are shared among primates. This secretion is under the control of sex hormones.

In another experiment, two male rats were put in a cage separated by a screen, and the levels of pregnenolone in their olfactory bulbs were measured (Corpéchot et al 1985). When one of the males was replaced by a female, the level of pregnenolone in the olfactory bulb of the male decreased. However, when the female was neutered, there was no change in the pregnenolone concentration in the olfactory bulb of the male. So the signal for pheromone seems to be under the control of female sex hormones. If the recipient male is castrated, it does not respond to the female, unless you give it testosterone. So, in mammals, there are certainly pheromonal signals that are under the control of sex steroids.

In fishes, there is one extraordinary example of a hormone also acting as a pheromone. It's a progestin, 17,20α-dihydroxy-pregna-4-en-3-one. This steroid, in addition to being a meiotic agent in females, is excreted into the water, received by the nasal epithelium of males, and triggers spermatogenesis and male behaviour. Olfactory receptors are G protein-linked and constitute a family of seven-transmembrane-domain receptors that several groups are working on. There are hundreds of these receptors, probably only one per specialized cell. Interestingly, human, dog and pig sperm have, in their membranes, three, four or five members of this family of odorant receptors, quasi-identical to those in the olfactory system (Parmentier et al 1992).

Besides the olfactory system there is the vomeronasal organ. This is under-developed in primates except during fetal development, after which it regresses. The vomeronasal organ has a structure which is different from the olfactory epithelium. It has connections to the accessory olfactory bulb. There are electrophysiological techniques for recording from the vomeronasal organ. There are steroids called vomeropherins, pheromones specifically directed to the vomeronasal system, which show sex specificity; that is, steroids 'recieved' only by men or only by women.

References

Black LJ, Jones CD, Falcone JF 1983 Antagonism of estrogen action with a new benzothiophene-derived antiestrogen. Life Sci 32:1031–1036

Black LJ, Masahiko S, Rowley ER et al 1994 Raloxifene (LY139481 HCl) prevents bone loss and reduces serum cholesterol without causing uterine hypertrophy in ovariectomized rats. J Clin Invest 93:63–69

Corpéchot C, Leclerc P, Baulieu EE, Brazeau P 1985 Neurosteroids: regulatory mechanisms in male rat brain during heterosexual exposure. Steroids 45:229–234

Evans G, Bryant HU, Magee D, Sato M, Turner RT 1994 The effects of raloxifene on tibia histomorphometry in ovariectomized rats. Endocrinology 134:2283–2288

Faye JC, Fargin A, Bayard F et al 1986 Dissimilarities between the uterine estrogen receptor in cytosol of castrated and estradiol treated rats. Endocrinology 118: 2276–2283

Manolagas SC, Bellido T, Jilka RL 1995 Sex steroids, cytokines and the bone marrow:
    new concepts on the pathogenesis of osteoporosis. In: Non-reproductive aspects of
    sex steroids. Wiley, Chichester (Ciba Found Symp 191) p 187–202

McClintock MK 1971 Menstrual synchrony and suppression. Nature 229:244–245

Parmentier M, Libert S, Schiffman S et al 1992 Expression of the putative olfactory
    receptor gene family in mammalian germ cells. Nature 355:453–455

Robel P, Baulieu EE 1994 Neurosteroids: biosynthesis and function. Trends Endocrinol
    Metab 5:1–8

Yang NN, Hardikar S, Kim J, Sato M 1993 Raloxifene, an 'anti-estrogen', stimulates
    the effects of estrogen on inhibiting bone resorption through regulating TGFβ3
    expression in bone. J Bone Miner Res 8 (suppl 1):7(abstr)

Symptoms related to the menopause and sex steroid treatments

T. Bäckström

Department of Obstetrics and Gynecology, University Hospital, S-901 85 Umeå, Sweden

Abstract. In the menopause transition, ovarian steroid production is gradually inhibited and around 35% of women will seek medical help for postmenopausal symptoms. The hot flush is a characteristic manifestation occurring in about 70% of women; it is associated with oestrogen withdrawal and disappears with oestrogen-based hormone replacement therapy. The exact mechanism behind it is still unclear but is probably related to heat loss mechanisms. The flush often occurs in parallel to changes in skin temperature, blood flow, pulse rate and pulses of luteinizing hormone (LH). These are probably secondary to a disturbance in the thermoregulatory centre of the CNS, which is anatomically close to neurons containing gonadotropin-releasing hormone. Depression is no more frequent in the menopausal transition than at other times in life. After surgical menopause, however, oestrogen improves low mood over placebo. In women with premenstrual syndrome, an increased feeling of well-being is associated with the pre-ovulatory oestrogen peak. Progestogens are associated with negative mood changes during the menstrual cycle, oral contraception and postmenopausal replacement therapy. Certain progesterone metabolites are anaesthetic and have anti-epileptic and anxiolytic properties, effects which are mediated via the type A γ-aminobutyric acid ($GABA_A$) receptor. Oestrogen is associated with increased sensory perception, locomotory activity, limb coordination and balance: this may help explain the increased frequency of bone fractures in the early postmenopausal period. Oestrogen improves memory and performance in patients with mild Alzheimer's dementia and increases epileptic activity in patients with partial epilepsy. These effects can be related to amplifying effects of oestrogen on excitatory amino acids in the CNS.

1995 Non-reproductive actions of sex steroids. Wiley, Chichester (Ciba Foundation Symposium 191) p 171–186

The ageing of the ovary has already begun by the time women reach the age of 40, and can be detected as an increase in early follicular and mid-cycle levels of follicle-stimulating hormone (FSH) (Sherman et al 1976). Around this time, menstrual cycles continue to be ovulatory, but the ovary gradually becomes less responsive to gonadotropin stimulation; this process begins several years before menses cease (Sherman & Korenman 1975). The decrease in production of oestradiol removes the negative feedback inhibition by oestrogen on the hypothalamic pituitary production of FSH. This is probably one reason for the gradual increase in the early-follicular-phase FSH levels (Monroe et al 1972). An increase in luteinizing hormone (LH) occurs later, although this is not as pronounced as the increase in FSH. Another reason for the increase in FSH production is probably a gradual reduction in the production of inhibin by the ovary. The mean follicular-phase levels of immunoreactive inhibin decrease with age, in parallel with an increase in FSH (MacNaughton et al 1992).

The periodic pulsatile pattern of LH and FSH secretion continues in postmenopausal women. The gonadotropin pulses occur with intervals of approximately 60–90 min (Yen et al 1972). As women approach menopausal age, menstrual cycles become irregular and the number of ovulatory cycles decreases (Metcalf 1983). When the menopause occurs the ovaries have few or no primordial follicles (Richardson et al 1987).

Postmenopausal ovaries usually produce very little oestrogen and their removal is not accompanied by any significant decrease in circulating oestrogen levels (Judd et al 1974). However, the ovary continues to be endocrinologically active, especially with regard to the production of androgens. In particular, testosterone and androstenedione are produced by the ovarian stromal and hilus cells, which retain their LH receptors and are endocrinologically active (Judd et al 1974). Androstenedione is also produced by the adrenal, which is responsible for a large proportion of the total androstenedione production. Testosterone, however, is mainly produced by the ovary in postmenopausal women. The androgens produced by both the ovaries and the adrenal are converted to oestrogen at extraglandular sites, i.e. outside either the ovary or adrenal gland. This extragonadal aromatization is an important concept in the production of oestrogen in postmenopausal women. All the sites of this aromatization have not yet been identified, but it is known to occur in adipose tissue, liver, kidney and brain. It mainly involves the conversion of androstenedione to oestrone (Grodin et al 1973).

The effects of the extragonadal aromatization are influenced by age and weight. Heavy women have higher conversion rates and higher circulating oestrogen concentrations than slender women. There is a close relationship between the body mass index and oestrogen concentration in blood (Boman et al 1990). This indicates that the weight of women is of great significance for the expression of postmenopausal symptoms.

The production of androgens by the adrenal decreases with age and has already begun to decline by the age of 30. The production of cortisol from the adrenal, however, stays the same throughout life. Besides androstenedione, the adrenal gland also produces dehydroepiandrosterone (DHEA) and DHEA sulphate (Vermeulen & Verdonck 1978). The changes in the production of androgens by the adrenal are of significance for later postmenopausal changes such as osteoporosis.

The level of oestrogen that causes menstrual bleeding and development of endometrium varies among individuals, but a continuous serum concentration of oestradiol above 100 pmol/l will usually have these effects.

The median age for menopause in western countries is 51 years. The median age for smokers is two years earlier than for non-smokers. For about 10% of women the menopause arrives abruptly whereas, for the remainder, the menopause is preceded by a period of up to four years of irregular menstrual bleeding (McKinlay et al 1992). Up to 70% of postmenopausal women will experience symptoms of the menopause—mainly hot flushes and perspiration. These symptoms occur during the first four to five years but decrease in severity during later years: only 20% still have hot flushes five years after the menopause (McKinlay et al 1992).

Around 35% of women will seek medical help for their postmenopausal symptoms. Hot flushes have been described in different races and cultures. They are more common in women who smoke. Women with postmenopausal symptoms weigh less than those without (Jaszman 1976).

The hot flush

The exact mechanism behind the hot flush is still not fully understood. The phenomenon is clearly related to a withdrawal of oestrogen; the symptoms disappear with oestrogen replacement therapy (Jaszman 1976). To treat the symptoms, an oestrogen dose high enough also to stimuate endometrial proliferation is required. This will, after some time, result in endometrial hyperplasy with increased risk of endometrial carcinoma (Ziel & Finkle 1975). Therefore, sequential oestrogen/progestogen replacement therapy or combined oestrogen/progestogen treatments are used. Because progesterone inhibits endometrial proliferation, this treatment induces regular endometrial sheddings or, in the case of combined therapy, produces an atrophic endometrium. The risk of endometrial cancer is therefore abolished (Persson et al 1989). This is the rationale behind the combined use of oestrogen and progesterone in postmenopausal hormone replacement therapy. The doses of oestrogen needed for osteoporosis prophylaxis also appear to be above the concentration that induces endometrial proliferation, hence progesterone is needed in this treatment also.

The hot flush is widely considered to be the characteristic clinical manifestation of the menopause. It is often described as a sensation of pressure in the head similar to a headache. This progresses in intensity until the actual flush occurs. The actual flush begins in the head and neck regions, and often passes in waves over the entire body: it is described as a sensation of heat or burning. This period is followed by an outbreak of sweating across the whole body, but particularly marked in the head, neck, chest and back. The entire episode can vary in duration from 30 s up to 30 min but, on average, lasts for about 10 min. Shortly after the onset of a flush, there is an increase in blood flow in the hand and forearm as well as an increased pulse rate, symptoms which persist for some minutes after the sensation of the flush disappears. At the same time, the temperature of the fingers begins to increase; this reaches a maximum about five minutes after the end of the flush (Ginsburg et al 1981, Molnar 1975). There have been many attempts to identify a trigger for the flush sensation. A relationship has been found between the LH pulses from the pituitary and the presence of a hot flush (Casper et al 1979). However, LH pulses occur without hot flushes and there are hot flushes which are not followed by LH pulses. In two studies, about 76–83% of LH pulses were associated with rises in skin temperature or were accompanied by hot flushes (Casper et al 1979, Tataryn et al 1979). Several studies have been undertaken to see whether the LH pulse induces the peripheral changes, but so far these have failed. If gonadotropin-releasing hormone (GnRH) agonist preparations are given, the LH pulses are abolished, but the flush continues. Therefore, the LH pulses and the flushes are now considered as parallel phenomena to a common disturbance in the CNS. It seems that the flush represents an alternation in the central thermoregulatory function and that the change in skin temperature is probably a mechanism promoting heat loss. The relationship to LH pulses could be due to the close anatomical occasion between the thermoregulatory centre and GnRH-containing neurons (Ginsburg & Hardiman 1991). An oestrogen replacement therapy will decrease both the production of LH and the frequency of hot flushes.

After the start of oestrogen therapy, the number of hot flushes decreases gradually over a period of up to 20 days. During the next two months the number of hot flushes will stabilize at a level that is related to the dose of the oestrogen replacement therapy, as well as the effect of decreasing FSH concentration. A sequential oestrogen/progestogen replacement therapy will be more effective at depressing FSH levels than oestrogen alone (Holst et al 1989).

Mood changes

Mood changes during the menopause are frequently discussed in the press and among patients. Several longitudinal studies have been unable to confirm an increase in clinical depression or severe mood change in relation to the menopausal transition (Kaufert et al 1992). However, a prospective study on a general population sample in Norway showed that the menopause was accompanied by an increase in less-severe depression of mood, but there was no link between menopause and severe depression (Holte 1992). This has been confirmed in longitudinal studies in North America and in Europe (Kaufert et al 1992, Hunter 1992). The effect of oestrogen replacement therapy on clinical depression has been tested in some randomized controlled clinical studies. In three studies in which oestrogen was used at a hormone replacement therapy dosage, no effect on clinical depression was noted in naturally postmenopausal women with clinical depression (Coope 1981, Campbell 1976, Thompson & Oswald 1977). In an open study, Klaiber et al (1979) gave a high dose of oestrogen to severely depressed psychiatric patients with good effect. It has not been possible to replicate these results in controlled studies. However, in a recent double-blind controlled study, Gregoirc et al (1994) found that oestrogen was beneficial at a higer dosage for treating severe post partum depression.

In depressed patients who are surgically postmenopausal (i.e. women who have been oophorectomized), the results were different. Three studies all show a significant improvement in depressive mood (using Beck's Depression Scale or Hamilton's Rating Scale) after oestrogen therapy. In these studies a higher oestrogen dose had a significantly better effect than a lower dose (Montgomery et al 1987, Sherwin & Gelfand 1985, Campbell 1976).

Some women show cyclical mood changes in relation to different phases of the menstrual cycle. In the so-called 'premenstrual syndrome', patients experience increased feelings of well-being in association with the pre-ovulatory oestrogen peak. After ovulation, these women experience a negative change in mood, increasing in severity until the last five premenstrual days. After the onset of the menstrual bleeding the symptoms decrease and are usually gone three to four days after the onset of bleeding (Bäckström et al 1983). The increased feeling of well-being during the pre-ovulatory peak in these women is not simply due to relief of the negative mood, because they have a reduced sense of well-being during the pre-ovulatory period in anovulatory cycles induced by GnRH agonists (Hammarbäck & Bäckström 1988).

During the menstrual cycle and luteal phase, progestogens are associated with negative changes in mood. Negative mood symptoms are a well-known side effect of oral contraceptives in women of fertile age. Women who experience negative mood changes whilst taking oral contraceptives are largely those who suffer from premenstrual syndrome in their natural menstrual cycles (Cullberg 1972). In patients taking hormone replacement therapy there is an increase in negative mood parallel with the addition of progestogens in sequential replacement therapy. In these women, symptoms begin shortly after the progestogen is added to the treatment and last for 2–3 days after the treatement ends (Hammarbäck et al 1985). There seems to be a dose-dependent increase in negative mood changes with increasing dosage of the progestogen (Magos et al 1986). These results have, however, not yet been confirmed. Negative change in mood is a well-known side effect of combined hormone replacement therapy and is one of the major problems with its acceptability to the patient. The exact mechanism behind this depression is still not understood, although it is clear that it is related to the addition of progestogen to the postmenopausal replacement therapy. The effect of the progestogen is probably directly on the CNS. Natural progesterone and certain progesterone metabolites given intravenously seem, however, to have opposite effects at pharmacological doses. The progesterone metabolites 3α-hydroxy-5α-pregnan-20-one and 3α-hydroxy-5β-pregnan-20-one have been shown to have anaesthetic (Carl et al 1990), anti-epileptic (Langren et al 1987) and anxiolytic properties (Finn & Gee 1994). These substances show cyclical changes in serum concentration during the menstrual cycle (Paul & Purdy 1992) and are probably produced by the corpus luteum (Bäckström et al 1986). When micronized oral progesterone is given it will be metabolized, mostly to allopregnanolone. High plasma concentrations result and patients experience sedation as a side effect (Freeman et al 1993). These effects have now been shown to be mediated by a type-A γ-aminobutyric acid (GABA$_A$) receptor in the CNS. These steroids act in a similar way to benzodiazepines and barbiturates, enhancing the effect of GABA on the GABA$_A$ receptor (Finn & Gee 1994).

Pregnenolone sulphate and some androgen metabolites have been shown *in vitro* to have effects that suggest they might be anxiogenic in humans (Paul & Purdy 1992). The negative effects on mood of progestogen may be related to the existence of such anxiogenic substances.

Sensory, motor and memory functions

Behavioural studies have suggested that the oestrogen-dominated follicular phase is associated with increased alertness and activity (Asso & Braier 1982). Increases in rapidly alternating tasks such as walking, finger tapping frequency, typing and word fluency have been reported (Broverman et al 1968). In general, these tasks involve movement and control over the limbs and digits. Increased levels of oestradiol are also associated with increased two-point discrimination, fine-touch perception, hearing, smell and vision (Zimmerman & Parlee 1973), and increased locomotory activity (Morris & Udry 1970, Beatty 1979) in both humans and animals. There are also reports that oestrogen is associated with improved limb coordination (Hampson & Kimura 1988). A further indication that oestrogen affects the control of movements is that one of its side effects is an increase in involuntary movements such as chorea and tardive dyskinesia in humans (Maggi & Peres 1985). These effects may be related to the amplifying effects of oestradiol on excitatory amino acids in the CNS (Smith 1989, Weiland 1992).

There is an increase in the incidence of bone fractures shortly after the menopause in women not receiving hormone replacement therapy; this increase is not related to osteoporosis. Instead, it has been suggested that it is a consequence of changes in balance and limb coordination, secondary to a decrease in oestrogen during the first few postmenopausal years (Naessén et al 1990).

There is also evidence that oestrogen enhances short-term memory and alertness following cyclical elevation of oestradiol during the menstrual cycle (Patkai et al 1974). Shorter reaction times have also been seen in women during the follicular phase of the menstrual cycle. Oestrogen has been shown to have a beneficial effect on memory function and performance in patients with mild Alzheimer's dementia, although in severely affected patients no effect was seen (Honjo et al 1989, Fillit et al 1986). The mechanisms behind these effects are not understood completely, but changes in neuronal cell development and number of synapses between neurons have been shown to occur alongside changes in oestrogen (Woolley & McEwen 1992). In rats, the cerebral circulation increases within 10 min of injection of oestradiol (Goldman et al 1976).

The results discussed above demonstrate that many of the symptoms of the menopause are related to the function of the CNS, and that both oestrogen and progestogens are involved in creation of these symptoms and their treatment.

Acknowledgements

Some of the studies described here were supported by the Swedish Medical Research Council, Project 4X-11198, and CoCensys Inc., Irvine, CA, USA.

References

Asso D, Braier JR 1982 Changes with the menstrual cycle in psychophysiological and self-report measures of activation. Biol Psychol 15:95–107

Bäckström T, Sanders D, Leask R, Davidson D, Warner P, Bancroft J 1983 Mood, sexuality, hormones and the menstrual cycle. II. Hormone levels and their relationship to the premenstrual syndrome. Psychosom Med 45:503–507

Bäckström T, Andersson A, Baird DT, Selstam G 1986 5-Alpha-pregnan-3,20-dione in peripheral and ovarian vein blood of women at different stages of the menstrual cycle. Acta Endocrinol 111:116–121

Beatty WW 1979 Gonadal hormones and sex differences in non-reproductive behaviors in rodents: organizational and activational influences. Horm Behav 12:112–163

Boman K, Bäckström T, Gerdes U, Stendahl U 1990 Oestrogens and clinical characteristics in endometrial carcinoma. Anticancer Res 10:247–252

Broverman DM, Klaiber EL, Kobayashi Y, Vogel W 1968 Roles of activation and inhibition in sex differences in cognitive abilities. Psychol Rev 75:23–50

Campbell S 1976 Double blind psychometric studies on the effects of natural oestrogens on postmenopausal women. In: Campbell S (ed) Management of the menopause and the postmenopausal years. MTP Press, Lancaster, p 149–158

Carl P, Hogskilde S, Nielsen JW et al 1990 Pregnanolone emulsion. A preliminary pharmacokinetic and pharmacodynamic study of a new intravenous anaesthetic agent. Anesthesia 45:189–197

Casper RF, Yen SSC, Wilkes MM 1979 Menopausal flushes: a neuroendocrine link with pulsatile luteinizing sectretion. Science 205:823–825

Coope J 1981 Is oestrogen effective in the treatment of menopausal depression? J R Coll Gen Pract 31:134–140

Cullberg J 1972 Mood changes and menstrual symptoms with different gestagen/estrogen combinations: a double blind comparison with a placebo. Acta Psychiatr Scand Suppl 236:1-86

Fillit H, Weinreb H, Cholst I et al 1986 Observations in a preliminary open trial of estradiol therapy for senile dementia—Alzheimer's type. Psychoneuroendocrinology 11:337–345

Finn DA, Gee KW 1994 The significance of steroid action on the GABA-A receptor complex. In: Berg G, Hammar M (eds) The modern management of the menopause. Parthenon, Park Ridge, NJ, p 301–313

Freeman E, Purdy R, Coutifaris C, Rickels K, Paus S 1993 Anxiolytic metabolites of progesterone: correlation with mood and performance measures following oral progesterone administration to healthy female volunteers. Neuroendocrinology 58:478–484

Ginsburg J, Hardiman P 1991 What do we know about the pathogenesis of the menopausal hot flush? In: Sitruk-Ware R, Utian WH (eds) The menopause and hormonal replacement therapy. Marcel Dekker, New York, 15–46

Ginsburg J, Swinhoe J, O'Reilly B 1981 Cardiovascular responses during the menopausal hot flush. Br J Obstet Gynaecol 88:925–930

Goldman H, Skelley EB, Sandman CA, Kastin AJ, Murphy S 1976 Hormones and regional brain blood flow. Pharmacol Biochem Behav (suppl 1) 5:165–169

Gregoirc A, Henderson A, Kumar R, Studd J 1994 A controlled trial of oestradiol therapy for post natal depression. Neuropsychopharmacology 10:9015(abstr)

Grodin JM, Siiteri PK, MacDonald PC 1973 Source of esrogen production in postmenopausal women. J Clin Endocrinol & Metab 36:207–214

Hammarbäck S, Bäckström T 1988 Induced anovulation as treatment of premenstrual tension syndrome—a double-blind crossover study with GnRH-agonist versus placebo. Acta Obstet Gynecol Scand 67:159–166

Hammarbäck S, Bäckström T, Holst J, von Schoultz B, Lyrenäs 1985 Cyclical mood changes as in the premenstrual tension syndrome during sequential estrogen–progestagen postmenopausal replacement therapy. Acta Obstet Gynecol Scand 64:515–518

Hampson E, Kimura D 1988 Reciprocal effects of hormonal fluctuations on human motor and perceptual–spatial skills. Behav Neurosci 102:456–459

Holst J, Bäckström T, Hammarbäck S, von Schoultz B 1989 Progestogen addition during estrogen replacement therapy—effects on vasomotor symptoms and mood. Maturitas 11:13–20

Holte A 1992 Influences of natural menopause on health complaints: a prospective study of healthy Norwegian women. Maturitas 14:127–141

Honjo H, Ogino Y, Naitoh K et al 1989 In vivo effects by estrone sulfate on the central nervous system—senile dementia (Alzheimer's type). J Steroid Biochem 34:521–525

Hunter M 1992 The south-east England longitudinal study of the climacteric and postmenopause. Maturitas 14:117–126

Jaszman L 1976 Epidemiology of the climacteric syndrome. In: Campbell S (ed) The management of the menopause and postmenopausal years. MTP Press, Lancaster, p 11–23

Judd HL, Judd GE, Lucas WE, Yen SSC 1974 Endocrine function of the postmenopausal ovary: concentration of androgens and estrogens in ovarian and peripheral vein blood. J Clin Endocrinol & Metab 39:1020–1024

Kaufert PA, Gilbert P, Tate R 1992 The Manitoba Project: a reexamination of the link between menopause and depression. Maturitas 14:143–155

Klaiber EL, Broverman DM, Vogel W, Kobayashi Y 1979 Estrogen replacement therapy for severe persistent depression in women. Arch Gen Psychiatr 36:550–554

Landgren S, Aasly J, Bäckström T, Dubrowsky B, Danielsson E 1987 The effect of progesterone and its metabolites on the interictal epileptiform discharge in the cat's cortex. Acta Physiol Scand 131:33–42

McKinlay SM, Brambilla DJ, Posner JG 1992 The normal menopause transition. Maturitas 14:103–115

MacNaughton J, Bangah M, McCloud P, Hee J, Burger HG 1992 Age related changes in follicle stimulating hormone, luteinizing hormone, estradiol and immunoreactive inhibin in women of reproductive age. Clin Endocrinol 36:339–345

Maggi A, Perez J 1985 Role of female gonadal hormones in the CNS: clinical and experimental aspects. Life Sci 37:893–906

Magos AL, Brewster E, Sing R, O'Dowd TM, Studd JWW 1986 The effect of norethisterone in postmenopausal women on oestrogen therapy: a model for the premenstrual syndrome. Br J Obstet Gynaecol 93:1290–1296

Metcalf MG 1983 Incidence of ovulation from the menarche to the menopause. N Z Med J 96:645–648

Molnar GW 1975 Body temperature during menopausal hot flashes. J Appl Physiol 1:499–503

Monroe SE, Jaffe RB, Midgley AR 1972 Regulation of human gonadotropins. XIII. Changes in serum gonadotropins in menstruating women in response to oophorectomy. J Clin Endocrinol & Metab 34:420

Montgomery JC, Appelby L, Brincat M et al 1987 Effect of estrogen and testosterone implants on psychological disorders in the climacteric. Lancet I:297–299

Morris NM, Udry JR 1970 Variations in pedometer activity during the menstrual cycle. Obstet Gynecol 35:199–201

Naessén T, Persson J, Adami H, Bergström R, Bergkvist L 1990 Hormone replacement therapy and the risk for first hip fracture. Ann Intern Med 113:95–103

Patkai P, Johannson G, Post B 1974 Mood, alertness and sympathetic–adrenal medullary activity during the menstrual cycle. Psychosom Med 36:503–512

Paul S, Purdy R 1992 Neuroactive steroids. FASEB J 6:2311–2322

Persson I, Adami HO, Bergkvist L et al 1989 Risk of endometrial cancer after treatment with oestrogens alone or in conjunction with progestogens: results of a prospective study. BMJ 298:147–151

Richardson SJ, Senikas V, Nelson JF 1987 Follicular depletion during the menopausal transition: evidence for accelerated loss and ultimate exhaustion. J Clin Endocrinol & Metab 65:1231–1237

Sherman BM, Korenman SG 1975 Hormonal characteristics of human menstrual cycle throughout reproductive life. J Clin Invest 55:699–706

Sherman BM, West JH, Korenman SG 1976 Menopausal transition: analysis of LH, FSH, estradiol, and progesterone concentrations during menstrual cycles of older women. J Clin Endocrinol & Metab 42:629–636

Sherwin BB, Gelfand MM 1985 Sex steroids and affect in the surgical menopause: a double-blind crossover study. Psychoneuroendocrinology 10:325–335

Smith SS 1989 Estrogen produces long-term increases in excitatory neuronal responses to NMDA and quisqualate. Brain Res 503:354–357

Tataryn IV, Meldrum DR, Lu KH, Frumar AM, Judd HL 1979 LH, FSH and skin temperature during the menopausal hot flash. J Clin Endocrinol & Metab 49:152–154

Thompson J, Oswald I 1977 Effect of oestrogen on the sleep, mood and anxiety of menopausal women. BMJ 2:1317–1319

Vermeulen A, Verdonck L 1978 Sex hormone concentrations in postmenopausal women: relation to obesity, fat mass, age and years post-menopause. Clin Endocrinol 9:59–66

Weiland NG 1992 Estradiol selectively regulates agonist binding sites on the N-methyl-D-aspartate receptor complex in the CA1 region of the hippocampus. Endocrinology 131:662–668

Woolley CS, McEwen BS 1992 Estradiol mediates fluctuation in hippocampal synapse density during the estrous cycle in the adult rat. J Neurosci 12:2549–2554

Yen SS, Tsai CC, Naftolin F, Vanderberg G, Ajabor L 1972 Pulsatile patterns of gonadotropin release in subjects with and without ovarian function. J Clin Endocrinol & Metab 34:671

Ziel HK, Finkle WD 1975 Increased risk of endometrial carcinoma among users of conjugated estrogens. N Engl J Med 293:1167–1170

Zimmerman E, Parlee MB 1973 Behavioral changes associated with the menstrual cycle: an experimental investigation. J Appl Soc Psychol 3:335–344

DISCUSSION

McEwen: You mentioned that on average there is a rather gradual fall in oestrogen levels after the menopause. Are there any data showing variations among individual women?

Bäckström: Yes, if you take the body mass index into consideration, some individuals maintain high levels of oestrogen even into old age. There are also certain conditions that increase oestrogen concentrations, for instance, congestive heart failure and liver disease (Pintor et al 1982).

McEwen: In relation to your point about surgically menopausal individuals being more responsive to oestrogens, are there individuals undergoing natural menopause who show a more abrupt decline in oestrogens and who respond most strongly in terms of postmenopausal symptoms?

Bäckström: It has been shown that the number and the severity of hot flushes are related to the steepness of the decline in oestrogen levels at menopause. But I've never seen any study relating the speed of oestrogen decline to mood changes.

McEwen: In relation to the earlier discussion about a randomized trial of oestrogen replacement in postmenopausal women (Vickers et al 1995, this volume), is it not important to have some measurement to predict which women have the greatest oestrogen deficiency? This would enable you to assess the response to hormone replacement and get around the problem of heterogeneity of the responding population that results in 'noise' in the data.

Bäckström: This is one of problems with doing a clinical trial. If one is not aware of and does not control for all these confounding factors, it might be impossible to sort out the effects at the end of the trial. That could well be the reason for the noise.

Baulieu: Is the decrease in oestrogen after menopause the consequence of a decrease in oestrogen secretion from the ovaries, or is it, for instance, linked to the decrease in dehydroepiandrosterone (DHEA) sulphate?

Backstrom: I believe that it's related to the decrease in both adrenal and ovarian androgen production.

Thijssen: I would like to introduce another androgen to this discussion: androstenedione is not only secreted by the adrenal, but in postmenopausal women it is also produced by the ovaries. So there's a rather basic difference between women who are surgically menopausal and those who still have their ovaries and who continue to produce androstenedione. You described differences between surgically and naturally menopausal women: has anyone looked for the effects of androstenedione? This is, of course, the major precursor of oestrone.

Bäckström: As far as I know no one has looked at androstenedione replacement therapy. Sherwin & Gelfand (1987) have done studies using testosterone, which increased the feeling of well-being compared with oestrogen-only replacement, especially in regard to sexuality. But about 40% of the androgens in postmenopausal women are produced by the ovary.

Thijssen: Some people have suggested that androgen production by the ovary increases a little with age (Thijssen & Longcope 1976).

Bäckström: I'm not completely convinced by this. I think that the ovary also ages, and that the production of androstenedione from the ovary decreases with age. But I agree that it's not completely clear.

Thijssen: Mortola & Yen (1990) carried out a study in which they used the androgen DHEA in postmenopausal women.

James: My concept is that the decline in oestrogen levels which occurs at the menopause is very much a reflection of the failure of ovarian oestrogen secretion. Then it is the production of androstenedione by the ovary and the adrenal, converted in the periphery, which provides the main source of oestrogen production. DHEA sulphate is in fact a very poor precursor of oestrogen, so the fact that DHEA sulphate production is falling bears no direct relation to the decline in the production of oestrogens.

Baulieu: You are probably right, but it has not been demonstrated.

James: Longcope et al (1982) carried out some infusion studies and showed that the conversion ratio of DHEA sulphate to oestrone was about $1:200$, which is low compared with the production of oestrone from androstenedione. I believe that ovarian oestrogen secretion gradually declines through the menopause, and it is compensated to some extent by the peripheral oestrogen production. I think that this is what you're seeing in these data.

Bäckström: I agree with you. DHEA sulphate was used as a long-term marker of DHEA production from the adrenal. It's quite clear that androstenedione is of much greater importance for the peripheral conversion to oestrogen than DHEA.

James: My point, though, is that DHEA sulphate is not a marker for androgens: it's produced by the adrenal independently of androstenedione and it's not sensitive to adrenocorticotrophic hormones in the way that adrenal androstenedione is. You're looking at the wrong steroid if you are looking at DHEA sulphate—if you measure androstenedione, you will see the precursor directly.

Robel: Concerning progestins: there might be a big difference between the progestins that are most often used and progesterone itself, because the latter is a precursor of metabolites (in particular, allopregnanolone) that have anxiolytic and sedative properties. I would like you to comment on the influence of progesterone versus other progestins.

Bäckström: Progesterone, taken orally in micronized preparations, for instance, is converted to a large extent to the 5α- and 5β-reduced metabolites, which are active on the $GABA_A$ receptor. As such, they are anaesthetic, anxiolytic and anti-epileptic. This can also be seen in their effects on epileptic seizure frequency during the menstrual cycle. I am not aware of any study comparing progesterone with progestins. However, if natural progesterone (mainly the vaginal progesterone) is used, we see similar mood changes to those that occur with progestins, although perhaps not as pronounced.

Thijssen: Nahoul et al (1993) claimed to show different effects of progesterone on mood depending on the means of administration. Oral administration could have different effects on mood compared with vaginal administration. This is due to the metabolic pattern of progesterone determined by the route of entry.

Bäckström: We have studied the pharmacokinetics of oral, vaginal and intramuscular administration. With vaginal application very few of these 5α- and 5β-reduced metabolites are produced. With the oral route, a large number of 5α-reduced metabolites are produced (T. Bäckström, L. Dennerstein, B. De Lingerre & P. MacGill, unpublished results).

Bonewald: I thought your comment on bone fracture occurring before osteoporosis in females was very intriguing, because we're told in the bone field that fractures would be prevented by developing a compound to prevent osteoporosis. Perhaps when we're developing compounds that maintain bone mass, we really need to be looking at the effect on neural function too. You're saying that the first effect of oestrogen decline after menopause is really on the nervous system, and then the effects on other systems are secondary.

Bäckström: The results point in that direction.

Manolagas: I take exception to your definition of osteoporosis. Osteoporosis is not a process that happens abruptly; it's a very gradual process. The disease osteoporosis is a combination of things, one of which is certainly low bone density: thin bones break more easily. However, as you very correctly pointed out, to break your bones you have to fall down, and if you are unstable, you're more likely to fall down. Osteoporosis does not happen at the age of 65—bone loss is a gradual process for both men and women. Of course, bone loss is accelerated after the menopause. I do not agree with the concept that you first have defects of the nervous system, and then you have osteoporosis.

Bäckström: My point was that the amount of bone loss that would cause fracture on its own arises later in life. I'm also aware of the fact that bone loss starts early—perhaps even before the age of 40—and it's gradual.

Manolagas: Is there an association between the mood changes and depression after the menopause with post partum depression?

Bäckström: This is a very interesting topic. It seems that there is a link between those women who have post partum depression, premenstrual syndrome, and who react negatively to progestins in postmenopausal replacement therapy. It might even be the same women in all three cases, but there has not yet been a detailed study on these issues. In a recent controlled study, Gregoirc et al (1994) showed an effect of oestradiol on post partum depression.

Uvnäs-Moberg: Until now we have been discussing direct effects of steroids on brain function. I would like to present some data showing that sex steroids may also indirectly influence behaviour by modulating effects of certain brain peptides. Drugs may also influence behaviour by modifying the metabolism or release of steroids.

We have been studying the behaviour of rats in a behavioural box. We study the amount of locomotion: if we see a decrease in locomotory activity, we classify this as a sedative effect. For instance, valium decreases locomotory activity. The second effect valium has is to increase the time the rat spends in the centre of the box; in pharmacological terms, this is called an anxiolytic effect.

Interestingly, an endogenous peptide, oxytocin, also has these two different effects. If you give oxytocin (nanogram quantities intracerebroventricularly, or microgram amounts subcutaneously), you get a shift in locomotory activity from the periphery to the centre; that is, an anxiolytic-like effect. Doses 100- to 1000-fold higher, administered intracerebroventricularly or subcutaneously, have a sedative effect. These experiments were performed on male rats (Uvnäs-Moberg et al 1994). If you study female rats in oestrus, they already show a maximal anxiolytic-like behaviour (i.e. they spend more time in the centre of the box) and exogenous oxytocin has no further effect. During the other stages of the rat cycle, i.e. pro-, di- and mid-oestrus, you get an effect of oxytocin similar to that found in the males. This indicates that oestrogen, high levels of which occur during oestrus, has facilitated the effect of endogenous oxytocin to cause an anxiolytic-like effect. In contrast, the effect of oxytocin on locomotory activity can be obtained with 10-fold lower doses of oxytocin in mid- and di-oestrus when the levels of oestrogen and progesterone are high, suggesting that these steroids together potentiate the sedative effects of oxytocin (K. Uvnäs-Moberg et al, unpublished results).

If oxytocin exerts similar effects in humans it follows that if the levels of natural steroids decrease as, for example, during menopause, then these kinds of oxytocin-regulated behaviours are likely to change, perhaps contributing to the increased anxiety and tension recorded during this period in women.

There is another effect of oxytocin that is also very much influenced by the stage of the oestrous cycle. A normal rat increases in weight during mid- and di-oestrus. During oestrus, rats lose weight. This is not so much due to differences in food intake, but rather due to efficiency of metabolism. If you give high doses of oxytocin to female rats, they increase disproportionately in weight without eating more. The effect occurs during oestrus, thus the loss of weight normally occurring in oestrus does not happen. It seems that high doses of exogenous oxytocin can overcome the inhibitory effect of oestrogen on weight gain during oestrus (Uvnäs-Moberg et al, unpublished results).

I want to show you another example of possible 'indirect' steroid activity, which is perhaps more surprising. We were studying the effects of alcohol on oxytocin levels in humans. There was an effect, but not much of one (Lindman et al 1994). However, another effect (which took us by surprise) was that alcohol increased testosterone levels in women. During a normal cycle, testosterone levels are elevated slightly after one glass of wine. Women on low-dose oral contraceptives showed a greater increase in their testosterone levels (Eriksson et al 1994). The question is: will this have any effect on their behaviour? What happens if these women on oral contraceptives drink alcohol frequently? Coincidentally, the same day I saw these results I read in a Swedish newspaper that we now have a new social phenomenon: gangs of teenage girls running about fighting each other. I wondered if these girls might be on oral contraceptives and drinking a lot of alcohol.

We do not know what is causing this increase in testosterone. It could be a consequence of changed liver metabolism or an increased release of testosterone from the ovaries.

Manolagas: Is there testosterone in the wine? There are testosterone receptors in yeast.

Uvnäs-Moberg: But then testosterone levels would rise in men, too, after they had drunk wine, and they don't.

Thijssen: I have a very basic question. It is extremely difficult to measure testosterone in women: how have you been doing it? The way people do it nowadays is by using commercial kits: these are not reliable in women.

Uvnäs-Moberg: These experiments and the testosterone assays have been performed by Peter Eriksson in Helsinki. He measured total testosterone (Eriksson et al 1994).

Thijssen: I'm amazed that you found that testosterone levels are going down in women on oral contraceptives: the usual effect of oral contraceptives is an increase in sex hormone binding globulin, which will increase total testosterone.

Uvnäs-Moberg: But most studies I've seen do show a decrease.

Thijssen: In free testosterone, yes, but not in total testosterone.

James: These are extraordinary and interesting results. What was your reason for choosing to measure testosterone? Have you measured any other steroids? Is this a metabolic phenomenon or is it selective for testosterone?

Uvnäs-Moberg: We measured testosterone to investigate the connection between alcohol and sexual behaviour. Peter Eriksson suspected that there might be a difference between men and women.

James: What other steroids did you measure?

Uvnäs-Moberg: Oestrogen and progesterone—nothing happens.

James: What happens to the behaviour in these women? Do they all become very aggressive when they drink their wine?

Uvnäs-Moberg: I don't think it has been studied. We have a generation in Sweden where teenagers are given oral contraceptives to prevent abortions and there has been a change in behaviour. I am not saying there is a connection, but I think these kinds of ideas and relationships have to be analysed, as well as underlying psychological and social factors.

References

Eriksson CJP, Fukunaga T, Lindman R 1994 Sex hormone response to alcohol. Nature 369:711

Gregoirc A, Henderson A, Kumar R, Studd J 1994 A controlled trial of oestradiol therapy for post natal depression. Neuropsychopharmacology 10:9015(abstr)

Lindman RE, Ahlund N, Hannuksela S, Holmström-Mellberg K, Eriksson CJP, Uvnäs-Moberg K 1994 Hormonal covariates of socioemotional communication as a function of assertiveness, gender, and alcohol. Seventh Congress of the International Society for Biomedical Research on Alcoholism, Queensland, Australia, June 26–July 1, 1994

Longcope C, Bourget C, Flood C 1982 The production and aromatization of dehydroepiandrosterone in post-menopausal women. Maturitas 4:325–332

Mortola JF, Yen SS 1990 The effects of oral dehydroepiandrosterone on endocrine metabolic parameters in postmenopausal women. J Clin Endocrinol & Metab 71: 696–704

Nahoul K, Dehennin L, Jondet M, Roger M 1993 Profiles of plasma estrogens, progesterone and their metabolites after oral and vaginal administration of estradiol or progesterone. Maturitas 19:185–202

Pintor C, Finoccharo G, Fanni T 1982 Steroid plasma levels in chronic liver disease. In: Langer M, Chiandussi L, Chopra I, Martini L (eds) The endocrines and the liver. Academic Press, London, p 427–429

Sherwin BB, Gelfand MM 1987 The role of androgen in the maintenance of sexual function in oophorectomized women. Psychoson Med 49:397–409

Thijssen JHH, Longcope C 1976 Postmenopausal oestrogen production. In: van Keep PA, Greenblatt RB, Albeaux-Fernet M (eds) Consensus on menopause research. MTP Press, Lancaster, p 25–29

Uvnäs-Moberg K, Ahlenius S, Hillegaart V, Alster P 1994 High doses of oxytocin cause sedation and low doses cause an anxiolytic-like effect in male rats. Pharmacol Biochem Behav 49:101–106

Vickers MR, Meade TW, Wilkes HC 1995 Hormone replacement therapy and cardiovascular disease: the case for a randomized controlled trial. In: Nonreproductive actions of sex steroids. Wiley, Chichester (Ciba Found Symp 191) p 150–164

Sex steroids, cytokines and the bone marrow: new concepts on the pathogenesis of osteoporosis

Stavros C. Manolagas, Teresita Bellido and Robert L. Jilka

Department of Medicine, Division of Endocrinology and Metabolism and the UAMS Center for Osteoporosis and Metabolic Bone Diseases, University of Arkansas for Medical Sciences, 4301 West Markham, Mail Slot 587, Little Rock, AR 72205, USA

Abstract. Osteoclasts and osteoblasts, originating in the bone marrow from haemopoietic progenitors and mesenchymal stromal cells, respectively, are responsible for the remodelling of the skeleton throughout adult life. Upon loss of sex steroids, the production of osteoclasts in the bone marrow is increased. This is mediated by an increase in the production of interleukin 6 (IL-6), as well as an increase in the sensitivity of the osteoclastic precursors to the action of cytokines such as IL-6, owing to an up-regulation of the gp130 signal transduction pathway. Consistent with this, oestrogens as well as androgens inhibit IL-6 production through an indirect effect of their specific receptors on the transcriptional activity of the IL-6 gene promoter, and inhibit the expression of the gp130 gene. With advancing age, the ability of the marrow to maintain the high rate of osteoclastogenesis caused by the acute loss of sex steroids is diminished. This is probably the result of the negative effect of senescence on the ability of the marrow to produce stromal/osteoblastic cells, which provide the essential support for osteoclastogenesis. These observations suggest that inappropriate production of osteoclasts or inadequate production of osteoblasts in the bone marrow are fundamental cellular changes in the pathogenesis of postmenopausal and senescence-associated osteoporosis, respectively.

1995 Non-reproductive actions of sex steroids. Wiley, Chichester (Ciba Foundation Symposium 191) p 187–202

Bone is remodelled continuously throughout adult life by the resorption of old bone by osteoclasts and the subsequent formation of new bone by osteoblasts. These two events are tightly coupled to each other and are responsible for the renewal of the skeleton while maintaining its anatomical and structural integrity. Sex hormone deficiency is a major cause of the disturbance of the balance between resorption and formation and the loss of bone mass characterizing the syndrome of osteoporosis. Indeed, at the menopause (or following castration in men and women), the rate of bone remodelling is highly accelerated and bone is lost rapidly. The phase of rapid bone loss which results from acute gonadal deficiency is characterized by excessive osteoclast activity and affects trabecular bone primarily (Parfitt et al 1983, Eriksen et al 1990). The phase of rapid bone loss seen in the early postmenopausal years, however, is transient; it is followed, after a few years, by a lower rate of remodelling and a phase of slow bone loss seen in elderly eugonadal males as well (Parfitt 1992, Gallagher et al 1987, Heaney 1990, Nordin et al 1992).

Consistent with the evidence from humans, evidence from experimentally induced loss of ovarian function (ovariectomy) in primates and rodents indicates that the loss of bone that follows is associated with an increase in bone resorption and formation rates, with the former exceeding the latter, thus resulting in bone loss (Balena et al 1993, Poli et al 1994, Kalu 1991) and an increase in the number of osteoclasts present in trabecular bone (Turner et al 1988, Jilka et al 1992).

A series of discoveries has established that osteoclasts and osteoblasts originate in the bone marrow from haemopoietic and mesenchymal cells, respectively, and that the development of these cells is under the control of an intricate interplay between locally produced cytokines and systemic bone-active hormones. Indeed, osteoclasts arise from multipotent haemopoietic cells, specifically the colony-forming unit for granulocytes and macrophages (CFU-GM). Osteoblasts, as well as the cells of the marrow stroma which support haemopoiesis, develop from multipotent mesenchymal progenitors, termed CFU-fibroblasts (CFU-F). Marrow stromal cells and osteoblastic cells exhibit an extensive overlap in their phenotypic properties. Cellular cross-talk between stromal/osteoblastic cells and osteoclast progenitors of the bone marrow is critical for the development of osteoclasts (Shevde et al 1994). Moreover, the ability of parathyroid hormone, 1,25-dihydroxyvitamin D_3 [1,25(OH)$_2$D$_3$], interleukin (IL)-1, and tumour necrosis factor (TNF) to stimulate osteoclast formation is mediated via cells of the stromal/osteoblastic lineage (Suda et al 1992), which synthesize a number of factors critical for the proliferation and differentiation of osteoclast progenitors. The list of these factors includes the interleukins IL-1, IL-6, IL-11, granulocyte/macrophage colony-stimulating factor, macrophage colony-stimulating factor, leukaemia inhibitory factor (LIF) and stem cell factor (Mundy 1992, Horowitz & Jilka 1992, Girasole et al 1994, Demulder et al 1992). Among these, IL-6 seems to play a prominent role in the pathogenesis of several disease states associated with excessive osteoclastic resorption.

IL-6 enhances osteoclast formation and bone resorption in cultures of fetal mouse bone *in vitro* (Ishimi et al 1990, Lowik et al 1989) and stimulates bone resorption cooperatively with IL-1 *in vivo* (Black et al 1991). More importantly, IL-6 seems to play a critical role in the focal osteolytic lesions associated with Paget's disease (Roodman et al 1992) and multiple myeloma (Klein et al 1991).

Both stromal cells and osteoblastic cells produce IL-6 in nanomolar quantities in response to stimulation by locally produced cytokines such as IL-1 and TNF (Ishimi et al 1990, Girasole et al 1992, Linkhart et al 1991), growth factors such as transforming growth factor (TGF)-β (Horowitz et al 1992) and systemic hormones such as parathyroid hormone (Lowik et al 1989, Feyen et al 1989, Passeri et al 1994).

The production of IL-6 by (and the abundance of the IL-6 mRNA in) cultured bone marrow stromal and osteoblastic cell lines, as well as its production by primary bone cell cultures from rodents and humans, is inhibited by 17β-oestradiol (Girasole et al 1992). This effect is mediated via the oestrogen receptor (Bellido et al 1993a). Consistent with the mediating role of the oestrogen receptor in the regulation of IL-6 by 17β-oestradiol, 17β-oestradiol completely suppresses stimulated transcription in HeLa cells co-transfected with constructs of the human IL-6 promoter linked to the reporter gene chloramphenicol acetyltransferase and a human oestrogen receptor expression plasmid, but has no effect on IL-6 transcription in HeLa cells not transfected with oestrogen receptor (Pottratz et al 1994). Similarly, oestradiol inhibits stimulated transcription from the human IL-6 promoter in transfected bone marrow stromal cells (which express the oestrogen receptor constitutively) without the requirement for co-transfection of the oestrogen receptor plasmid. These hormonal effects are indistinguishable between constructs containing either a 1.2 kb fragment of the 5′ flanking region of the human IL-6 gene promoter or only the proximal 225 bp segment. Nevertheless, yeast-derived recombinant oestrogen receptor does not bind, neither does it compete with the binding of nuclear extracts, to the 225 bp segment in DNA shift assays. This evidence indicates that oestradiol inhibits IL-6 production by inhibiting the stimulated expression of the IL-6 gene through an oestrogen receptor-mediated indirect effect on the transcriptional activity of the proximal 225 bp sequence of the promoter, perhaps through an interference with events along the signalling pathways initiated by the IL-6 stimulating agents.

Because loss of testicular function is also associated with bone loss, we have recently examined whether androgens, acting through the androgen-specific receptor, exert similar effects to those of oestrogens on the IL-6 gene. We have found that murine bone marrow-derived stromal cells contain specific high-affinity androgen receptors, and that both testosterone and dihydrotestosterone inhibit IL-6 production by these cells. Moreover, testosterone and dihydrotestosterone, but not oestradiol, inhibit the stimulated transcription from the human IL-6 promoter in HeLa cells co-transfected with an androgen receptor expression plasmid, whereas they do not when the cells are co-transfected with the oestrogen receptor expression plasmid (Bellido et al 1993b). These findings indicate that the IL-6 gene is regulated by both classes of sex steroids through their respective receptors.

On the basis of the *in vitro* inhibition of IL-6 production by sex steroids, we have tested *in vivo* the hypothesis that oestrogen loss up-regulates osteoclastogenesis through an increase in the production of IL-6 in the microenvironment of the marrow, by using the ovariectomized mouse model (Jilka et al 1992). We found that the number of CFU-GM colonies and the number of osteoclasts was greater in short-term cultures of marrow cells from ovariectomized mice than in cultures from sham-operated animals. The increased osteoclast formation in the marrow cultures in our studies was mirrored by an increase in the number of osteoclasts present in sections of trabecular bone. More importantly, administration of 17β-oestradiol or injections of an IL-6-neutralizing antibody (but not administration of an IgG isotype control antibody) to the ovariectomized animals prevented all these cellular changes. Consistent with these findings, we have found that oestrogen loss causes an up-regulation of IL-6 production by *ex vivo* bone marrow cell cultures in response to either $1,25(OH)_2D_3$ or parathyroid hormone, and that a similar phenomenon can be elicited *in vitro* by withdrawal of 17β-oestradiol from primary cultures of calvarial cells (Passeri et al 1993).

Similar to ovariectomy, orchidectomy in mice causes increased replication of CFU-GM, as well as up-regulation of osteoclast formation (Girasole et al 1992). These effects are prevented by administration of either testosterone or an IL-6-neutralizing antibody. On the basis of this evidence, we have proposed that the loss of bone that follows the loss of gonadal function in either sex is caused by similar mechanisms (Bellido et al 1993b).

Although the results discussed above strongly suggest that IL-6 is an important mediator of the up-regulation of the early stages of osteoclastogenesis precipitated by loss of oestrogens or androgens, IL-6 does not appear to play a role in the development of osteoclast precursors in the marrow or affect the number of osteoclasts present in trabecular bone of sham-operated or ovariectomized animals receiving oestrogen replacement (Jilka et al 1992). This suggests that under physiological conditions, i.e. the oestrogen-replete state, IL-6 is either redundant or its levels are kept below a critical threshold relative to the sensitivity of osteoclastogenesis to this cytokine. Consistent with the contention that other cytokines are responsible for osteoclast development in the oestrogen-replete state, we have found that IL-11 induces the formation of osteoclasts in cultures of murine bone marrow and calvarial cells, and stimulates bone resorption (Girasole et al 1994). In addition, an antibody neutralizing IL-11 suppresses osteoclast development induced by either $1,25(OH)_2D_3$, parathyroid hormone, IL-1 or TNF, whereas antibodies to IL-1 or TNF have no effect on the ability of IL-11 to induce osteoclast development. More importantly, in contrast to IL-6, the effects of IL-11 are independent of the oestrogen status of the marrow donors, indicating that, as opposed to IL-6 (which attains its importance for osteoclastogenesis in pathological states only), IL-11 is involved in osteoclastogenesis under physiological circumstances.

Strong support for the hypothesis that IL-6 mediates the ovariectomy-induced up-regulation of CFU-GM, osteoclastogenesis and the bone loss associated with this condition, was recently provided by studies in IL-6 knockout mice (Poli et al 1994). It was demonstrated that the genetically engineered IL-6-deficient mice fail to upregulate CFU-GM formation and osteoclastogenesis following loss of ovarian function, and they are indeed protected from the loss of bone that follows loss of ovarian function in the IL-6-replete mice.

Although the observations discussed above provide an explanation for the up-regulation of osteoclastogenesis and loss of bone that follows the loss of gonadal steroids, they do not explain the fact that following menopause in women, as well as ovariectomy in rodents and primates, there ensues not only increased osteoclast activity, but also increased osteoblast activity (Parfitt 1992, Kalu 1991, Balena et al 1993). A potential mechanistic explanation for the relationship between loss of ovarian function and increased osteoblastic activity has emerged from studies examining the effects of ovariectomy on the development of osteoblast progenitors in *ex vivo* cultures of the murine bone marrow. Specifically, we have found that marrow cell cultures from ovariectomized mice exhibit an increase in the number of CFU-F colonies (presumed to be the precursors of osteoblasts), as well as the number of CFU-F colonies exhibiting the ability to mineralize (which we have designated as CFU-osteoblasts [CFU-OB]) (Jilka et al 1994). Hence, loss of ovarian function seems to enhance not only the development of osteoclast progenitors, but also the development of osteoblast progenitors, in the bone marrow.

The concordant up-regulation of osteoclast and osteoblast progenitor formation in the marrow and the high rate of remodelling which follows the loss of gonadal function (and probably other states, such as hyperpara-thyroidism) may be due to enhanced sensitivity of bone marrow progenitors to both osteoclastogenic and osteoblastogenic signals, caused by an up-regulation of the signal transduction pathway which mediates the effects of the subfamily of cytokines that includes IL-6, IL-11, LIF, oncostatin M (OSM) and ciliary neurotropic factor (CNTF) (Kishimoto et al 1994).

The receptors for IL-6, IL-11, LIF, OSM and CNTF are composed of complexes of two to three subunits. Upon binding of the ligand to the cytokine-specific receptor, the ligand–receptor complex binds to either one or two additional transmembrane proteins (β components) with phosphotyrosine activation properties. One of the β subunits, namely the glycoprotein gp130, is shared by all these receptors. The dimerization of the β subunits of the receptor complex after cytokine stimulation activates a series of kinases, termed JAKs (Janus tyrosine kinases), which, in turn, are responsible for protein tyrosine phosphorylation and, hence, the transduction of the cytokine signal (Stahl et al

1994). Elucidation of the fact that gp130 is used by all the members of this cytokine family explains to a large extent the functional pleiotropy and redundancy of these cytokines.

Recent studies of ours indicate that oestrogens, as well as androgens, strongly down-regulate the expression of the mRNA encoding gp130 in stromal and osteoblastic cells. In contrast to sex steroids, the systemic hormones parathyroid hormone and $1,25(OH_2)D_3$ dramatically up-regulate the expression of gp130 mRNA (as well as the expression of the protein and its phosphorylation in response to cytokines) in rodent and human osteoblasts and in primary cultures of murine bone marrow cells (Bellido et al 1994). More importantly, and in line with the *in vitro* inhibitory effect of oestrogens on gp130 mRNA, ovariectomy in mice causes a several-fold increase in the abundance of this transcript in the bone marrow.

This evidence raises the possibility that loss of sex steroids not only up-regulates IL-6 production, but may also render IL-6-responsive cells in the marrow more sensitive to this and other cytokines that share the gp130 signal transduction pathway. In other words, gonadal loss may potentiate not only the effects of IL-6 on osteoclastogenesis, but also sensitize the system to the actions of osteoblastogenic signals that act through gp130. One potential candidate for the initiation of such osteoblastogenic signals is LIF. LIF stimulates osteoblast proliferation and differentiation (Rodan et al 1990, Evans et al 1994) and overexpression of LIF causes ectopic bone formation (Metcalf & Gearing 1989). In view of the stimulatory effects of parathyroid hormone on gp130, up-regulation of gp130 might explain not only the high bone turnover associated with the loss of gonadal function, but also the high bone turnover state associated with primary hyperparathyroidism.

Senescence in both sexes is characterized by loss of bone. However, the characteristics of this so-called involutional osteoporosis are distinct from those associated with the bone loss that accompanies loss of gonadal function. Indeed, the pathogenetic hallmark of the former is low-turnover bone remodelling with decreased osteoblast activity; whereas that of the latter is high-turnover bone remodelling with excessive osteoclast activity. In view of this, we have proceeded to investigate whether, in the mouse model, the increased osteoclastogenesis and osteoblastogenesis that follows oestrogen deficiency persists with advancing age or whether it is a transient phenomenon. Ten months after ovariectomy, there was no difference between ovariectomized and sham-operated mice in CFU-GM or CFU-F in the bone marrow, indicating that the changes we had observed at the early stages following ovariectomy (two to four weeks) had subsided. From this evidence, it appears that the mouse model mimics the situation in women whereby the high-turnover bone remodelling with increased osteoclast activity of the early postmenopausal state changes to a lower-turnover remodelling with decreased osteoblastic activity with advancing age. These observations support the hypothesis that the transition of rapid bone loss associated with oestrogen loss to the slow bone loss associated with advancing age is due to a negative effect of ageing on the ability of the bone marrow to maintain the high rate of osteoclastogenesis.

To test this hypothesis, we proceeded to dissect the effects of the loss of ovarian function from the effects of ageing on the bone marrow using a mouse model of accelerated ageing, namely the SAM-P/6 mouse (Matsushita et al 1986). SAM-P/6 is a sub-strain of AKR/J, which has been selected by inbreeding and exhibits early senescence associated with low-turnover osteopenia. However, these mice exhibit intact ovarian function, making them a unique model whereby the effects of ageing can be dissected from the effects of sex steroids.

We found that the number of osteoclasts formed in cultures of marrow cells from SAM-P/6 mice was approximately 10-fold lower than those formed in marrow cultures from senescence-resistant mice (SAM-R/1) which exhibit normal bone mass (negative controls). Furthermore, whereas ovariectomy caused significant up-regulation of osteoclast formation (three- to fourfold) in cultures of marrow cells from SAM-R/1 mice, there was no such response in the case of SAM-P/6 mice. The absence of an osteoclastogenic response of the SAM-P/6 mice following ovariectomy was apparently secondary to a decrease in the activity and/or the number of the stromal/osteoblastic cells that support osteoclastogenesis, rather than to a decrease in the number of haemopoietic osteoclast progenitors. Indeed, when osteoblasts from normal mice were co-cultured with bone marrow cells from SAM-P/6 mice, the formation of osteoclastic cells was indistinguishable from that of cultures using normal osteoblastic cells and normal bone marrow cells, suggesting that the defect in these mice lies with cells of the mesenchymal lineage (Manolagas et al 1993).

To confirm the suggestion that the bone marrow of SAM-P/6 mice is defective in its ability to support osteoclastogenesis owing to a defect in stromal/osteoblastic support cells, we compared the number of CFU-F colonies (the presumed progenitor of osteoblastic cells) in bone marrow cultures from SAM-P/6 mice with cultures from the normal control. We found that the former cultures exhibited a dramatically decreased ability to form CFU-F compared with the control mice. Moreover, there was an even larger decrease in the number of CFU-F colonies exhibiting mineralization (a characteristic of osteoblastic progenitors) in cultures from SAM-P/6 mice compared with controls. The association of decreased osteoclastogenic capability of the bone marrow with low-turnover osteopenia in an animal model of senescence suggests, for the first time, a cellular mechanism explaining the transition of the rapid bone loss associated with oestrogen loss to the slow bone loss associated with ageing.

In conclusion, results from *in vitro* studies in animal and human cells, as well as *in vivo* studies in animal models of sex steroid deficiency and senescence, suggest that both states exert their adverse effects on skeletal homeostasis by altering the normal process of bone cell development from their progenitors in the marrow, and that osteoblast and osteoclast formation in the bone marrow (and hence the rate of bone remodelling) may be coordinated by the convergence of osteoclastogenic and osteoblastogenic signals on the common signal transduction pathway mediated by gp130. On the basis of this, we have formulated the following general hypothesis. The high rate of bone remodelling and the loss of bone caused by loss of gonadal function may be explained by an increase in the sensitivity of bone marrow progenitors of both osteoclasts and osteoblasts to osteoclastogenic and osteoblastogenic signals mediated by up-regulation of the gp130 signal transduction pathway and an imbalance in favour of osteoclast formation due to increased production of the osteoclastogenic cytokine IL-6. With advancing age, the rate of remodelling slows owing to a relative decrease in the ability of the marrow to respond to osteoblastogenic and osteoclastogenic signals resulting from a decrese in gp130 alone, or in combination with a decrease in the production of cytokines promoting osteoblast formation. Under the latter conditions, the increased production of IL-6, which is associated with ageing in both sexes (Daynes et al 1993, Ershler et al 1993), could still be responsible for the imbalance between osteoblastogenesis and osteoclastogenesis.

Acknowledgements

This work was supported by the National Institutes of Health (AR 41313, AR43003) and the Department of Veterans Affairs. The authors thank Mrs Barker for typing this manuscript.

References

Balena R, Toolan BC, Shea M et al 1993 The effects of 2-year treatment with the aminobisphosphonate alendronate on bone metabolism, bone histomorphometry, and bone strength in ovariectomized nonhuman primates. J Clin Invest 92: 2577–2586

Bellido T, Girasole G, Passeri G et al 1993a Demonstration of estrogen and vitamin D receptors in bone marrow-derived stromal cells: up-regulation of the estrogen receptor by 1,25-dihydroxyvitamin-D_3. Endocrinology 133:553–562

Bellido T, Girasole G, Jilka RL, Crabb D, Manolagas SC 1993b Demonstration of androgen receptors in bone marrow stromal cells and their role in the regulation of transcription from the human interleukin-6 (IL-6) gene promoter. J Bone Miner Res (suppl 1) 8:S131(abstr)

Bellido T, Girasole G, Passeri G, Jilka RL, Manolagas SC 1994 Sex steroids regulate the expression of the gp130 transduction pathway by bone marrow and bone cells. Bone Miner (suppl 1) 25:39(abstr)

Black K, Garrett IR, Mundy GR 1991 Chinese hamster ovarian cells transfected with the murine interleukin-6 gene cause hypercalcemia as well as cachexia, leukocytosis and thrombocytosis in tumor-bearing nude mice. Endocrinology 128:2657–2659

Daynes RA, Araneo BA, Ershler WB, Maloney C, Li G-Z, Ryu S-Y 1993 Altered regulation of IL-6 production with normal aging: possible linkage to the age-associated decline in dehydroepiandrosterone and its sulfated derivative. J Immunol 150:5219–5230

Demulder A, Suggs SV, Zsebo KM, Scarcez T, Roodman GD 1992 Effects of stem cell factor on osteoclast-like cell formation in long-term human marrow cultures. J Bone Miner Res 7:1337–1344

Eriksen EF, Hodgson SF, Eastell R, Cedel SL, O'Fallon WM, Riggs BL 1990 Cancellous bone remodeling in type I (postmenopausal) osteoporosis: quantitative assessment of rates of formation, resorption, and bone loss at tissue and cellular levels. J Bone Miner Res 5:311–319

Ershler WB, Sun WH, Binkley N et al 1993 Interleukin-6 and aging: blood levels and mononuclear cell production increase with advancing age and *in vitro* production is modifiable by dietary restriction. Lymphokine Cytokine Res 12:225–230

Evans DB, Gerber B, Feyen JHM 1994 Recombinant human leukemia inhibitory factor is mitogenic for human bone-derived osteoblast-like cells. Biochem Biophys Res Commun 199:220–226

Feyen JHM, Elford P, Dipadova RE, Trechsel U 1989 Interleukin-6 is produced by bone and modulated by parathyroid hormone. J Bone Miner Res 4:633–638

Gallagher JC, Goldgar D, Moy A 1987 Total bone calcium in normal women: effect of age and menopause status. J Bone Miner Res 2:491–496

Girasole G, Jilka RL, Passeri G et al 1992 17β-estradiol inhibits interleukin-6 production by bone marrow-derived stromal cells and osteoblasts *in-vitro*: a potential mechanism for the antiosteoporotic effect of estrogens. J Clin Invest 89:883–891

Girasole G, Passeri G, Knutson S, Manolagas SC, Jilka RL 1992 Up-regulation of osteoclastogenic potential of the marrow is induced by orchiectomy and is reversed by testosterone replacement in the mouse. J Bone Miner Res (suppl 1) 7:S96(abstr)

Girasole G, Passeri G, Jilka RL, Manolagas SC 1994 Interleukin-11: a new cytokine critical for osteoclast development. J Clin Invest 93:1516–1524

Heaney RP 1990 Estrogen–calcium interactions in the postmenopause: a quantitative description. Bone Miner 11:67–84

Horowitz MC, Jilka RL 1992 Colony stimulating factors and bone remodeling. In: Gowen M (ed) Cytokines and bone metabolism. CRC Press, Boca Raton, FL, p 185–227

Horowitz M, Phillips J, Centrella M 1992 TGF-β regulates interleukin-6 secretion by osteoblasts. In: Cohn DV, Gennari C, Tashjian AH Jr (eds) Calcium regulating hormones and bone metabolism. Excerpta Medica, Amsterdam, vol 11:275–280

Ishimi Y, Miyaura C, Jin CH et al 1990 IL-6 is produced by osteoblasts and induces bone resorption. J Immunol 145:3297–3303

Jilka RL, Hangoc G, Girasole G et al 1992 Increased osteoclast development after estrogen loss: mediation by interleukin-6. Science 257:88–91

Jilka RL, Williams D, Manolagas SC 1994 Ovariectomy in mice increases osteocalcin levels in the serum and enhances osteoblast progenitor formation in the bone marrow. Bone Miner (suppl 1) 25:15(abstr)

Kalu DN 1991 The ovariectomized rat model of postmenopausal bone loss. Bone Miner 15:175–192

Kishimoto T, Taga T, Akira S 1994 Cytokine signal transduction. Cell 76:253–262

Klein B, Wijdenes J, Zhang X-G et al 1991 Murine anti-interleukin-6 monoclonal antibody therapy for a patient with plasma cell leukemia. Blood 78:1198–1204

Linkhart TA, Linkhart SG, MacCharles DC, Long DL, Strong DD 1991 Interleukin-6 messenger RNA expression and interleukin-6 protein secretion in cells isolated from normal human bone: regulation by interleukin-1. J Bone Miner Res 6:1285–1294

Lowik CWGM, van der Pluijm G, Bloys H et al 1989 Parathyroid hormone (PTH) and PTH-like protein (PLP) stimulate interleukin-6 production by osteogenic cells: a possible role of interleukin-6 in osteoclastogenesis. Biochem Biophys Res Commun 162:1546–1552

Manolagas SC, Knutson S, Jilka RL 1993 The senescence accelerated mouse (SAM-P/6), a model of senile osteoporosis, exhibits decreased osteoclastogenesis and fails to up-regulate this process following ovariectomy. J Bone Miner Res (suppl 1) 8:S141(abstr)

Matsushita M, Tsuboyama T, Kasai R et al 1986 Age-related changes in bone mass in the senescence-accelerated mouse (SAM). SAM-R/3 and SAM-P/6 as new murine models for senile osteoporosis. Am J Pathol 125:276–283

Metcalf D, Gearing DP 1989 Fatal syndrome in mice engrafted with cells producing high levels of the leukemia inhibitory factor. Proc Natl Acad Sci USA 127: 5948–5952

Mundy GR 1992 Local factors regulating osteoclast function. In: Rifkin BR, Gay CV (eds) Biology and physiology of the osteoclast. CRC Press, Boca Raton, FL, p 171–185

Nordin BEC, Need AG, Bridges A, Horowitz M 1992 Relative contributions of years since menopause, age, and weight to vertebral density in postmenopausal women. J Clin Endocrinol & Metab 74:20–23

Parfitt AM 1992 The two-stage concept of bone loss revisited. Triangle 31:99–110

Parfitt AM, Mathews CHE, Villanueva AR, Kleerekoper M, Frame B, Rao DS 1983 Relationships between surface, volume, and thickness of iliac trabecular bone in aging and in osteoporosis: implications for the microanatomic and cellular mechanisms of bone loss. J Clin Invest 72:1396–1409

Passeri G, Girasole G, Jilka RL, Manolagas SC 1993 Increased interleukin-6 production by murine bone marrow and bone cells after estrogen withdrawal. Endocrinology 133:822–828

Passeri G, Girasole G, Manolagas SC, Jilka RL 1994 Endogenous production of tumor necrosis factor by primary cultures of murine calvarial cells: influence on IL-6 production and osteoclast development. Bone Miner 24:109–126

Poli V, Balena R, Fattori E et al 1994 Interleukin-6 deficient mice are protected from bone loss caused by estrogen depletion. EMBO J 13:1189–1196

Pottratz S, Bellido T, Mocharla H, Crabb D, Manolagas SC 1994 17β-estradiol inhibits expression of human interleukin-6 promoter–reporter constructs by a receptor-dependent mechanism. J Clin Invest 93:944–950

Rodan SB, Wesolowski G, Hilton DJ, Nicola NA, Rodan GA 1990 Leukemia inhibitory factor binds with high affinity to preosteoblastic RCT-1 cells and potentiates the retinoic acid induction of alkaline phosphatase. Endocrinology 127:1602–1608

Roodman GD, Kurihara N, Ohsaki Y et al 1992 Interleukin 6: a potential autocrine/ paracrine factor in Paget's disease of bone. J Clin Invest 89:46–52

Shevde N, Anklesaria P, Greenberger JS, Bleiberg I, Glowacki J 1994 Stromal cell-mediated stimulation of osteoclastogenesis. Proc Soc Exp Biol Med 205:306–315

Stahl N, Boulton TG, Farruggella T et al 1994 Association and activation of Jak-Tyk kinases by CNTF-LIF-OSM-IL-6 β receptor components. Science 263:92–95

Suda T, Takahashi N, Martin TJ 1992 Modulation of osteoclast differentiation. Endocr Rev 13:66–80

Turner RT, Wakley GK, Hannon KS, Bell NH 1988 Tamoxifen inhibits osteoclast-mediated resorption of trabecular bone in ovarian hormone-deficient rats. Endocrinology 122:1146–1150

DISCUSSION

Baulieu: What sort of protein is gp130?

Manolagas: It is a signal-transducing transmembrane glycoprotein which is physically attached to the JAKs. In the case of interleukin 6 (IL-6), binding of the ligand to its cell-surface receptor leads to the assembly of the receptor complex with gp130 and homodimerizaton of gp130. Binding of other cytokines to their receptors leads to the association of gp130 with different β-subunits. In other words, bipartite or tripartite complexes containing gp130 are essential for JAK activation in response to several cytokines. gp130 itself is not a protein kinase.

Baulieu: Do osteoblasts possess intracellular oestrogen receptors?

Manolagas: Yes, this evidence was initially reported five or six years ago. Our group has more recently sequenced the bone marrow stromal cell oestrogen receptor and we have reported that it is identical to the classical oestrogen receptor (Bellido et al 1993).

Baulieu: Did you not imply that the mechanism of oestrogen action is a direct (non-genomic) effect upon the osteoblasts?

Manolagas: I have only discussed here genomic actions of oestrogens. Osteoclastogenesis is supported by the stromal cells of the bone marrow. Therefore, effects of oestrogens on stromal cells can fully account for the modulation of osteoclast development. None the less, there are reports that oestrogen receptors are present in osteoclasts, and that osteoclasts themselves make IL-6. It is therefore possible that there is yet another level of control for IL-6 production by sex steroids. In the latter case, osteoclast-derived IL-6 might be acting as an autocrine factor to stimulate more osteoclasts.

Although there might be other effects involved here, the evidence that by knocking out the IL-6 gene you can stop the bone loss induced by loss of sex steroids suggests strongly that IL-6 is an essential factor.

Baulieu: Would it therefore be true to say that you see a sort of triple intervention of sex steroids? On the bone marrow cells indirectly, on the intracellular receptor, and direct (non-genomic) effects at the membrane level?

Manolagas: No, I have not implied non-genomic effects.

Parker: Did you look at the effects of tamoxifen on either IL-6 or gp130 expression?

Manolagas: We have only looked for effects of tamoxifen on bioassayable IL-6 production in human cells, and we found none. Hermann et al (1994) from Ligand Pharmaceuticals (San Diego, CA) have examined the effect of tamoxifen on IL-6 transcription in transfection assays; they also found no effect.

Parker: Their results might have been complicated by the use of heterologous cells such as HeLa cells.

Manolagas: The San Diego group used murine +/+ LDA11 cells (Hermann et al 1994).

Parker: So you cannot demonstrate that tamoxifen acts as an oestrogen in terms of its effect on IL-6 production and gp130 expression.

Manolagas: We have not looked at the effects of tamoxifen on gp130. We didn't see any effect on IL-6 production. I think that production is the smaller component of the story: I think gp130 up-regulation and, therefore, increased sensitivity to IL-6 and related cytokines might be much more significant.

Parker: Is gp130 expressed in both osteoblasts and osteoclasts?

Manolagas: I don't think that anyone has looked for gp130 expression in osteoclasts so far. Our experiments were carried out in a number of cell lines representing the stromal osteoblastic lineage. However, I fully expect cells of the osteoblast lineage to express gp130. I would also suggest that the observation reported here that the IL-6 effect could be blocked by oestradiol in endothelial cells indicates that endothelial cells must express gp130. I would therefore like to add oestrogen to these cells and see what happens to gp130. If oestrogens down-regulate gp130, you might have an explanation of how oestrogens down-regulate the effects of IL-6.

Bonewald: IL-6 is a central factor as far as bone resorption is concerned, because it seems to play a crucial role in myeloma, disappearing bone disease (Gorham–Stout syndrome), osteoporosis and Paget's disease. But those diseases are quite different from each other in the type of resorption that occurs and the time course, so there have to be other factors and other mechanisms involved in interactions with IL-6. It probably is one of the central cytokines playing a role in bone resorption, but IL-6 has to be interacting with other cytokines and be regulated by the presence or absence of IL-6 soluble and membrane receptors.

Baulieu: What do you think the role of sex steroids is?

Bonewald: Sex steroids are having multiple effects on bone cells, including direct effects by inhibiting IL-6 production. I think they have other direct effects on bone cells and indirect effects on processes such as matrix formation. I also think sex steroids are having non-genomic effects on bone cells. That's what I was alluding to earlier when I mentioned our matrix vesicle studies. We have found non-genomic effects of vitamin D and my colleague has found non-genomic effects of oestrogen (Gates et al 1994).

Manolagas: There are several reports that cytokines other than IL-6 are involved in the pathological bone resorption that follows loss of oestrogens. Everyone in our field has had a favourite cytokine. For example, it has been reported that if you block both IL-1 and TNF you can block bone loss (Kimble et al 1994). We have published evidence that IL-1 and TNF stimulate IL-6 production and probably affect osteoclastogenesis through IL-11. Hence, it is clear that there is a whole cascade of interactive cytokines affecting bone. I therefore believe that you can interfere with the pathological bone resorption process not only by blocking IL-6, but also by blocking other cytokines that stimulate IL-6. However, to date, evidence that sex steroids directly regulate cytokine production is only available for IL-6.

Baulieu: When osteoporosis develops, who is the conductor of the orchestra?

Manolagas: I think the trigger is IL-6.

Bonewald: It's probably playing a central role in the resorption process.

Manolagas: This is a very important concept: IL-6 has no effect in normal osteoclastic development. We can block IL-6 completely in mice with normal levels of sex steroids and nothing will happen: it is only when the mice lose oestrogens or androgens that IL-6 becomes important. This is a critical distinction.

Beato: I was interested in the mechanism of the inhibition of the 225 bp fragment of the IL-6 promoter by oestrogen in the transfection studies. It's not mediated by binding to the oestrogen-responsive element; it's probably due to some interference by another factor. I wonder whether there is an NF-κB binding site there, because there's evidence that oestrogens can inhibit NF-κB.

Manolagas: We did experiments looking at this. Oestrogens exert some change in NF-κB activity, but in our hands it's not an inhibition. Recently, Ray et al (1994) provided indirect evidence that NF-κB is the mediator of the inhibitory effect of oestrogen on the transcriptional activity of the IL-6 promoter. They suggested this because when they co-transfected the p50 subunit of NF-κB, they were able to show down-regulation of the activity of the IL-6 promoter. However, they did not perform experiments with oestrogens plus the promoter plus NF-κB. We think that if it is not NF-κB, it could be another member of the NF-κB family, but we haven't demonstrated this yet.

Beato: Certainly, within this 225 bp part of the promoter there is an NF-κB binding site.

Manolagas: There are indeed NF-IL-6 and NF-κB response elements, as there is a glucocorticoid response element in this region of the promoter.

Beato: I was referring to the results of Michael Karin, which showed that steroid hormones (I'm not sure whether oestrogens were included) enhance I-κB gene expression and in this way inhibit NF-κB action.

Horwitz: Your results suggest an analogy to a negative glucocorticoid response element, because you're looking at inhibition of a gene and not stimulation. So perhaps you are dealing with a negative oestrogen response element, rather than the classical consensus positive oestrogen response element.

Manolagas: If this were the case, we should be able to supershift the receptor with anti-oestrogen receptor antibody.

Parker: And you should be able to compete with a positive oestrogen response element.

Manolagas: In our studies, we found that the oestrogen-responsive element did not compete with the binding of the 225 bp fragment of the promoter to nuclear extracts.

Beato: NF-κB is unlikely to be involved because most of the other interleukins are also regulated by NF-κB. Such a mechanism would affect all of them, and yet you see selectivity for IL-6.

Bonewald: It is now clear that IL-6 is an important player in bone resorption; it will obviously be important to develop an inhibitor of this cytokine. Has anyone treated any bone conditions or diseases with a resorption component (except for osteoporosis) with sex steroids?

Manolagas: Glowacki et al (1993) and Bismar et al (1994) have demonstrated that in human bone marrow there is increased production of IL-6 after the menopause. There are also studies which suggest that hyperthyroidism-induced bone loss might be mediated by IL-6 (Lakatos et al 1994). Furthermore, a group from Japan showed that the local bone loss that occurs in rheumatoid arthritis is also due to over-production of IL-6 (Kotake et al 1994). In terms of treatment, there have been a couple of studies from Bataille's group in France on patients with multiple myeloma (which is another condition associated with destruction of bone), showing very nicely that the hypercalcaemia associated with this condition is decreased by injection of anti-IL-6 antibodies.

Parker: Could you just clarify the set of experiments where you talked about the effects of ovariectomy that were not rescued by oestrogen?

Manolagas: In these experiments we measured the amount of gp130 RNA in bone marrow aspirates following sham operation, ovariectomy, or ovarietcomy plus oestrogen administration. Three weeks after each operation we aspirated the bone marrow and measured the gp130 mRNA immediately or, in the second experiment, we measured it after culture of the marrow cells for nine days. We found a big increase in the ovariectomized mice compared with the sham-operated animals. However, the ovariectomized mice that received oestrogen replacement therapy also had elevated gp130 mRNA levels. Our current explanation for this apparent discrepancy is that there might be another steroid or factor besides oestrogen that is eliminated by the removal of ovaries and that is important for gp130 suppression.

Parker: But in the cell lines, oestrogen by itself was sufficient to down-regulate the levels of gp130.

Manolagas: Yes.

Baulieu: Which cell types express IL-6?

Manolagas: Several cell types in the bone marrow and other tissues produce IL-6. Our interest in the relationship between sex steroids and IL-6 was stimulated by a report by Tabibzadeh et al (1989), who demonstrated that oestrogen inhibits IL-6 in the stromal cells of the endometrium. In fact, the authors postulated that the menstrual cycle is a consequence of indirect effects of oestrogens not on the endothelial cells but on the production of IL-6 by stromal cells of the endometrium. They thought that IL-6 production was responsible for the proliferative changes that characterize the menstrual cycle.

Bonewald: It's interesting that almost every cell type studied makes IL-6. Osteoblasts make femtograms of IL-6 per cell (Linkhart et al 1991), but the one cell that makes huge amounts of IL-6 is the osteoclast—it makes nanograms of IL-6 per cell (Ohsaki et al 1992). The only other cells that come close are endometrial cells (Tabibzadeh et al 1989).

Manolagas: Hepatocytes also make large quantities of IL-6.

Baulieu: Constitutively?

Manolagas: No, this is an inducible gene.

Baulieu: What induces IL-6 in the liver, for instance?

Manolagas: IL-1, TNF, phorbol esters, fibroblast growth factor, insulin-like growth factor and several other factors.

Bonewald: Our preliminary data suggest that peptido-leukotrienes inhibit and the leukotriene LTB4 stimulates IL-6 production by osteoclasts.

References

Bellido T, Girasole G, Passeri G et al 1993 Demonstration of estrogen and vitamin D receptors in bone marrow-derived stromal cells: up-regulation of the estrogen receptor by 1,25-dihydroxyvitamin-D₃. Endocrinology 133:553–562

Bismar H, Diel I, Ziegler R, Pfeilschifter J 1994 Increased cytokine secretion by human bone marrow cells after menopause or discontinuation of estrogen replacement. J Bone Miner Res (suppl) 9:S158(abstr)

Gates PA, Mendez J, Schwartz Z et al 1994 Sex-dependent effects of 17β-estradiol on resting zone chondrocyte membrane fluidity. J Dent Res 73 (suppl 1):376(abstr)

Glowacki J, Girasole G, Lycette C, Kilander K, Manolagas S 1993 Osteoclast precursors and interleukin-6 production by human bone marrow: modulation by estrogen and age. J Bone Miner Res (suppl) 7:S136(abstr)

Hermann T, Chang YC, Esty A, Pike W 1994 Inhibition of interleukin-6 production by estrogen is mediated through NFκB. Calcif Tissue Int 54:340(abstr)

Kimble RB, Matayoshi AB, Vannice JL, Kung V, Pacifici R 1994 Simultaneous inhibition of TNF and IL-1 is required to block bone loss in the early postovariectomy period. J Bone Miner Res (suppl) 9:S159(abstr)

Kotake S, Sato K, Kim KJ et al 1994 Interleukin-6 and soluble interleukin-6 receptors found in synovial fluids are responsible for inducing joint disruption in patients with rheumatoid arthritis. J Bone Miner Res (suppl) 9:S140(abstr)

Lakatos P, Kiss L, Tatrau A et al 1994 Serum IL-6 levels and bone mineral content in patients with hyper- and hypothyroidism. J Bone Miner Res (suppl) 9:S139(abstr)

Linkhart TA, Linkhart SG, MacCharles DC, Long DL, Strong DD 1991 Interleukin-6 messenger RNA expression and interleukin-6 protein secretion in cells isolated from normal human bone: regulation by interleukin-1. J Bone Miner Res 6:1285–1294

Ohsaki Y, Takahashi S, Scarcez T et al 1992 Evidence for an autocrine/paracrine role for inteleukin-6 in bone resorption by giant cells from giant tumours of bone. Endocrinology 131:2229–2234

Ray A, Prefontaine KE, Ray P 1994 Down-modulation of interleukin-6 gene expression by 17β-estradiol in the absence of high-affinity DNA binding by the estrogen receptor. J Biol Chem 269:12940–12946

Tabibzadeh SS, Santhanam U, Sehgal PB, May LT J 1989 Cytokine-induced production of IFN-β2/IL-6 by freshly explanted human endometrial stromal cells. Modulation by estradiol-17β. Immunology 142:3134–3139

Sex steroids and the immune system

Howard S. Fox

Department of Neuropharmacology, CVN-8, The Scripps Research Institute,
10666 N. Torrey Pines Road, La Jolla, CA 92037, USA

Abstract. Autoimmune diseases afflict women at a much higher rate than they do men. Although the differences in sex steroids probably play a role in this sexual dimorphism, the effects of sex steroids on the immune system and the mechanisms by which those effects occur are uncertain. We have begun studies to examine systematically these processes by examining the cells responsive to sex steroids, the genes regulated by these hormones, and the effect on the immune system and autoimmunity.

1995 Non-reproductive actions of sex steroids. Wiley, Chichester (Ciba Foundation Symposium 191) p 203–217

Women suffer from an increased (up to ninefold) prevalence of many autoimmune diseases, such as systemic lupus erythematosus (SLE), rheumatoid arthritis, primary biliary cirrhosis and Graves' disease, although in men some diseases, such as ankylosing spondylitis, are more prevalent. Additionally, women appear to have a more vigorous immune system, manifested by higher immunoglobulin levels, increased response to immunization, increased resistance to a variety of infections, and decreased graft rejection time (Schuurs & Verheul 1990, Ansar Ahmed et al 1985, Grossman 1985). It is likely that the differences in sex steroids—continual androgen in males and cyclic oestrogen and progesterone in females— contribute to this difference, but the mechanisms by which these hormones alter immune responses is unknown. Although the reported experimental effects of sex steroids on immune responses vary, oestrogens in general increase and androgens decrease autoimmune reactions.

The testicular hormones responsible for the male phenotype are the androgens testosterone and its active metabolite, 5α-dihydrotestosterone (DHT). In males, testosterone is synthesized primarily by the gonads, whereas DHT is formed largely within the target cells for this hormone by the enzyme 5α-reductase. The effects of these hormones are first noted during embryonic development, and production rises sharply at puberty. A single androgen receptor binds testosterone and DHT, and mediates the effects of the hormones within cells. In females, both oestrogen and progesterone are synthesized primarily by the gonads, where production begins in a cyclical fashion at puberty. Separate specific intracellular receptors are present for oestrogen and progesterone.

Hormones can affect the immune system by direct actions on immune cells, or by indirect means. In the best studied examples, the presence of the specific hormone and receptor induces steroid hormone/receptor complexes that can bind to particular DNA response elements in regulated genes. Steroid hormone/receptor complexes may also interact with other nuclear transcription factor proteins. These protein–DNA and protein–protein interactions change the level of expression of regulated genes, thus altering cellular functions (Beato 1991). Other mechanisms of actions of sex steroid hormones are also possible.

Sex steroid effects on cells of the immune system

The immune system consists of sites of precursor production and differentiation (the thymus and bone marrow), sites of organized lymphoid tissues (such as the lymph nodes and spleen), and numerous circulating and organ-resident cells. Immune interactions are divided into two large classes, those of cellular immunity mediated by T lymphocytes, and humoral immunity mediated by B lymphocytes. Interactions between these two systems are broad, and it is clear that both helper T cells and specialized antigen-presenting cells, such as macrophages, are required for most immune reactions.

Examination of cells of the immune system for sex steroid receptors is incomplete, but both oestrogen and androgen receptors have been detected in the thymus. We have examined lymphoid organs for the expression of the sex steroid receptor mRNAs using a sensitive and specific RNase protection technique. In adult mice, relatively high levels of oestrogen and androgen receptor mRNAs are found in lymph nodes and thymus, whereas much lower levels are seen in spleen. Analysis of individual cell populations has revealed that for androgen receptor mRNA, the highest levels of expression are found in T cells, whereas for oestrogen receptor, macrophages contain increased amounts of mRNA relative to T and B cells.

Both *in vitro* and *in vivo* studies have shown that sex steroids have a number of effects on the immune system, some of which parallel the reported immune differences between females and males. In studies on humoral immunity, oestrogen at physiological levels increased the number of immunoglobulin-secreting cells (Paavonen et al 1981). *In vivo* treatment of rodents with oestrogen increased antibody responses to a variety of antigens (Brik et al 1985). Conflicting data have been obtained regarding the effect of oestrogen on lymphocyte responses to mitogens and their effects on lymphocyte sub-populations.

Examination of the role of androgens has also been performed. Castration of male mice has long been known to increase thymic size and delay its involution, and androgen treatment can decrease thymic mass in organisms as diverse as mice and turtles (Fuji et al 1975, Saad et al 1991). In the enlarged thymus of castrated male mice, the absolute number of all thymocytes is increased, but the proportions of certain subpopulations vary. In functional studies on immune cells, androgens appear to be suppressive on both the humoral and cell-mediated arms of the immune system. Administration of testosterone to castrated male mice can reduce the antibody titre to a specific antigen (Carsten et al 1989), and *in vitro* testosterone inhibits pokeweed mitogen-induced human B cell differentiation (Sthoeger et al 1988). A genotype-dependent immune suppression by testosterone has been mapped to the H-2 complex (Ansar Ahmed et al 1987). Responses to infectious disease are also suppressed by androgens, as is response to histoincompatible grafts.

Sex steroid effects on genes of the immune system

Cytokines are proteins, such as the interleukins (IL) and interferons (IFN), which are produced by immune and non-immune cell types. Within the immune system, cytokines mediate many of the immune interactions between immune cells and between the immune system and the host. Sex steroid modulation of cytokine expression has been examined by a number of investigators. When lymph node cells are stimulated by anti-CD3 antibody, DHT treatment does not change IL-2 production, but instead decreases IFN-γ, IL-4 and IL-5 (Araneo et al 1991). This effect was seen on isolated CD4$^+$ and CD8$^+$ cells and could not be induced by treatment with testosterone. 5α-reduction to DHT was necessary, a task thought to be carried out by adjacent macrophages. In contrast, others have found a stimulatory effect of androgen on IL-5 production (Wang et al 1993).

It has been shown that in both *in vivo* and *in vitro* experimental systems lymphoid cells from females can produce higher levels of IFN-γ than those from males. We have examined oestrogen regulation of IFN-γ expression (Fox et al 1991). We found that *in vitro* exposure of isolated spleen cells decreases the amount of IFN-γ mRNA and protein following mitogen stimulation. IFN-γ was not detectable in unstimulated cells, regardless of exposure to oestrogen. Mitogen stimulation produced the expected rise in IFN-γ, whereas mitogen plus oestrogen led to enhanced IFN-γ mRNA and protein expression.

This increase in both protein and mRNA suggested that oestrogen may serve to enhance expression of the IFN-γ promoter. To test this hypothesis, we constructed a plasmid in which 3.5 kb containing the promoter and 5′-flanking region from the murine IFN-γ gene directs expression of chloramphenicol acetyltransferase (CAT), and used this construct to perform transient transfection assays in cultured human T lymphoid lines. Expression of the human oestrogen receptor was simultaneously achieved by transfection of an expression construct. Only a low level of CAT activity was produced in the transfected cells in the absence of oestrogen. The activity of the hybrid plasmid was markedly increased (approximately ninefold), however, in cells that had been exposed to oestrogen. In the absence of oestrogen, chemical activation of the transfected cells increased CAT expression approximately fivefold above basal levels when oestrogen was absent; in the presence of oestrogen and activation, more than a 20-fold increase was seen.

The DNA sequences in the IFN-γ regulatory region responsible for this effect were mapped by producing and testing a series of deletion mutants of the IFN-γ construct as described above. For the maximal response to oestrogen, at least 3.2 kb of 5'-flanking sequence was necessary, and the enhancing effect of oestrogen, but not activation, was entirely lost when only 0.58 kb remained. The data in their entirety suggest that the full oestrogen response may require the concerted actions of two or more functional elements lying 0.5–3.2 kb upstream from the start of transcription.

Therefore, there is a direct effect of oestrogen in enhancing expression of the IFN-γ gene. IFN-γ has pleiotropic effects in the immune system, largely serves a pro-inflammatory role in immune interactions, and is a mediator of many effects of immune cells on the host. Thus, differences in its level of production induced by sex steroids can have significant biological consequences. The regulation by oestrogen of other cytokines in cells of the immune system is currently under investigation. Given our findings (above) on the highest levels of oestrogen receptor mRNA in macrophages, we have been examining cytokine expression in cultured macrophage lines. Our studies on the P388D1 line have revealed that IL-6 is decreased by oestrogen, but no effect is seen on IL-1 or tumour necrosis factor-α. The mechanism involved in this specific effect is unknown, but decrease in IL-6 production by oestrogen has been seen in bone, as reported by Manolagas et al (1995, this volume).

Sex steroid effects on autoimmunity

A number of animal models for autoimmune diseases exist. In mice, it is clear that there are certain genetic backgrounds that will allow the development of autoimmune disease, as well as some defined genes or chromosomes that can accelerate the disease on those backgrounds. In some of these models a distinct sexual dimorphism exists in disease severity. In one case, the BXSB model for lupus, the difference is due to a gene(s) on the Y chromosome; X-chromosome genes that affect the immune system (such as *xid*) are also known. Still, most differences found appear to be due to sex steroids.

One of the best characterized systems for immunological sex steroid effects is the spontaneous lupus-like disease suffered by F_1 hybrids between New Zealand black and New Zealand white mice. Female mice have a dramatically more severe disease than males do, and it has been shown that oestrogens accelerate whereas androgens ameliorate the disease (Roubinian et al 1977, 1978). The cellular mechanisms involved in this effect remain obscure. The oestrogen-enhanced autoantibodies included ones reactive to DNA, which are strongly linked to SLE in humans. Other rodent models further support the autoimmune-enhancing effects of oestrogen, manifested mostly in the humoral branch of the immune system.

In human SLE, the effects of sex steroids are less clear owing to the ethical limitations inherent in the investigation of human disease. In most reports, oestrogen levels alone cannot be identified as the sole differential agent, and some studies are based on small numbers of patients or single case reports. Pregnancy (Mund et al 1963) and oestrogen-containing contraceptive pills (Jungers et al 1982) were both associated with increased risk of SLE. Oopherectomy may lower the risk of developing SLE (Yocum et al 1975). However, rheumatoid arthritis, although female-predominant, may show a negative correlation with oestrogen levels. In rheumatoid arthritis, symptoms have been reported to be decreased in association with increased oestrogen levels, such as those which occur during pregnancy (Oka & Vainio 1966) and during the post-ovulatory phase of the menstrual cycle (Latman 1983).

Although many experiments have shown that the defect leading to autoimmunity in these and many other models of autoimmunity can be attributed to bone-marrow-derived cells, few data exist regarding the level of action of sex steroids. In the obese strain chicken model for thyroiditis, protection from autoimmunity by androgens is linked to effects on two non-bone-marrow-derived systems: degeneration of bursal epithelial cells, and decreased hepatic corticosteroid-binding globulin production (Fassler et al 1988). The level of action of androgens in protecting mammals from autoimmunity remains unexplored.

In another animal model for autoimmunity, MRL–*lpr/lpr* mice, animals suffer from both lupus and a rheumatoid arthritis-like disease. Both male and female mice suffer from disease, but androgen treatment of gonadally intact mice of either sex lowers anti-DNA antibodies, lessens proteinuria and prolongs survival (Steinberg et al 1980). The recessive lymphoproliferation *lpr* gene accelerates the disease on the MRL genetic background. The *lpr* mutation is a defect in the *Fas* gene, the product of which is involved in apoptosis (Watanabe-Fukunaga et al 1992). When the *lpr* mutation is inbred on other genetic backgrounds, autoantibody production and lymphoproliferation occur but tissue pathology is minimal. On one non-disease-inducing background, C57BL6/J (B6), introduction of the *lpr* mutation causes anti-DNA and anti-immunoglobulin (Ig) autoantibody production, which is greatly elevated in female mice compared with males of the same congenic line (Warren et al 1984). We have confirmed that this effect is probably due to gonadal steroids, as castration of males and oopherectomy of females abrogates the sex difference in autoantibody levels.

Both polyclonal activation and antigen-driven responses in the B cell production of autoantibodies may have a pathogenic role in lupus, and specifically in the *lpr* mouse models. Autoreactive B cells in young MRL–*lpr*/*lpr* mice appear to be polyclonally activated (Klinman & Steinberg 1987). Genetic analyses of IgG and IgA anti-IgG hybridomas derived from MRL–*lpr*/*lpr* and C3H–*lpr*/*lpr* mice reveal clonal relatedness (Schlomchik et al 1987). Others have shown that lipopolysaccharide immunization of B6–*lpr*/*lpr* mice, which should lead to polyclonal activation, stimulated IgM, but not IgG, anti-DNA levels, but with no difference between the sexes; anti-IgG autoantibodies were unchanged by lipopolysaccharide immunization (Warren et al 1984).

However, it had not been determined whether female B6–*lpr*/*lpr* mice are merely capable of mounting a more vigorous humoral immune response than males. To examine whether any sexually dimorphic IgG immune response is found in response to antigenic immunization, we immunized both male and female B6–*lpr*/*lpr* mice with either rabbit IgG or tetanus toxoid, and performed serial bleeds. Females responded with significantly higher levels of IgG anti-IgG antibodies than the response achieved in male mice, whereas the response to tetanus toxoid was equivalent between the sexes. In the case of IgG anti-IgG antibodies, we found that immunization with IgG leads to increased response in females, whereas immunization with tetanus toxoid does not. This autoantigen specificity in immune response provides support for specific antigen-driven activation of the humoral autoimmune response, leading to autoantibody production and disease.

Another mouse strain, NOD (non-obese diabetic), provides an animal model of human type 1 (juvenile-type, or insulin-dependent) diabetes mellitus. The spontaneous inflammatory diabetes that develops in NOD mice is due to cell-mediated immunity (Kikutani & Makino 1992). Although inflammation of the pancreatic islets of Langerhans (insulitis) is present in male NOD mice, less than half of these mice progress to diabetes. Overt diabetes occurs in most female mice beginning at three months of age. A role for sex steroids in this sexual dimorphism is likely. Castration of male mice raises their incidence of diabetes, and oophorectomy of females decreases the progression to diabetes (Makino et al 1981, Fitzpatrick et al 1991).

Although both male and female NOD mice develop insulitis, only females reproducibly develop clinical diabetes. To discern the effect of androgens on this sexual dimorphism, we examined the effect of treating female NOD mice with DHT (Fox 1992). Beginning at approximately eight weeks of age, we performed continual treatment with DHT, or control treatment in female NOD mice. Control mice, as expected, developed clinical diabetes. However, the DHT-treated mice were completely protected from diabetes. Average blood glucose values from this treated group did not differ significantly from non-NOD mice maintained under similar conditions.

In NOD mice, insulitis precedes the onset of diabetes, and immune cells, mainly T cells, are responsible for the islet destruction. Therefore, it is possible that DHT prevents diabetes by affecting the inflammatory infiltrate in the islets. Examination of NOD mice at eight weeks of age (when the hormone treatment began) by histological analysis of the pancreas revealed that islets were present containing inflammatory infiltrates and with evidence of destructive lesions. When pancreatic tissues from 31-week-old DHT-treated mice were examined, insulitis remained present, with lesions showing a spectrum of severity, ranging from lesions with minimal inflammation to islets with destructive lesions. Still, the islets of Langerhans remained present, and immunohistochemical analysis revealed abundant insulin-containing beta cells within the islets. In contrast, in the control group, killed after the onset of diabetes, islets could not be identified. Structures probably representing the end stage of islet destruction, with only a few remaining endocrine cells, were occasionally noted. Immunohistochemical analysis did not identify any remaining beta cells in pancreatic tissues of the control mice.

We next questioned how DHT protects the beta cell from destruction. Was the effect on the immune system, preventing the killing of the beta cells by physically or functionally eliminating the cells necessary for the self-destruction; or could the effect be mediated by non-immune cell types, such as endothelial cells, by preventing lymphocyte trafficking; or on the beta cell itself by making it resistant to immune attack?

We performed adoptive transfer experiments to attempt to distinguish between these possibilities. In NOD mice, the injection of spleen cells from donor animals, either newly diabetic or capable of progressing to diabetes, results in reproducible diabetes in the recipients. In these experiments DHT-treated female mice received spleen cells from diabetic female mice, and the recipients were then monitored for the onset of diabetes. All of the DHT-treated recipient mice receiving donor spleen cells from diabetic mice themselves became diabetic within four weeks after transfer. Similarly treated control recipient female DHT-treated mice, receiving donor spleen cells from young, non-diabetic female NOD mice, did not become diabetic.

Thus, immune-mediated diabetes can indeed be induced in the DHT-treated NOD mice. The hormone cannot be acting to prevent trafficking of pathogenic immune cells into the pancreas or islets, or by protecting the beta cell from immune damage. The combination of the histopathological data and the results of the adoptive transfer experiments indicate that androgen treatment blocks the progression of the immune cell attack.

Perspectives

The effects of sex steroids in the immune system and their role in human immune responses and autoimmune diseases remain a subject of investigation.

Through a definition of the cells responsive to sex steroids, the molecules in these cells altered by sex steroids, and the *in vitro* and *in vivo* changes resulting from sex steroid actions, a better understanding of the hormone–immune interactions can be achieved, and potential mechanisms to prevent or treat autoimmune disease envisaged.

References

Ansar Ahmed S, Penhale WJ, Talal N 1985 Sex hormones, immune responses, and autoimmune diseases. Am J Pathol 121:531–551

Ansar Ahmed S, Talal N, Christadoss P 1987 Genetic regulation of testosterone-induced immune suppression. Cell Immunol 104:91–98

Araneo BA, Dowell T, Diegel M, Daynes RA 1991 Dihydrotestosterone exerts a depressive influence on the production of interleukin-4 (IL-4), IL-5, and γ-interferon, but not IL-2 by activated murine T cells. Blood 78:688–699

Beato M 1991 Transcriptional control by nuclear receptors. FASEB J 5:2044–2051

Brick JE, Wilson DA, Walker SE 1985 Hormonal modulation of responses to thymus-independent and thymus-dependent antigens in autoimmune NZB/W mice. J Immunol 134:3693–3698

Carsten H, Holmdahl R, Tarkowski A, Nilsson LA 1989 Oestradiol- and testosterone-mediated effects on the immune system in normal and autoimmune mice are genetically linked and inherited as dominant traits. Immunology 68:209–214

Fassler R, Dietrich H, Kromer G, Bock G, Brezinschek HP, Wick G 1988 The role of testosterone in spontaneous autoimmune thyroiditis of Obese strain (OS) chickens. J Autoimmun 1:97–108

Fitzpatrick F, Lepault F, Homo-Delarch F, Bach JF, Dardenne M 1991 Influence of castration, alone or combined with thymectomy, on the development of diabetes in the nonobese diabetic mouse. Endocrinology 129:1382–1390

Fox HS 1992 Androgen treatment prevents diabetes in nonobese diabetic mice. J Exp Med 175:1409–1412

Fox HS, Bond BL, Parslow TG 1991 Estrogen regulates the IFN-γ promoter. J Immunol 146:4362–4367

Fuji H, Nawa Y, Tsuchiya H et al 1975 Effect of a single administration of testosterone on the immune and lymphoid tissues in mice. Cell Immunol 20:315–326

Grossman CJ 1985 Interactions between the gonadal steroids and the immune system. Science 227:257–261

Jungers P, Dougados M, Pelissier C et al 1982 Influence of oral contraceptive therapy on the activity of systemic lupus erythematosus. Arthritis Rheum 25:618–623

Kikutani H, Makino S 1992 The murine autoimmune diabetes model: NOD and related strains. Adv Immunol 51:285–322

Klinman DM, Steinberg AD 1987 Systemic autoimmune disease arises from polyclonal B cell activation. J Exp Med 165:1755–1760

Latman NS 1983 Relation of menstrual cycle phase to symptoms of rheumatoid arthritis. Am J Med 74:957–960

Manolagas SC, Bellido T, Jilka RL 1995 Sex steroids, cytokines and the bone marrow: new concepts on the pathogenesis of osteoporosis. In: Non-reproductive actions of sex steroids. Wiley, Chichester (Ciba Found Symp 191) p 187–202

Makino S, Kunimoto K, Muraoka Y, Katagiri K 1981 Effect of castration on the appearance of diabetes in NOD mouse. Exp Anim (Tokyo) 30:137–140

Mund A, Swison J, Rothfield N 1963 Effect of pregnancy on course of systemic lupus erythematosus. JAMA 183:917–920

Oka M, Vainio U 1966 Effect of pregnancy on the prognosis and serology of rheumatoid arthritis. Acta Rheum Scand 12:47–52

Paavonen T, Andersson LC, Adlercreutz H 1981 Sex hormone regulation of *in vitro* immune response: estradiol enhances human B cell maturation via inhibition of suppressor T cells in pokeweed mitogen-stimulated cultures. J Exp Med 154: 1935–1945

Roubinian JR, Papoian R, Talal N 1977 Androgenic hormones modulate autoantibody responses and improve survival in murine lupus. J Clin Invest 59:1066–1070

Roubinian JR, Talal N, Greenspan JS, Goodman JR, Siiteri PK 1978 Effect of castration and sex hormone tretment on survival, anti-nucleic acid antibodies, and glomerulonephritis in NZB/NZW F_1 mice. J Exp Med 147:1568–1583

Saad AH, Torroba M, Varas A, Zapata A 1991 Testosterone induces lymphopenia in turtles. Vet Immunol Immunopathol 28:173–180

Schuurs AHWM, Verheul HAM 1990 Effects of gender and sex steroids on the immune response. J Steroid Biochem 35:157–172

Shlomchik MJ, Aucoin AH, Pisetsky DS, Weigert MG 1987 The role of clonal selection and somatic mutation in autoimmunity. Nature 328:805–811

Steinberg AD, Roths JB, Murphy ED, Steinberg RT, Raveche ES 1980 Effects of thymectomy or androgen administration upon the autoimmune disease of MRL/Mp-*lpr/lpr* mice. J Immunol 125:871–873

Sthoeger ZM, Chiorazzi N, Lahita RG 1988 Regulation of the immune response by sex hormones. I. *In vitro* effects of estrodiol and testosterone on pokeweed mitogen-induced human B cell differentiation. J Immunol 141:91–88

Wang Y, Campbell HD, Young IG 1993 Sex hormones and dexamethasone modulate interleukin-5 gene expression in T lymphocytes. J Steroid Biochem Mol Biol 44: 203–210

Warren RW, Roths JB, Murphy ED, Pisetsky DS 1984 Mechanisms of polyclonal B-cell activation in autoimmune B6-*lpr/lpr* mice. Cell Immunol 84:22–31

Watanabe-Fukunaga R, Brannan CI, Copeland NG, Jenkins NA, Nagata S 1992 Lymphoproliferation disorder in mice explained by defects in Fas antigen that mediates apoptosis. Nature 356:314–317

Yocum MW, Grossman J, Waterhouse C, Abraham GN, May AG, Condemi JJ 1975 Monozygotic twins discordant for systemic lupus erythematosus: comparison of immune response, autoantibodies, viral antibody titres, gamma globulin and light chain metabolism. Arthritis Rheum 18:193–199

DISCUSSION

Baulieu: You didn't say anything about progesterone.

Fox: No, I did not. We didn't find reasonably detectable levels anywhere other than in the mature thymocytes. Even there, the levels were still much lower than androgens and oestrogen. If we are going to study effects, we would rather study cells with higher levels of receptors. But that's not to say that there aren't good effects of progesterones to be seen. I could not induce progesterone receptor in the macrophages—I have tried that.

Baulieu: Do you mean that you tried to find progesterone receptor and you couldn't?

Fox: We got a very low level of progesterone receptor in macrophages determined only by reverse transcriptase polymerase chain reaction, but never a signal that I could believe by RNase protection. So we tried treating the macrophages with oestrogens to see whether we could induce the progesterone receptor, and we could not.

Baulieu: One of the reasons I asked this question was because there has been a long series of studies which probably haven't even begun to give a better understanding of the phenomenon concerning progesterone activity mediating immunological changes affecting the presence of the fetus in the uterus. There is, first of all, non-rejection of the embryo. Have you looked at this?

Fox: I haven't, but it is a fascinating problem because it is relevant to the issue of how pregnancy fits into autoimmunity—the symptoms of rheumatoid arthritis decrease during pregnancy and those of lupus get worse. They are both autoimmune diseases, yet there is this pronounced difference in pregnancy.

Sarnyai: You suggested that the androgens have a counterproductive effect in autoimmune function. Is there any clinical evidence available on the autoimmune functions in testicular feminization and hypoandrogen states?

Fox: In testicular feminization there are some case reports of autoimmunity, but the numbers are not very large. In Klinefelter's syndrome, there are increased incidences of lupus as well as scleroderma (Valloton & Forbes 1967, De Keyser et al 1989, Lahita & Bradlow 1987). Klinefelter's is a hypogonadal state, with less androgen and autoimmunity is significantly increased from that of normal males.

Sarnyai: Have you looked at other factors of autoimmune activation processes, like the complement system? Actual damage is caused by the complement activation locally.

Fox: Especially in lupus and diseases like that. There is certainly evidence for increased production of complement components and their receptors by oestrogen and during the menstrual cycle (Hasty et al 1994).

Bäckström: There is an increase in the incidence of rheumatoid arthritis in menopausal women (Fox 1895, Da Silva & Hall 1992). How does this fit in with your data?

Fox: My argument for that is the same as my argument for Sjogren's syndrome, which arises post-menopausally with an increased incidence in women: I think these diseases take a long time to develop. We don't do mass screening for autoantibodies unless there are symptoms and I think the cascade takes a while to get going. I wonder whether it's really an increased incidence at menopause or rather an increased incidence with age.

Bäckström: If you give gonadotropin-releasing hormone agonists (which block the ovarian function of steroids) to women who have a family history of rheumatoid arthritis, you can sometimes see symptoms within a month.

Fox: I'm perplexed. Does one reduce androgens when one gives that therapy?

Bäckström: Not really. A bit, but there is not a big change.

Fox: I don't have a good explanation for this, just as I lack an explanation for the different incidences of autoimmune diseases in pregnancy.

Manolagas: You showed that the oestrogen receptor was present in bone marrow-derived and peritoneal macrophages. Have you looked for the oestrogen receptor in undifferentiated circulating monocytes?

Fox: No, but it would be reasonably simple to do that.

Manolagas: How many receptors were present per macrophage?

Fox: I didn't do it per cell, I did it per milligram protein. My control was MCF-7 cells and it was approximately 25-fold less than binding per milligram protein.

Beato: Is this a critical point?

Manolagas: I don't know whether it's critical. In the bone field we've been trying to find out how many oestrogen receptors are enough. I have come to the conclusion that it doesn't really matter. The difference between 1000 and 20 000 sites per cell might not be relevant; you don't need all 20 000 copies of the receptor to induce an effect.

Beato: Be careful about that. When glucocorticoid-resistant variants of lymphoid cells are selected, most mutants that are labelled receptor negative still have a tenth of the number of receptor molecules found in normal cells. Instead of 30 000 molecules per cell they have 3000 and do not respond to hormone.

Manolagas: Most of these instances are caused by defective DNA binding.

Beato: No, they are perfectly normal receptors, there is just one-tenth the number of them.

Fox: I was looking for the cells that had the highest levels of receptor, thinking that in those I would see the most profound effects.

Beato: But the fact that you had to use such a sensitive procedure as RNase mapping to see it suggests already that there's very little receptor present.

Manolagas: *In vitro* estimations of receptor levels might not accurately reflect the *in vivo* situation. *In vivo* you have a number of factors that might affect receptor levels. For example, we found that in the presence of 1,25-dihydroxyvitamin D_3, the oestrogen receptor is up-regulated two- or threefold (Bellido et al 1993). It is therefore possible that there are factors *in vivo*, that you are missing *in vitro*, which keep the levels of receptor higher.

There was a recent report that 1,25-dihydroxyvitamin D_3 administration to NOD mice dramatically inhibits the expression of the disease (Mathieu et al 1994). Do you understand how either sex steroids or vitamin D might influence autoimmunity in these animal models?

Fox: The one criticism I have of the NOD model is that many things you do to it can inhibit the disease (other than my placebo!). If you give NOD mice Freund's adjuvant or irradiate them, this inhibits disease. One has to examine the level of action of these treatments; specifically, whether they affect immune cells. I think those effects are effects on the macrophage. I am not sure any parameters have been worked out concerning the effect of vitamin D. We have looked extensively at how androgens affect T cell subsets and did not find significant changes.

Manolagas: Zac Lemire at the University of California at San Diego has done a lot of this work. You should be talking to him.

Jensen: If I might go back to ancient history for a minute, it was shown very early on that target tissues, such as the immature rat uterus, are provided with a rather large reserve supply of oestrogen receptors. A physiological dose of oestrogen causes only a small fraction of the total receptors to bind strongly in the nucleus and even hyperphysiological amounts do not use all the receptor (Jensen et al 1968). Large amounts of oestrogen can be taken up by the receptors in uterine cells, but the excess hormone is not retained in the nucleus (Jensen et al 1967, Clark & Peck 1976).

Fox: Do you see any different physiological effects by filling all the receptors?

Jensen: Not really. One may have secondary effects on the pituitary, but as far as the uterotrophic response goes, once a certain level is reached, more hormone does not lead to any more growth. There is some evidence that massive amounts of oestrogen, which essentially deplete the entire receptor content, inhibit the replenishment of receptor that normally follows a physiological or mildly hyperphysiological dose of hormone.

Thijssen: However, at very high doses there is a difference: when you treat postmenopausal women with 100–1000 times higher doses of oestrogens for breast cancer, you get an atrophic endometrium.

McEwen: I want to ask you about T cell selection, because you've mentioned it a couple of times, and yet the data with the NOD mouse suggest that it was mainly androgen suppression of an already established problem. Given what you know about the distribution of androgen and oestrogen receptors in the immature thymus, would you expect the effects of androgen and oestrogen to occur at the stage when the positive and negative selection is taking place?

Fox: I would; it's just a matter of whether we currently have the tools to detect it. I've looked at the proportion of T cells expressing different β chains of the T cell receptor (TCR). This is typically selected in mice, including NOD mice, with some endogenous superantigens that react with several TCR βs and delete them intrathymically; there is no measurable effect of oestrogen or androgen on this. I would expect that there may be a response on certain self antigens that can get by either without androgen or with oestrogen. But I don't have the tools to detect that. The other place that there are sex steroid receptors is the stromal network of the thymus. The thymus does not only contain the developing T cells but there are macrophages and dendritic cells that are presenting the antigen. Those cells also have a level of sex steroid receptor mRNA equivalent to the double-negative thymocytes that I showed. This is very different from glucocorticoids, where the receptors are on the double-positive T cells. If you add glucocorticoids or oestrogen or androgens at high levels *in vivo*, you get thymic atrophy and apoptosis of those cells. If you do it *in vitro*, glucocorticoids are very effective in causing apoptosis but oestrogens and androgens are inactive. So I think a lot of the effect is coming from the presenting cells—the cells that are presenting the self antigens in the context of major histocompatibility complex to the immature T cells.

McEwen: I was wondering about the natural hormone secretion pattern around the time of birth, with testosterone secretion in the male, whereas the female mouse or rat has an α fetoprotein that binds oestrogen and tends to reduce oestrogens unless there's a local aromatizing source. Has anyone looked for aromatase in the thymus?

Fox: I have not looked extensively, but I have seen no evidence for aromatization in the thymus.

Manolagas: There is extensive evidence for aromatization in the bone marrow.

Bäckström: Is there any lasting effect of testosterone? If you stop treatment, how long does it take until the T cell is activated?

Fox: It takes about two to three months. If we start treating at four to six weeks, most of the animals will develop disease by 20–24 weeks. It takes about two to three months for the cells to build up enough or cause enough destruction to cause disease. In that model you need about 90–95% β cell destruction to cause diabetes. If we kill them at various time points, we see the peri-insulitis developing into infiltrating insulitis with loss of β cells, and eventually disease.

Bäckström: Are there any clinical trials using androgens for the treatment of diabetes?

Fox: Not that I know of.

Horwitz: Related to this, type 1 insulin-dependent diabetes develops, to a large extent, in pre-pubertal children. I would therefore guess that in type 1 diabetes the sex steroids are not involved in the pathogenesis of the disease. What happens at puberty to the sex ratio of patients who develop diabetes?

Fox: The NOD mouse is a wonderful model for autoimmunity, but it has many differences from humans including the fact that there is no sex difference of diabetes in humans. It's been one of the most extensively immunologically studied strains of mice, with numerous reagents available to examine the immunological consequence of almost anything you do to them. It provides a wonderful model system, but in some cases the relevance to human diabetes is questionable. Also, human diabetes isn't as 'clean' in its development. But the relevance of this model to general T cell autoimmune-mediated processes is high.

Thijssen: Some years ago Schuurs & Verheul (1990) reviewed the effect of sex steroids on the immune system. They concluded that there was a discrepancy between the biological activity regarding sex effects and the effect on the immune system. They claimed that a lot of the non-reproductively active steroids would have an effect on the immune system.

Fox: I read that review. It was a very nice examination and compilation of the literature. In the context of the discussions we have had on cardiovascular-active steroids and bone-active steroids, I would be quite curious about specific immune system-active hormones. Interestingly, in the discussion about these analogues that are active on organs other than the reproductive organs, for most of the targets there are relatively lower levels of receptors; I wonder whether that is somehow behind this selectivity.

We have discussed the mechanisms behind tissue specificity: could receptor number have something to do with this?

Robel: In that context, there are recent reports claiming that dehydroepiandrosterone (DHEA) can stimulate immunity. Protection against lethal viral infection of mice was obtained with enormous amounts of DHEA given by gavage or intraperitoneal injection (Loria et al 1988). In similar conditions, DHEA and pregnenolone increased the production of antibodies against exogenous proteins (Morfin & Courchay 1994). Their 7α-hydroxylated derivatives are, however, about 100-fold more potent, and might be the active metabolites. The presence of a DHEA-specific receptor in murine T cells is highly controversial (Meikle et al 1992).

Fox: Healthy mice don't normally secrete DHEA, and most of those experiments were done on rodents, so it really is quite an abnormal situation. Rather than replenishing it, as in humans, it is taking it from a null level to a higher level.

References

Bellido T, Girasole G, Passeri G et al 1993 Demonstration of estrogen and vitamin D receptors in bone marrow-derived stromal cells: up-regulation of the estrogen receptor by 1,25-dihydroxyvitamin D_3. Endocrinology 133:553–562

Clark JH, Peck EJ Jr 1976 Nuclear retention of receptor–oestrogen complex and nuclear acceptor sites. Nature 260:635–637

Da Silva JA, Hall GM 1992 The effect of gender and sex hormones on outcome in rheumatoid arthritis. Clin Rheumatol 6:196–219

De Keyser F, Mielants H, Veys EM 1989 Klinefelter's syndrome and scleroderma. J Rheumatol 16:1613–1614

Fox RF 1895 The varieties of rheumatoid arthritis. Lancet II:79–84

Hasty LA, Lambris JD, Lessey BA, Pruksananonda K, Lyttle CR 1994 Hormonal regulation of complement components and receptors throughout the menstrual cycle. Am J Obstet Gynecol 170:168–175

Jensen EV, DeSombre ER, Jungblut PW 1967 Interaction of oestrogens with receptor sites *in vivo* and *in vitro*. In: Proc 2nd Int Congr Hormonal Steroids, Milan, 1966. Excerpta Medica, Amsterdam, p 492–500

Jensen EV, Suzuki T, Kawashima T, Stumpf WE, Jungblut PW, DeSombre ER 1968 A two-step mechanism for the interaction of oestradiol with rat uterus. Proc Natl Acad Sci USA 59:632–638

Lahita RG, Bradlow HL 1987 Klinefelter's syndrome: hormone metabolism in hypogonadal males with systemic lupus erythematosus. J Rheumatol 14:154–157

Loria RM, Inge T, Cook SS, Szakai AK, Regelson W 1988 Protection against acute lethal viral infections with the native steroid dehydroepiandrosterone (DHEA). J Med Virol 26:301–314

Mathieu C, Waer M, Laureys J, Rutgeerts O, Bouillon R 1994 Prevention of autimmune diabetes in NOD mice by 1,25-dihydroxyvitamin D_3. Diabetologia 37:552–558

Meikle AW, Dorchuck RW, Araeno BA et al 1992 The presence of a dehydroepiandrosterone-specific receptor binding complex in murine T cells. J Steroid Biochem Mol Biol 42:293–304

Morfin R, Courchay G 1994 Pregnenolone and dehydroepiandrosterone as precursors of native 7-hydroxylated metabolites which increase the immune response in mice. J Steroid Biochem Mol Biol 50:91–100

Schuurs AH, Verheul HA 1990 Effects of gender and sex steroids on the immune system. J Steroid Biochem 25:157–172

Valloton MB, Forbes AP 1967 Autoimmunity in gonadal dysgenesis and Klinefelter's syndrome. Lancet I:648–651

Steroidal regulation of cell cycle progression

Robert L. Sutherland, Jenny A. Hamilton, Kimberley J. E. Sweeney, Colin K. W. Watts and Elizabeth A. Musgrove

Cancer Biology Division, Garvan Institute of Medical Research, St Vincent's Hospital, Darlinghurst, Sydney, NSW 2010, Australia

Abstract. Sex steroid hormones and their antagonists have well-defined mitogenic and growth-inhibitory effects on target cells including cancer cells. These effects are mediated by cell cycle phase-specific actions, implying that steroids control rates of cell cycle progression by regulating the expression of key cell cycle regulatory genes. An emerging model of cell cycle control involves transcriptional induction of cyclin genes and consequent activation of cyclin-dependent kinases, which initiate cellular events necessary to complete checkpoints within the cell cycle. Our recent studies have focused on the roles of G1 cyclins, particularly cyclin D1, in the control of cell cycle progression in human breast cancer cells. These studies show that cyclin D1 induction is an early response to mitogenic stimulation by oestrogens and progestins, is rate-limiting for G1 progression and is sufficient for completion of the cell cycle in cells arrested in early G1 phase by serum deprivation. Furthermore, inhibition of cyclin D1 expression is an early response to growth-inhibitory anti-oestrogens. These results suggest that cyclin D1 is a target for regulation of cell cycle progression by sex steroids and their antagonists.

1995 Non-reproductive actions of sex steroids. Wiley, Chichester (Ciba Foundation Symposium 191) p 218–234

Sex steroid hormones are essential for the programmed development and function of reproductive tissues during fetal development, at puberty and in adult life. These roles entail precise regulation of cellular proliferation and differentiation, often in a cell- and tissue-specific manner. Sex steroids have also been implicated in the pathogenesis of several important human cancers. For example, oestrogens are involved in the development of carcinoma of the breast, endometrium and ovary, whereas androgens are implicated in the aetiology of prostate cancer (Henderson et al 1991). Because of the high incidence of steroid hormone-responsive cancers (Henderson et al 1991), knowledge of the mechanisms underlying control of proliferation by steroids is relevant not only to a deeper understanding of normal physiological function,

TABLE 1 Activation of cyclin/cyclin-dependent kinase (CDK) complexes throughout the cell cycle

Cyclin	Associated CDK	Time of activation
A	CDC2	G2/M
	CDK2	S/G2
B	CDC2	M
D	CDK4, CDK6	G1
E	CDK2	G1/S

effects resulted from progestin agonist or antagonist activity. However, when we compared the changes in cell cycle phase distribution, we saw temporal differences between growth inhibition by progestins and antiprogestins (Musgrove & Sutherland 1993). Such data are supportive of the view that antiprogestin inhibition occurs by a process which opposes the stimulatory effect of progestin. Interestingly, antiprogestins and anti-oestrogens led to temporally similar changes in cell cycle phase distribution, suggesting that their actions occurred at similar points within G1 phase (Fig. 1) (Musgrove & Sutherland 1993).

Mechanisms of cell cycle control

The demonstration of steroidal control of cell cycle progression at defined points within G1 phase indicates that steroids, acting via their respective receptors, are controlling the expression of key cell cycle regulatory genes, the products of which determine the rate of G1 progression. Such an interpretation is compatible with current concepts of mammalian cell cycle control where environmental signals act within G1 phase to regulate rates of cell proliferation (Baserga 1990). It is further supported by the demonstration of steroidal regulation of some immediate-early genes also activated in response to mitogenic peptide growth factors (e.g. Musgrove et al 1991). Significant advances in our understanding of the molecular basis of cell cycle control in mammalian cells, which have emerged in recent years owing to the discovery and functional analysis of the cell cycle-regulatory cyclins, CDKs and inhibitors of these kinases (Motokura & Arnold 1993, Pines 1994), suggest additional targets for steroid action.

Cyclins and CDKs are the regulatory and catalytic subunits, respectively, of cell cycle-regulated kinases. Mammalian cells contain multiple cyclins and CDKs (Motokura & Arnold 1993) (Table 1). The members of each family share sequence homology within specific motifs which are thought to have functional significance. Some cyclins are particularly closely related, e.g. cyclins D1, D2 and D3, and thus form subgroups within the cyclin family. The

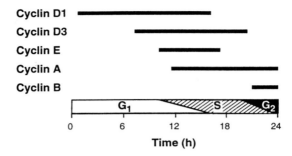

FIG. 2. Sequential induction of cyclin mRNA following growth factor stimulation of breast cells. Bars indicate approximate times of maximum expression of individual cyclins after growth factor stimulation at 0 h.

existence of three closely related D-type cyclins raises questions of the functional significance of apparently similar molecules. However, there is now clear evidence for tissue-specific expression and distinct roles in the control of differentiation as well as proliferation, indicating that these genes are not redundant but may have complementary functions (Inaba et al 1992, Kato & Sherr 1993, Sherr 1993). Some cyclins are capable of binding to multiple CDKs (e.g. cyclin D1 can activate CDK4 and CDK6, presumably to mediate multiple functions [Pines 1993]); differential expression of the CDKs allows further scope for cell- and tissue-specific roles for particular cyclin/CDK complexes. In T-47D human breast cancer cells, the experimental model used in the majority of the studies presented here, CDK4 is the predominant partner for cyclin D1.

The sequential transcriptional activation of cyclin genes and consequent transient accumulation and activation of different cyclin/CDK complexes is thought to be the central mechanism for a series of control points in the mammalian cell cycle. In growth factor-stimulated cells (for example, breast cancer cells), cyclins D1, D3, E and A are sequentially induced during G1 phase progression and entry into S phase (Musgrove et al 1993) (Fig. 2). Passage through G1 control points is regulated by cyclins D1 and E. Ectopic expression of cyclin E shortens the duration of G1, indicating that it is rate-limiting for G1 transit (Ohtsubo & Roberts 1993). However, the duration of S phase is extended so the overall population doubling time is only slightly affected.

Microinjection studies with anti-cyclin D1 antibodies or anti-sense oligonucleotides have shown cyclin D1 to be necessary for entry into S phase (Baldin et al 1993, Quelle et al 1993, Lukas et al 1994). Since cyclin D1 was the first cyclin to increase in abundance following growth factor stimulation of breast cancer cells, we assessed the effects of alterations in cyclin D1 expression in breast cancer cells by generating T-47D cells expressing human cyclin D1 under the control of a metal-inducible metallothionein promoter. In cycling cells, induction of cyclin D1 following zinc treatment resulted in an increase in

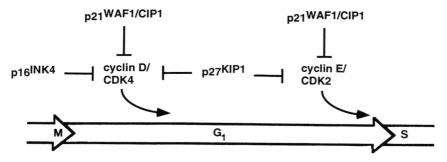

FIG. 3. Actions of cyclin-dependent kinase (CDK) inhibitors. Effects of CDK inhibitors on cyclin D/CDK4 and cyclin E/CDK2 are illustrated. Blunt arrow indicates inhibition.

the number of cells progressing through G1 and in the rate of transition from G1 to S phase, indicating that cyclin D1 is rate-limiting for progress through G1 phase (Musgrove et al 1994). Similar results obtained using rodent fibroblasts (Quelle et al 1993) indicate that this function is likely to be universal in cells which express cyclin D1. In addition, in T-47D breast cancer cells arrested in early G1 phase after growth factor deprivation, zinc induction of cyclin D1 was sufficient for completion of the cell cycle, a process requiring growth factor stimulation in control cells (Musgrove et al 1994).

Together, these observations provide evidence for a central role for cyclin D1 in breast cancer cell proliferation and suggest that molecules which regulate cyclin D1 expression or the function of associated kinases, particularly CDK4, also play a critical role in the control of cell cycle progression in these cells. Recently, a number of endogenous inhibitors of CDK activity have been identified (Fig. 3). These include: p16[INK4], which specifically inhibits the catalytic activitiy of cyclin D/CDK4 complexes (Serrano et al 1993); p27[KIP1], which inhibits both cyclin D/CDK4 and cyclin E/CDK2 complexes and appears to link the functions of these kinases (Polyak et al 1994); and p21[WAF1/CIP1], a general inhibitor of cyclin/CDK complexes (Hunter 1993). The potential down-stream targets for mediating environmental effects on rates of cell proliferation, including the effects of sex steroids and their antagonists, thus include cyclins, CDKs and CDK inhibitors, as well as the immediate-early genes activated in response to mitogenic peptide growth factors.

Steroidal regulation of cyclin gene expression

To test the hypothesis that these genes are targets for steroid hormone action, we examined the effects of sex steroids and their antagonists on the expression of cyclin, CDK and CDK inhibitor mRNA. Preliminary data do not suggest

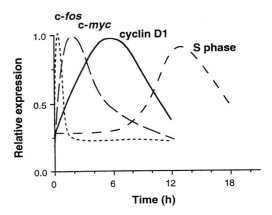

FIG. 4. Effects of progestin stimulation on cell cycle-regulatory genes. Temporal changes in c-*fos*, c-*myc* and cyclin D1 mRNA following progestin stimulation of T-47D breast cancer cells are shown.

regulation of mRNA for CDK4, p16^{INK4} or p21$^{WAF1/CIP1}$ as a likely mechanism for the mitogenic effects of oestrogens or progestins. However, cyclin D1 is induced by both progestin and oestrogen treatment, although the response to progestins is more marked (Musgrove et al 1993, E. A. Musgrove & R. L. Sutherland, unpublished results). The induction occurs later than the induction of either of the immediate-early genes c-*fos* or c-*myc*, but is apparent within one to two hours (Fig. 4). No data are yet available on the mechanism of transcriptional activation. However, the region for which a sequence is available, extending approximately 2.0 kb upstream of the cyclin D1 transcription start site, does not contain regions homologous to consensus steroid response elements. This raises the possibility that the regulation of cyclin D1 by steroid hormones occurs via intermediary proteins, either alone or complexed with the steroid receptor.

Since simultaneous treatment with steroid and steroid antagonist prevents the effects of the steroid on cell cycle progression, it would also be expected to prevent the modulation of genes involved in the regulation of cell cycle progression. In experiments we carried out to test this proposition, the effects of antiprogestin treatment on progestin induction of c-*fos* were equivocal, but simultaneous antiprogestin treatment prevented progestin induction of c-*myc* (Musgrove et al 1991). In addition, progestin induction of cyclin D1 and stimulation of cell cycle progression were both prevented by antiprogestin, added either simultaneously or after a three-hour delay (Musgrove et al 1993). Induction of c-*myc* expression by progestins is transient, and by three hours c-*myc* mRNA levels begin to decline. It might be expected, therefore, that addition of antiprogestin three hours after progestin treatment would be too late to prevent the consequences of c-*myc* induction. Since antiprogestin addition three hours

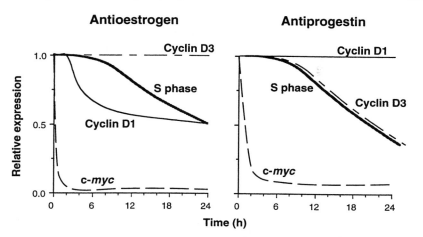

FIG. 5.   Effect of anti-oestrogen or antiprogestin treatment on cell cycle-regulatory genes. Changes in the expression of c-*myc*, cyclin D1 and cyclin D3 mRNAs in breast cancer cells following anti-oestrogen or antiprogestin treatment are presented schematically.

after progestin treatment prevents stimulation of cell cycle progression, regulation of cyclin D1 appears to be more closely connected to progestin regulation of proliferation than does regulation of c-*myc*.

Anti-oestrogens reduce the abundance of cyclin D1 mRNA by approximately 50% (Musgrove et al 1993, Watts et al 1994), i.e. the effects of anti-oestrogens on cyclin D1 and S phase are of similar magnitude (Fig. 5). Simultaneous oestrogen treatment prevented anti-oestrogen down-regulation of cyclin D1 mRNA (Watts et al 1994). Cyclin gene expression is to some degree dependent on the growth state of the cell and therefore might be expected to decrease in growth-arrested cells. However, the decrease in cyclin D1 expression occurs more than six hours before any changes in cell cycle phase distribution are apparent, and is therefore not merely a consequence of growth inhibition (Musgrove et al 1993, Watts et al 1994). The similar kinetics of the changes in cell cycle-phase distribution after anti-oestrogen or antiprogestin treatment argues that these compounds might inhibit cell cycle progression by convergent pathways (Musgrove & Sutherland 1993), and both rapidly and profoundly decrease c-*myc* expression (Fig. 5). However, there are clear differences between the changes in cyclin gene expression resulting from anti-oestrogen and anti-progestin treatments (Fig. 5) (Musgrove et al 1993). Antiprogestins decrease expression of cyclin D3 but not cyclin D1, whereas anti-oestrogens decrease cyclin D1 but not cyclin D3 expression. In the experiments we have carried out to date, treatment with anti-oestrogens or antiprogestins had no significant effect on the abundance of CDK2, CDK4, p16[INK4] or p21[WAF1/CIP1] mRNA before changes in cell cycle phase distribution were apparent, suggesting that regulation of the expression of these genes is unlikely to be a cause of growth inhibition after anti-oestrogen or antiprogestin treatment. It remains possible that regulation of the function of these molecules contributes to growth inhibition by steroid antagonists, but although CDK2 activity (measured using histone H1 as a substrate for immunoprecipitated kinase) decreased after anti-oestrogen treatment, the response did not precede growth inhibition.

Although the cyclin D1 gene is clearly a cell cycle regulatory gene in breast cancer cells, it should not be construed that it is the only gene responsible for regulation of cell cycle progression in these cells. Growth inhibition by antiprogestins without accompanying decrease in cyclin D1 expression clearly identifies that this is not the case. The processes which actually control cell proliferation are fewer than those which are necessary for cell proliferation. Whereas stimulation of cell cycle progression requires action on a controlling process, inhibition will occur after interference with any necessary process. Evidence that inhibition of the expression or function of either cyclin D1 or c-*myc* alone (for example, by the use of antibodies or anti-sense techniques) inhibits breast cancer cell entry into S phase (Watson et al 1991, Lukas et al 1994) suggests that the down-regulation of either c-*myc* or cyclin D1 could be sufficient for growth arrest after anti-oestrogen or antiprogestin treatment. The data summarized above are consistent with the interpretation that stimulation of breast cancer cell cycle progression is controlled by the level of cyclin D1, but that c-*myc* is also necessary.

Conclusions

Sex steroid hormones and their antagonists exert effects on cell proliferation by affecting cell cycle control points in G1 phase. The demonstration that some nuclear proto-oncogenes and G1 cyclins with key regulatory roles in cell cycle control are transcriptionally regulated by steroids and their antagonists implicates these genes as down-stream targets in steroid receptor-mediated control of cell proliferation. However, the complexity of cell cycle control, as exemplified by the recent identification of inhibitors of cyclin/CDK complexes and the well-documented cell- and tissue-specific effects of steroids, provides strong evidence that other genes are likely to be involved in steroidal regulation of cell proliferation. Steroid-responsive breast cancer cell lines are likely to provide valuable model systems for further elucidating the complexities of cell cycle control by steroids.

Acknowledgements

This work is currently supported by the National Health and Medical Research Council of Australia, the New South Wales State Cancer Council and the St Vincent's Hospital Research Fund.

References

Baldin V, Lukas J, Marcote MJ, Pagano M, Draetta G 1993 Cyclin D1 is a nuclear protein required for cell cycle progression in G_1. Genes Dev 7:812–821

Baserga R 1990 The cell cycle: myths and realities. Cancer Res 50:6769–6771

Clarke CL, Sutherland RL 1990 Progestin regulation of cellular proliferation. Endocr Rev 11:266–302

Early Breast Cancer Trialists' Collaborative Group 1992 Systemic treatment of early breast cancer by hormonal, cytotoxic, or immune therapy. Lancet 339:1–15, 71–85

Henderson BE, Ross RK, Pike MC 1991 Toward the primary prevention of cancer. Science 254:1131–1138

Horwitz KB 1992 The molecular biology of RU486. Is there a role for antiprogestins in the treatment of breast cancer? Endocr Rev 13:146–163

Hunter T 1993 Braking the cycle. Cell 75:839–841

Inaba T, Matsushime H, Valentine M, Roussel MF, Sherr CJ, Look AT 1992 Genomic organization, chromosomal localization, and independent expression of human cyclin D genes. Genomics 13:565–574

Kato J-y, Sherr CJ 1993 Inhibition of granulocyte differentiation by G_1 cyclins D2 and D3 but not D1. Proc Natl Acad Sci USA 90:11513–11517

Leung BS, Potter AH 1987 Mode of estrogen action on cell proliferation in CAMA-1 cells. II. Sensitivity of G1 phase population. J Cell Biochem 34:213–225

Lukas J, Pagano M, Staskova Z, Draetta G, Bartek J 1994 Cyclin D1 protein oscillates and is essential for cell cycle progression in human tumour cell lines. Oncogene 9: 707–718

Motokura T, Arnold A 1993 Cyclins and oncogenesis. Biochim Biophys Acta 1155: 63–78

Musgrove EA, Sutherland RL 1993 Effects of the progestin antagonist RU486 on T-47D cell cycle kinetics and cell cycle regulatory genes. Biochem Biophys Res Commun 195:1184–1190

Musgrove EA, Wakeling AE, Sutherland RL 1989 Points of action of estrogen antagonists and a calmodulin antagonist within the MCF-7 human breast cancer cell cycle. Cancer Res 49:2398–2404

Musgrove EA, Lee CSL, Sutherland RL 1991 Progestins both stimulate and inhibit breast cancer cell cycle progression while increasing expression of transforming growth factor α, epidermal growth factor receptor, c-*fos* and c-*myc* genes. Mol Cell Biol 11:5032–5043

Musgrove EA, Hamilton JA, Lee CSL, Sweeney KJE, Watts CKW, Sutherland RL 1993 Growth factor, steroid and steroid antagonist regulation of cyclin gene expression associated with changes in T-47D human breast cancer cell cycle progression. Mol Cell Biol 13:3577–3587

Musgrove EA, Lee CSL, Buckley MF, Sutherland RL 1994 Cyclin D1 induction in breast cancer cells shortens G_1 and is sufficient for cells arrested in G_1 to complete the cell cycle. Proc Natl Acad Sci USA 91:8022–8026

Ohtsubo M, Roberts JM 1993 Cyclin-dependent regulation of G_1 in mammalian fibroblasts. Science 259:1908–1912

Pines J 1993 Cyclin and cyclin-dependent kinases: take your partners. Trends Biochem Sci 18:195–197

Pines J 1994 Arresting developments in cell-cycle control. Trends Biochem Sci 19: 143–145

Polyak K, Kato J-Y, Solomon MJ et al 1993 p27[Kip1], a cyclin–Cdk inhibitor, links transforming growth factor-β and contact inhibition to cell cycle arrest. Genes & Dev 8:9–22

Quelle DE, Ashmun RA, Shurtleff SA et al 1993 Overexpression of mouse D-type cyclins accelerates G_1 phase in rodent fibroblasts. Genes & Dev 7:1559–1571

Serrano M, Hannon GJ, Beach D 1993 A new regulatory motif in cell cycle control causing specific inhibition of cyclin D/CDK4. Nature 366:704–707

Sherr CJ 1993 Mammalian G_1 cyclins. Cell 73:1059–1065

Sutherland RL, Reddel RR, Green MD 1983a Effects of oestrogens on cell proliferation and cell cycle kinetics. A hypothesis on the cell cycle effects of antioestrogens. Eur J Cancer & Clin Oncol 19:307–318

Sutherland RL, Green MD, Hall RE, Reddel RR, Taylor IW 1983b Tamoxifen induces accumulation of MCF 7 human mammary carcinoma cells in the G_0/G_1 phase of the cell cycle. Eur J Cancer Clin Oncol 19:615–621

Sutherland RL, Hall RE, Taylor IW 1983c Cell proliferation kinetics of MCF-7 human mammary carcinoma cells in culture and effects of tamoxifen on exponentially growing and plateau-phase cells. Cancer Res 43:3998–4006

Taylor IW, Hodson PJ, Green MD, Sutherland RL 1983 Effects of tamoxifen on cell cycle progression of synchronous MCF-7 human mammary carcinoma cells. Cancer Res 43:4007–4010

Watson PH, Pon RT, Shiu RPC 1991 Inhibition of c-*myc* expression by phosphorothioate antisense oligonucleotide identifies a critical role for c-*myc* in the growth of human breast cancer. Cancer Res 51:3996–4000

Watts CKW, Sweeney KJE, Warlters A, Musgrove EA, Sutherland RL 1994 Antiestrogen regulation of cell cycle progression and cyclin D1 gene expression in MCF-7 human breast cancer cells. Breast Cancer Res Treat 31:95–105

DISCUSSION

Baulieu: Was there oestrogen in the medium in the anti-oestrogen experiments you presented in Fig. 5?

Sutherland: No. The experimental design involved taking cells that are in a serum-free, chemically defined environment and stimulating them with a single growth factor only—normally insulin—and then treating with anti-oestrogen. There will always be some debate about how much residual oestrogen is present in this system.

Baulieu: Did you use tamoxifen as the anti-oestrogen?

Sutherland: No, we used ICI 164 384 in this experiment, but tamoxifen or 4-hydroxytamoxifen have the same effect (Watts et al 1994).

Beato: You didn't show any effect of oestrogens.

Sutherland: In this experimental system (i.e. in a serum-free environment), oestrogens alone have very little effect. Therefore it is very difficult to distinguish the effect of oestrogen independently of a growth factor. However, in the presence of insulin, oestrogen stimulates cell cycle progression, c-*myc* and cyclin D1 gene expression in these T-47D cells.

Baulieu: Also, the cells you are using do not have much classical oestrogen receptor.

Sutherland: Although some T-47D lines have been reported to have low levels of oestrogen receptor, our cells have significant levels (Reddel et al 1985). We have also done similar experiments in MCF-7 cells, which are particularly sensitive to oestrogen. What one sees in MCF-7 cells is similar to what one sees in T-47D cells. So, if you treat with oestrogen alone, you get a small induction of c-*myc* and cyclin D1 mRNA with a small proportion of cells progressing to S phase, but if you treat the cells with oestrogen plus insulin or insulin-like growth factor 1 (IGF-1), you see the sort of effects described before by others (e.g. van der Burg et al 1988), where there is a synergistic effect on cell proliferation.

Beato: Do you think this may be an explanation for the effect of progestins? You also said that only a small proportion of cells go into S phase when you add progesterone in low-nutrient conditions.

Sutherland: This issue was addressed in our earlier publication (Musgrove et al 1991) where we showed that this effect of progestin is due to a shortening of G1 phase. In insulin-supplemented serum-free medium the cells are growing suboptimally; that is, they have a doubling time of 48–60 h, whereas when they are growing maximally in the presence of 5–10% fetal calf serum they have a doubling time of 24 h. Only the cycling cells are affected by progestin. If you deprive the cells of growth factor and allow them to arrest, progestins have a much smaller effect. Our explanation for this is that progestin is able to take cells that are committed to the cell cycle and shorten G1 phase, presumably by inducing cyclin D1. That's why one sees that cohort of cells entering S phase (Fig. 4).

Beato: Presumably, every cell will go at some point through the G1 phase, which will make them responsive to progesterone.

Sutherland: The problem is that there's another progestin control point just after mitosis, which is a stop signal, i.e. cell cycle progression is arrested at this point. Any cell that reaches this point cannot progress further and will not reach mid-G1 phase where the cells are sensitive to the stimulatory effect of progestin. What happens in our model system is that the cells in early mid-G1 phase are accelerated by the progestin, complete a round of replication, mitose and then arrest. It's not all that different from many experimental differentiation systems where cohorts of cells are required to complete DNA synthesis before they can enter a differentiation pathway. In many ways, progestin is a differentiation-inducing hormone.

Bäckström: Is this stop signal after mitosis always seen in cancer cells?

Sutherland: This is an important point. The stop signal seems to be predominant in endometrial epithelium. There are not many good model systems in which to study progestin effects on cell proliferation *in vitro*, so it's a difficult question to address. When we did these experiments some years ago, we were keen to try to distinguish the stimulatory effect from the inhibitory effect. We used different combinations and concentrations of antiprogestin and progestin and different timings of administration, but this was ineffective in differentiating the two effects.

Manolagas: Is it fair for us to conclude that oestrogens and anti-oestrogens, although they compete for the same receptor, affect different parameters?

Sutherland: No, I couldn't conclude that from the studies that we've done. In the presence of insulin/IGF-1, oestrogens induce responses that are inhibited by anti-oestrogens

Manolagas: Oestrogens do not have effects on any of the genes involved in cycling, and the anti-oestrogen works by blocking these genes.

Sutherland: This issue is complicated by the need for growth factors as well as oestrogen. As indicated earlier, in circumstances where oestrogens are strongly mitogenic, they induce the genes that anti-oestrogens inhibit. Additionally, one can rescue all the effects of anti-oestrogens in this system with oestrogens.

Manolagas: Can you demonstrate reversal of the effects on these particular genes?

Sutherland: Yes (Watts et al 1994).

Muramatsu: Do you have any idea of the genes that link oestrogen to cyclin D1 expression?

Sutherland: No, we don't. We are doing some work on the cyclin D1 promoter. However, there are no steroid response elements in the 2 kb of the promoter that have been characterized to date.

Castagnetta: I've noticed that you use ICI 164 384 at concentrations as high as 500 nM. This seems to be much higher than physiological concentrations, because the affinity for the oestrogen receptor commonly ranges from 10^{-11} to 10^{-10} M. Have you tested different concentrations of ICI 164 384?

Sutherland: The reason that we've used this concentration is that it gives a maximal response. We have done experiments at other concentrations, and we see the same effects, but they are of course concentration dependent (Watts et al 1994).

Castagnetta: Is there any difference between the pure anti-oestrogen and 4-hydroxytamoxifen, which may act as an antagonist or a weak oestrogen?

Sutherland: We have done these experiments with the non-steroidal anti-oestrogens tamoxifen and 4-hydroxytamoxifen and with the steroidal compounds ICI 164 384 and ICI 182 780. Essentially you get the same answers—the effects are dose dependent and reversible with oestrogen.

Castagnetta: It is surprising that the oestrogen had little effect on its own. Perhaps your results also depend more on the cell type used than the culture conditions. T-47D cells appear to be less oestrogen-responsive than MCF-7 cells. In our experience, they are sometimes lacking in oestrogen receptor and show a limited response to oestradiol. Why did you select this system? Do you think that some of your results might be attributable to this?

Sutherland: I think the system we use, like all others, has its limitations. The reason we have persisted with it is that Liz Musgrove (who did a large amount of this work when she was a PhD student in the group) has fully validated a system with which one can generate viable growth-arrested cells that can be stimulated to re-initiate cell cycle progression. T-47D cells seemed to be better for this than the MCF-7 cells, particularly because they respond to a broader spectrum of mitogens, including progestin. I wouldn't want to be dogmatic about the oestrogen effects, because we haven't looked at them in the detail that we have looked at those of progestins, antiprogestins and anti-oestrogens. All I can say is what I've said before: this cell line, which is responsive to oestrogen in terms of other parameters (like induction of progesterone receptor or other oestrogen-induced genes), does not respond to oestrogen alone with a major increase in cell cycle progression. However, in the presence of insulin/IGF-1, oestrogen increases the rate of cell cycle progression.

Horwitz: I think these are very elegant studies. At least for the progestins, I think the system is relevant. These cells clearly express a lot of progesterone receptors. My question deals with the very interesting description of the transient stimulatory effects of progestins, which are then followed by inhibitory effects: have you looked at the regulation by cyclin D1 of these two processes? How does inhibition by the progesterone agonist compare with inhibition by the anti-progestin in relation to cyclin D1 at this stage?

Sutherland: We haven't looked at this in detail. In a sense I've been slightly selective in the data that I've presented: cyclin D1 is certainly not the only gene involved. The antiprogestins have effects that are almost identical to the anti-oestrogens in terms of timing, which led us to hypothesize that they probably have effects on the same point in cell cycle progression. As it turns out, they have identical effects on c-*myc* mRNA (Fig. 5), but antiprogestin, which is able to inhibit cells to the same degree as anti-oestrogen, has no effect on cyclin D1 mRNA. We're currently doing experiments to determine whether there are effects of steroids and their antagonists on cyclin/CDK interactions and activity, but I have no new data to report. Clearly the situation is very complex and the activity of cyclin/CDK complexes can be modulated at a number of levels in addition to effects on cyclin gene expression.

Rochefort: I agree with you that you should not exclude the oestrogen as inducing cyclin D1, for instance. Our T-47D cells, in contrast to yours, are stimulated to grow quite well by oestrogen, as are our MCF-7 cells. I assume that you would find the same results in other cells with oestrogen as those you have found with progestin in your T-47D cells. Moreover, in patients, under physiological conditions progestin is rarely secreted and active in the absence of oestrogen. In the presence of oestrogen *in vitro*, progestin apparently has an anti-oestrogenic effect (Vignon et al 1983). When we tested the effects of progestin on cell proliferation in MCF-7 or T-47D cell lines, in the presence of oestrogen, progestin clearly inhibited cell growth. We showed that even in mammary cells, from fine-needle aspirates, the oestrogen receptor was down-regulated after progestin treatment, suggesting that progestin also has an anti-oestrogenic effect *in vivo* (Maudelonde et al 1991). When progestin is tested without oestrogen, it can have a stimulatory effect, but it might be relatively artificial.

Sutherland: Well it's not a stimulation in terms of being a continuous stimulation of cell proliferation. If you take a culture of T-47D cells, treat them with progestins and monitor cell proliferation over a number of cell cycles, there is inhibition of cell growth, because once the cells have gone through the first round of replication they arrest. However, if you look very closely during the first 24 h of progestin treatment, where you get progestin stimulation of cell cycle progression, cell numbers actually increase relative to control as approximately 50% of the cell population has progressed through the cell cycle and divided.

Rochefort: When you stimulate with insulin and observe induction of cyclin D1, do you assume that cyclin D1 is an immediate-early gene, or that this effect would be indirect and be mediated by another factor? Have you done experiments with cycloheximide?

Sutherland: No, cyclin D1 is not considered to be an immediate-early gene. In the situations where we see stimulation of cyclin D1, induction of c-*fos* and c-*myc* precedes it. A very important issue therefore is whether or not cyclin D1 induction is a consequence of the induction of c-*myc*; we've constructed the cell lines to address this question.

Parker: Is CDK4 the only kinase complexed with the D1 cyclin during G1?

Sutherland: We've screened all the breast cancer cell lines that we've worked with (there are about 20 of them); they express variable levels of CDK5 and CDK6, as well as CDK4.

Parker: Is CDK5 or CDK6 ever found in a complex with cyclin D1 in breast cancer cell lines?

Sutherland: We haven't done this with all the cell lines, but we've seen in the T-47D cells that CDK4 is the predominant CDK immunoprecipitated with cyclin D1.

Parker: Can the expression of cyclin D1 mRNA and/or protein be increased by any hormone without the cells progressing into S phase?

Sutherland: We haven't seen that.

These are preliminary data—I wouldn't want to be quoted in great detail. In the experiments that we've done, if you turn on cyclin D1, then you later see the cells progress into S phase.

Parker: Part of the problem with our MCF-7 cells is that even after serum deprivation they still continue to proliferate for some time.

Sutherland: It's another limitation of the system. You've got to get these questions in context. The reason that we are doing these experiments in these cell lines is twofold—first, we're interested in how these compounds exert their effects on breast cancer cells as opposed to other cell types unrelated to breast cancer; secondly, amplification and overexpression of cyclin D1 is one of the most common molecular lesions in human breast cancer. Consequently, in doing these experiments we are asking the question: what happens if you take a transformed breast cancer cell on a background of normal expression of cyclin D1 and then express the gene? How does this change the phenotype? That's a different sort of question to those that others are asking in fibroblasts, for instance, about what cyclin D1 does to cell cycle progression in general.

Bonewald: What is the relevance of the anti-oestrogen binding site? Does it have any relevance to this system at all?

Sutherland: Since we discovered anti-oestrogen binding sites I feel no guilt in saying that I don't think they have any major role. There are some ligands that inhibit cell proliferation and bind to those sites preferentially which could be put into this experimental system. But since the effects we see are reversed by oestrogen, it's unlikely that a specific anti-oestrogen binding site is involved.

Bonewald: You said that even the ICI compounds work the same way in this system. Do they also bind to the anti-oestrogen binding sites? Do the binding sites recognize all equally, with the same affinity?

Sutherland: The anti-oestrogen binding site we described is a specific binding site for triphenylethylene anti-oestrogens (Sutherland et al 1980). The steroidal anti-oestrogens do not bind to that particular binding site. Having said this, we've recently been collaborating with another group in Australia who have described a specific non-oestrogen-competable binding site for the steroidal anti-oestrogens (Parisot et al 1995). The function of this molecule is also unknown.

Beato: Although the endometrium is a reproductive organ and this symposium is on non-reproductive actions, it is interesting to think about this progestin effect in the endometrium, because there are indications that in endometrial cells you can stimulate cell growth by adding progestin.

Sutherland: In old experiments done 20–25 years ago, Len Martin and others looked very carefully at cell cycle progression and mitosis in mature and adult rats and mice using tritiated thymidine incorporation and autoradiography. They showed quite clearly that in the endometrium, progestin is an antagonist of cell cycle progression and will arrest oestrogen-stimulated cells by mechanisms that are similar in their cell cycle phase specificity to the inhibitory effects we have described in breast cancer cells (Sutherland et al 1988, Musgrove et al 1991).

Beato: In the presence of oestrogens, progestins act as anti-oestrogens and block proliferation. I'm talking about cells in culture that do not require oestrogen, but respond to progesterone. We have endometrial cell lines that grow much better with progesterone.

Baulieu: This can be seen in many cells.

Manolagas: There are suggestions that oestrogens, at least in haemopoietic cells, have effects on apoptosis. Have you seen any effects of anti-oestrogens on apoptosis?

Sutherland: There are two components to the steroid/steroid antagonist effects on cell proliferation: there are effects on cell cycle progression (i.e. rates of cell birth) and there are effects on apoptosis (i.e. rates of cell death). We have not, to date, addressed the second issue, which is obviously a very important one. As you know, in reproductive tissues in particular, the withdrawal of steroid hormones induces some of the most overt apoptosis that is seen in any normal tissue. We should and will address this important issue as it relates to steroids and their antagonists.

References

Maudelonde T, Lavaud P, Salazar G, Laffargue F, Rochefort H 1991 Progestin treatment depresses estrogen receptor but not cathepsin D levels in needle aspirates of benign breast disease. Breast Cancer Res Treat 19:95–102

Musgrove EA, Lee CSL, Sutherland RL 1991 Progestins both stimulate and inhibit breast cancer cell cycle progression while increasing expression of transforming growth factor α, epidermal growth factor receptor, c-*fos* and c-*myc* genes. Mol Cell Biol 11:5032–5043

Parisot JP, Hu XF, Sutherland RL, Wakeling A, Zalcberg JR, DeLuise M 1995 The pure antioestrogen ICI 182,780 binds to a high affinity binding site distinct from the oestrogen receptor. Submitted

Reddel RR, Murphy LC, Hall RE, Sutherland RL 1985 Differential sensitivity of human breast cancer cell lines to the growth inhibitory effects of tamoxifen. Cancer Res 45:1525–1531

Sutherland RL, Murphy LC, Foo MS, Green MD, Whybourne AM, Krozowski ZS 1980 High-affinity anti-oestrogen binding site distinct from the oestrogen receptor. Nature 288:273–275

Sutherland RL, Hall RE, Pang GYN, Musgrove EA, Clarke CL 1988 Effect of medroxyprogesterone acetate on proliferation and cell cycle kinetics of human mammary carcinoma cells. Cancer Res 48:5084–5091

van der Burg B, Rutteman GR, Blankenstein MA, de Laat SW, van Zoelen EJ 1988 Mitogenic stimulation of human breast cancer cells in a growth factor-defined medium: synergistic action of insulin and estrogen. J Cell Physiol 134:101–108

Vignon F, Bardon S, Chalbos D, Rochefort H 1983 Antiestrogenic effect of R5020, a synthetic progestin, in human breast cancer cells in culture. J Clin Endocrinol & Metab 56:1124–1130

Watts CKW, Sweeney KJE, Warlters A, Musgrove EA, Sutherland RL 1994 Antiestrogen regulation of cell cycle progression and cyclin D1 gene expression in MCF-7 human breast cancer cells. Breast Cancer Res Treat 31:95–105

Surprises with antiprogestins: novel mechanisms of progesterone receptor action

Kathryn B. Horwitz, Carol A. Sartorius, Alicia R. Hovland, Twila A. Jackson, Steve D. Groshong, Lin Tung and Glenn S. Takimoto

University of Colorado Health Sciences Center, Departments of Medicine and Pathology, and the Molecular Biology Program, Division of Endocrinology, 4200 East Ninth Avenue, Campus Box B151, Denver, CO 80262, USA

Abstract. When hormone antagonists have inappropriate agonist-like effects, the clinical consequences are grave. We describe novel molecular mechanisms by which antiprogestin-occupied progesterone receptors behave like agonists. These mechanisms include agonist-like transcriptional effects that do not require receptor binding to DNA at progesterone response elements, or that result from cross-talk between progesterone receptors and other signalling pathways. We discuss the complex structural organization of progesterone receptors, and demonstrate that the B receptor isoform has a unique third activation domain that may confer agonist-like properties in the presence of antiprogestins, whereas the A receptor isoform is a dominant-negative inhibitor. We argue that these novel mechanisms play a role in the apparent hormone resistance of breast cancers and the variable tissue-specific responses to antagonists.

1995 Non-reproductive actions of sex steroids. Wiley, Chichester (Ciba Foundation Symposium 191) p 235–253

The mechanisms by which steroid hormone antagonists produce unexpected agonist-like effects, or different tissue-specific effects, are unknown, but have important clinical implications. For example, tamoxifen, the oestrogen antagonist used widely to treat breast cancers, is an agonist in bone and uterus and has oestrogenic effects on lipid and lipoprotein levels. Tamoxifen can even be an agonist in breast cancers, producing undesirable side effects which exacerbate the disease. Thus, at the start of tamoxifen therapy patients often experience an oestrogenic tumour flare, and after long-term tamoxifen therapy inappropriate proliferative effects camouflage as 'resistance' (Horwitz 1993). Antiprogestins may also prove to be useful hormonal agents for the treatment of breast cancer (Horwitz 1992). However, as with tamoxifen, agonist-like proliferative effects have been reported with the progesterone antagonist RU486 in cultured breast cancer cell lines and in postmenopausal women, under conditions where inhibition would be expected (Gravanis et al

1985, Meyer et al 1990). Recent studies in our laboratory have addressed the molecular mechanisms by which these occur. They focus on the two natural isoforms of human progesterone receptors: B receptors, which are 933 amino acids in length, and A receptors, which lack the N-terminal 164 amino acids. When both A and B isoforms are present in equimolar amounts in wild-type progesterone receptor-positive cells, or are transiently co-expressed in progesterone receptor-negative cells, they dimerize and bind DNA as three species: A/A and B/B homodimers, and A/B heterodimers (Horwitz 1993 and references therein). This heterogeneity has complicated the study of antiprogestins.

This paper reviews our recent work with antiprogestins and human progesterone receptors (Tung et al 1993, Sartorius et al 1993, 1994a,b, Mohamed et al 1994). It demonstrates the extraordinary complexity of antiprogestin action and emphasizes the fact that steroid receptors regulate transcription through multiple mechanisms, some of which may not require direct interaction of the receptors with DNA at canonical hormone response elements, as previously thought.

Conventional actions of antiprogestins

Two fundamentally different mechanisms underlie the actions of antagonist–human progesterone receptor complexes. First is the classical effect of antagonists; namely, their ability to inhibit agonist actions directly. In this scenario, agonist-occupied progesterone receptors regulate transcription by binding as dimers to progesterone response elements present on the regulated gene. Antagonist-occupied progesterone receptor complexes also bind to progesterone response elements but are non-productive. Thus, by this mechanism, antagonist inhibition involves competition between the two ligands, agonist versus antagonist, for human progesterone receptor occupancy, followed by the competition between the two ligand–human progesterone receptor classes for binding to progesterone response elements. With agonists, DNA binding leads to a specific transcriptional response, whereas with antagonists, the DNA binding is non-productive. In the latter case, the non-productive or inhibitory potency of an antagonist is controlled by numerous factors, which include its affinity for the receptors, the affinity of antagonist-occupied progesterone receptor complexes for progesterone response elements, the number and occupancy of progesterone response elements on a promoter, and probably other factors (Horwitz 1993 and references therein).

Novel transactivation by antagonist-occupied B receptors without binding a progesterone response element

Additional data suggest that alternative mechanisms exist by which antagonist effects are mediated. By one of these, antagonist-occupied progesterone receptor complexes have inadvertent transcriptional stimulatory actions through DNA-binding sites or DNA-binding proteins that do not involve the canonical progesterone response elements. These novel mechanisms could, in theory, affect not only genes that contain progesterone response elements, but also genes that were never meant to be regulated by progesterone receptors and upon which agonists have no effects. Such mechanisms could explain how an antiprogestin can have effects on genes that are not targets of progesterone. There are several experimental models that demonstrate these unusual mechanisms.

We have studied the transient transcriptional activity of antagonist-occupied human progesterone receptors using a chloramphenicol acetyltransferase (CAT) reporter driven by a progesterone response element cloned upstream of the thymidine kinase (tk) gene promoter (Tung et al 1993). Treatment of HeLa cells expressing A receptors with the agonist R5020 leads to a 20-fold increase in transcription compared with basal levels, whereas none of three antiprogestins, RU486, ZK112993 or ZK98299, stimulate transcription. Instead, the antiprogestins typically suppress basal levels of transcription. However, in cells expressing B receptors, not only the agonist, but also all three antagonists strongly stimulate transcription. We were surprised that ZK98299 was a transcriptional activator because, in our hands, receptors occupied by this antagonist did not bind to DNA *in vivo* or *in vitro*. This suggested that transcriptional activation by antagonist-occupied B receptors was independent of binding to progesterone response elements. We therefore removed the progesterone response element from the promoter–reporter constructs. As expected, this eliminated the agonist-dependent transcription but, to our surprise, the anomalous antagonist-dependent transactivation was retained. Similar results were observed with a DNA-binding domain progesterone receptor mutant, whose specificity was altered so that it would no longer recognize a progesterone response element but would instead bind an oestrogen response element. When occupied by antiprogestins, this mutant still activated transcription of the progesterone response element-containing reporter.

Recent data show that other members of the steroid receptor superfamily can have effects that are independent of the canonical hormone response elements. Potential mechanisms fall into two broad categories: either the receptors bind to novel DNA sites that differ substantially from the consensus hormone response elements, or the receptors do not bind DNA at all, but interact with other DNA-binding proteins instead (Oro et al 1988, Sakai et al 1988, Diamond et al 1990a, Yang-Yen et al 1990, Jonat et al 1990, Schüle et al 1990, Miner & Yamamoto 1991, Kutoh et al 1992). By the latter mechanism, termed factor tethering, two factors establish protein–protein contacts on the DNA, but only one of the two actually binds DNA. However, both the DNA-bound protein and its tethered partner contain a DNA-binding domain. This model is of particular significance for transcription mediated by antagonist-occupied B receptors, because here we also find a requirement for an intact DNA-binding domain. Thus, a B receptor mutant lacking an ordered first zinc finger fails to stimulate transcription when occupied by RU486. In addition to its DNA-binding function, the DNA-binding domain of steroid receptors is implicated in mediating protein–protein interactions (Diamond et al 1990a, Yang-Yen et al 1990, Schüle et al 1990), perhaps through conserved surfaces that face away from the DNA. Indeed, several recent studies show that gluco-corticoid receptors and c-Jun repress one another's activity by protein–protein binding mechanisms that are independent of DNA binding. Nevertheless, to produce repression, an intact glucocorticoid receptor DNA-binding domain is required. Additionally, a dimerization function has been assigned to the second zinc finger (Härd et al 1990), providing further evidence that the DNA-binding domain mediates protein–protein interactions. We speculate that induction of transcription by antagonist-occupied B receptors can proceed through a mechanism in which the receptors are tethered to a DNA-bound protein partner, but do not bind DNA themselves. Alternatively, progesterone B receptors could function by linking an activator protein, bound to the tk promoter, to the basal transcriptional machinery.

Antagonist-occupied human progesterone B receptors bound to DNA are functionally switched to transcriptional agonists by cAMP

In contrast, we have described an entirely different antagonist-mediated activation mechanism, in which human progesterone receptors do have to be bound to DNA. Because we have a specific interest in the actions of steroid antagonists in breast cancer, we studied antiprogestins in a derivative line of T-47D human breast cancer cells which express high endogenous levels of A and B receptors and which stably express the mouse mammary tumour virus (MMTV) promoter cloned upstream of the CAT gene (Sartorius et al 1993). Treatment of these cells with the agonist R5020 produces high levels of CAT. When tested alone, the three antiprogestins RU486, ZK98299 and ZK112993 are unable to stimulate transcription, and all three inhibit R5020-mediated transcription. Thus, in this model, all three antiprogestins are good antagonists.

However, when cellular cAMP levels are raised, two of the antagonists demonstrate a surprisingly strong agonist activity: when present alone, RU486 and ZK112993 are transcriptionally inactive, but in the presence of 8-bromocyclic AMP, their transcription is agonist-like. Of interest is the fact that ZK98299 is entirely different, and despite elevated cAMP levels, this antagonist does not function as an agonist. Recall that ZK98299-occupied progesterone receptors either do not bind to DNA at progesterone response elements or have anomalous DNA-binding properties. From this and other controls we deduce that in order for antagonist-occupied progesterone receptors to become transcriptional activators under cAMP control, the receptors have to be bound to DNA.

The amplification of steroid-mediated responses in the presence of cAMP is not limited to progesterone receptors. Progesterone receptors are overexpressed in T-47D cells but the levels of glucocorticoid, androgen and oestrogen receptors are extremely low. In addition to progesterone receptors, the progesterone response elements of the MMTV promoter can be regulated by androgen and glucocorticoid receptors (Cato et al 1986, 1987). However, the MMTV promoter lacks an oestrogen response element and is not regulated by oestrogen receptors. In T-47D cells expressing the MMTV-CAT reporter, neither dexamethasone nor dihydrotestosterone stimulate CAT transcription, suggesting that levels of glucocorticoid and androgen receptors are too low in these cells to activate this promoter in the absence of other influences. However, when cAMP concentrations are raised, the cells acquire sensitivity to the steroid hormones, resulting in strong transcription. Thus, 8-bromocyclic AMP sensitizes the MMTV promoter to the actions of glucocorticoids and androgens. In contrast, no transcriptional amplification is seen with oestradiol, consistent with the inability of oestrogen receptors to bind the MMTV promoter. This again suggests that the cooperative effects of 8-bromocyclic AMP require that the receptors be bound to DNA.

Signal transduction pathways ultimately converge at the level of transcription to produce patterns of gene regulation that are specific to the gene and cell in question. Composite promoters may be regulated by multiple independent and interacting factors. In extreme cases, a transcription factor can yield opposite regulatory effects from one DNA-binding site owing to modulation by a second factor. A case in point is glucocorticoid receptors, which regulate proliferin gene transcription either positively or negatively. The direction of transcription by glucocorticoids is selected by DNA-bound Jun and Fos, which are postulated to interact with glucocorticoid receptors at progesterone response elements. cAMP-responsive signal transduction pathways are often involved in such cooperative interactions. These models suggest that on complex promoters, non-receptor factors (among which are cAMP-regulated proteins), can interact with steroid receptors to select the direction of transcription (Diamond et al 1990b, Gruol et al 1986, O'Shea et al 1989).

Our studies demonstrate that cAMP can both amplify the transcriptional signals of agonist-occupied steroid receptors and switch the transcriptional direction of some antiprogestins to render them potent agonists; an effect that can have unintended clinical consequences. We believe that this functional reversal requires that progesterone receptors bind to DNA, and that it is not due to ligand-independent phosphorylation or direct activation of the receptors by protein kinase A-dependent pathways. We find that elevated cAMP levels do not enhance phosphorylation of human progesterone receptors in breast cancer cells, neither do they modulate the hormone-dependent phosphorylation induced by progestins (Sartorius et al 1993). We therefore conclude that cAMP does not directly influence the activity of human progesterone receptors by phosphorylating them. Instead, our data are consistent with a model in which the direction of transcription by DNA-bound progesterone receptors is indirectly regulated by coactivator proteins whose activity is perhaps controlled by cAMP-dependent phosphorylation. This cooperativity between two signal transduction pathways, one involving steroid receptors, the other involving cAMP-regulated proteins, requires that the steroid receptors bind to DNA. It therefore does not occur on the MMTV promoter with oestrogen receptors, or when progesterone receptors are occupied by ZK98299.

New T-47D breast cancer cell lines for the independent study of progesterone A and B receptors: only antiprogestin-occupied B receptors are switched to transcriptional agonists with cAMP

The studies with wild-type T-47D cells described above do not permit analysis of the relative contributions of the A and B receptors to the synergism observed with cAMP, since these cells contain mixtures of the two receptors. However, their constitutive high level of production of progesterone receptors have made T-47D cells the major model used for the study of progesterone in breast cancer, unencumbered by the need for oestradiol priming. Because of several special phenotypic properties of T-47D cells and because factors other than receptors may be missing in persistently receptor-negative cells, we thought it prudent to retain the T-47D cellular milieu in developing new models to study the independent actions of the two progesterone receptor isoforms (Sartorius et al 1994a). First, we needed a progesterone receptor-negative T-47D subline. We developed a monoclonal progesterone receptor-negative cell line, called T-47D-Y, by selecting a progesterone receptor-negative subpopulation from a parental T-47D line that contained mixed positive and negative cells, as identified by flow cytometry. T-47D-Y cells are progesterone receptor-negative immunologically and by ligand binding assays, by growth resistance to progestins, by failure to bind a progesterone response element *in vitro*, and by failure to transactivate progesterone response element-regulated promoters.

T-47D-Y cells were then stably transfected with expression vectors encoding one or the other progesterone receptor isoform, and two monoclonal cell lines were selected that express only B receptors (called T-47D-YB) or only A receptors (called T-47D-YA). The ectopically expressed receptors are properly phosphorylated and, as do endogenously expressed receptors, they undergo ligand-dependent down-regulation. The expected B/B or A/A homodimers are present in cell extracts from each cell line, but A/B heterodimers are missing in both (Sartorius et al 1994a).

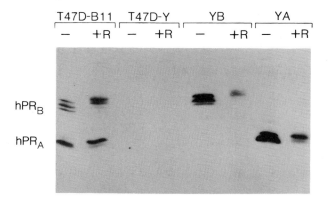

FIG. 1. Progesterone receptor content and structural analysis of wild-type T-47D cells (B11) and the three new T-47D cell lines (T-47D-Y, -YB and -YA) by immunoblotting. The cell lines indicated were treated (+ R) or not treated (−) with 100 nM of the agonist R5020 for one hour, and then harvested. Total cellular receptors were extracted by freeze-thawing with 0.4 M KCl, de-salted, resolved by SDS–PAGE, and electroblotted to nitrocellulose. The nitrocellulose sheet was probed with the anti-progesterone receptor antibody AB-52, and protein bands were detected by enhanced chemiluminescence and autoradiography. hPR$_A$, human progesterone A receptor; hPR$_B$, human progesterone B receptor. (Reproduced with permission from Sartorius et al 1994a.)

An immunoblot (Fig. 1) demonstrates the equimolar mixture of B and A receptors that are present in wild-type T-47D cells (Fig. 1, lanes 1 and 2), the absence of either receptor isoform in T-47D-Y cells (Fig. 1, lanes 3 and 4) and the unique presence of one or the other receptor isoform in T-47D-YB and T-47D-YA cells (Fig. 1, lanes 5–8). The levels of each isoform in YB and YA cells are approximately the same as the levels of that isoform in wild-type T-47D cells. The structure of the receptors in YB and YA cells is also analogous to that of the wild-type receptors. The triplet banding pattern of wild-type B receptors is retained in the ectopically expressed B receptors of hormone untreated (− R) YB cells, and the characteristic upwards shift in molecular weight of B and A receptors produced by R5020 treatment (+ R) is also retained in the new cell lines. Both of these structural features are due to receptor phosphorylation.

The new cell lines were used to study isoform-specific effects of agonists and antagonists when cAMP levels are raised. In the experiment shown in Fig. 2, YA or YB cells were transiently transfected with the MMTV-CAT reporter and treated with R5020 or the three antiprogestins in the presence or absence of 8-bromocyclic AMP. 8-Bromocyclic AMP alone does not stimulate CAT synthesis in either cell line. In YA cells, R5020 alone moderately stimulates CAT transcription from MMTV-CAT, and the agonist effect is synergistically

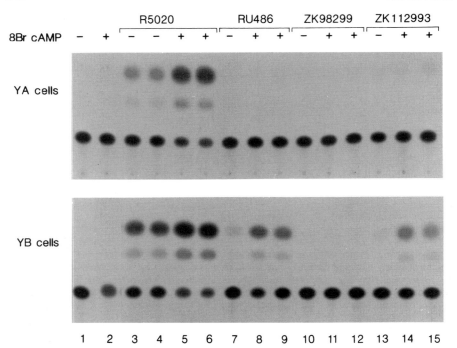

FIG. 2. When cyclic AMP levels are elevated, RU486 and ZK112993 but not ZK98299 are strong transactivators in T-47D-YB cells and not in T-47D-YA cells. T-47D-YA (*top*) and T-47D-YB (*bottom*) cells were transiently transfected with the MMTV-CAT (mouse mammary tumour virus–chloramphenicol acetyltransferase) reporter construct and treated with the steroid hormone indicated in the presence (+) or absence (−) of 1 mM 8-bromocyclic AMP (8Br cAMP); CAT activity levels normalized to β-galactosidase activity were measured by thin-layer chromatography. Hormone concentrations were 50 nM for R5020, RU486 and ZK112993, and 100 nM for ZK98299. YA cells contain progesterone A receptors only; YB cells contain B receptors only. (Reproduced with permission from Sartorius et al 1994a.)

enhanced when cAMP levels are raised. Thus, agonist-occupied A receptors are relatively weak transactivators whose activity is strongly enhanced by cAMP. When only A receptors are available (as in YA cells), the three antiprogestins RU486, ZK98299 and ZK112993 (Fig. 2, lanes 7–15) have no intrinsic agonist-like activity and 8-bromocyclic AMP does not alter this.

However, the agonist and antagonists have quite different effects in the B receptor-containing YB cells. R5020-regulated transcription from the MMTV-CAT reporter is very strong in these cells, making the cAMP-mediated synergism more difficult to observe than in YA cells. In contrast, in YB cells, 8-bromocyclic AMP strongly enhances the transcriptional phenotype of the antagonists RU486 and ZK112993. Both of these antiprogestins are weak agonists on the MMTV-CAT promoter, but become strong agonists when 8-bromocyclic AMP is added (Fig. 2, lanes 8, 9, 14 and 15). The antagonist ZK98299 is entirely different since on its own it has no intrinsic agonist activity and no enhancement is produced by 8-bromocyclic AMP (Fig. 2, lanes 12 and 13). This resistance of ZK98299 in YB cells to the activating effects of cAMP is similar to that which we described in wild-type T-47D cells that contain the natural mixture of both receptors.

These studies show that the two progesterone receptor isoforms behave differently in their cooperativity with cAMP. With regard to R5020, the synergism between cAMP and agonist-occupied receptors is most pronounced in YA cells. We speculate that in YA cells, cAMP sensitizes the MMTV promoter to the weak signal transmitted by R5020-occupied A receptors; this is similar to the manner in which cAMP amplifies the weak signals transmitted by hormone-occupied glucocorticoid and androgen receptors in wild-type T-47D cells. Because agonist-occupied B receptors in YB cells are already strong transactivators, cAMP has only modest further effects on this isoform.

With regard to progesterone antagonists, the isoform specificity of the cAMP effect is more interesting. We find that cAMP has absolutely no effect in YA cells, perhaps because the antagonists (specifically RU486 and ZK112993) exhibit no agonist-like activity on A receptors. Is there no minimal signal for cAMP to amplify? In contrast, the two antagonists appear to have some weak agonist-like activity in YB cells; hence, cAMP strongly amplifies this signal, converting the antagonist-occupied B receptors to potent transactivators. Therefore, it is significant that B receptors occupied by the antiprogestin ZK98299 are not subject to this functional modulation by cAMP. We speculate that ZK98299-occupied progesterone receptors are physically removed from cAMP control by their failure to bind DNA. This again implies that cooperativity between DNA-bound progesterone receptors and a cAMP-regulated coactivator can account for the transcriptional synergism.

A receptors are transdominant repressors of B receptors
without binding a progesterone response element

If B and A receptors are so different, what happens when the two are mixed? To determine the effects of A receptors on antagonist-stimulated transcription by B receptors, we co-transfected expression vectors encoding B receptors and increasing levels of A receptors into HeLa cells together with the PRE-tk-CAT reporter, and the cells were treated with either R5020 or RU486. B receptor isoforms alone stimulate CAT transcription in this model, whether the receptors are occupied by agonist or antagonist, whereas A receptors alone are stimulatory only when thay are agonist occupied. In fact, when RU486 is bound to the progesterone A receptor, transcription is always suppressed below basal levels. When the two receptor isoforms are coexpressed, strong transcription is maintained in the presence of the agonist, regardless of the B : A ratio. However, in the presence of the antagonist and at approximately equimolar amounts of the two receptors, the transcriptional phenotype of A receptors predominates, so that B receptor-stimulated transcription is almost entirely extinguished (Tung et al 1993).

When A and B receptors are equimolar, a 1 : 2 : 1 ratio of A/A, A/B and B/B dimers is expected. The extensive inhibition by A receptors suggested that A/B heterodimers have the same inhibitory transcriptional activity as A/A homodimers, and that only B/B homodimers are stimulatory. However, the presence of the two competing homodimeric species complicates functional analysis of the heterodimers, and B/B homodimers probably account for the incomplete suppression of transcription seen when A and B receptors are co-expressed. We therefore decided to construct receptors in which the heterodimeric species was the only class present.

When they are mixed, c-Jun and c-Fos preferentially form heterodimers over homodimers by at least 1000-fold (O'Shea et al 1989). Therefore, to force heterodimerization of progesterone receptors, we fused the leucine zippers of either c-Fos or c-Jun to the C-terminus of A or B receptors (Mohamed et al 1994). These chimeric progesterone receptors retain agonist and antagonist binding capacity. When agonist or antagonist-occupied A receptor–Jun and B receptor–Fos are each expressed alone, they have the same transcriptional phenotype as the wild-type receptors. However, when the two are co-transfected, the weak residual transcription seen with wild-type RU486-occupied B/A receptor mixtures is entirely eliminated. Thus, CAT levels are reproducibly below control values with B–Fos/A–Jun. These data confirm the A dominance hypothesis and show that antagonist-occupied pure A/B heterodimers exhibit exclusively the inhibitory transcriptional phenotype of antagonist-occupied A/A homodimers.

The dominance of A receptors is observed even when the antagonist used is ZK98299. The strong progesterone response element binding-independent transcriptional stimulation imparted by ZK98299-occupied B receptors in the tk promoter model is 80% suppressed by approximately equimolar concentrations of A receptors, and fully suppressed by a twofold molar excess of A receptor. Because ZK98299-occupied A receptors do not bind to a progesterone response element, these data imply that inhibitory effects of antagonist-occupied A receptors, similarly to the stimulatory effects of antagonist-occupied B receptors, are mediated by novel mechanisms that are not dependent on progesterone response elements. This was confirmed by experiments in which the antagonist-occupied A receptor DNA-binding domain specificity mutant, which cannot bind a progesterone response element, was used as the competing receptor species. On the PRE-tk-CAT construct, activation of CAT transcription by RU486-occupied wild-type B receptors was completely inhibited by this mutant.

Our experiments demonstrate that A receptors can inhibit the activity of B receptors (Tung et al 1993). In related studies it has been shown that A receptors inhibit not only B receptors, but also the activities of other members of the steroid receptor family, including oestrogen receptors (Vegeto et al 1993, McDonnell & Goldman 1994). Thus, the dominant inhibitory effects of A receptors are extensive, and may explain some of the 'anti-oestrogenic' actions reported for antiprogestins.

A third transactivation function (AF3) of human progesterone receptors located in the unique N-terminal segment of the B isoform—the B-upstream segment

Why do B receptors differ from A receptors? We postulated that the unique 164 amino acid B-upstream segment (BUS) is in part responsible for the functional differences between the two isoforms. With this in mind, we constructed a series of human progesterone receptor expression vectors encoding BUS fused to individual downstream functional domains of the receptors (Fig. 3A) (Sartorius et al 1994b). These include the two transactivation domains, AF1 located in a 90 amino acid segment just upstream of the DNA-binding domain (DBD) and nuclear localization signal (NLS); and AF2 located in the hormone-binding domain. BUS is a highly phosphorylated domain and contains the serine residues responsible for the B receptor triplet protein structure. The construct containing BUS-DBD-NLS binds tightly to DNA when aided by accessory nuclear factors. In HeLa cells, BUS-DBD-NLS strongly and autonomously activates CAT transcription from a promoter containing two progesterone response elements (PRE_2-$TATA_{tk}$-CAT) (Fig. 3B). This experiment shows that the empty expression vector is inactive (lane 1), and that B receptors are inactive in the absence of hormone (lane 4), but are strongly active in the presence of R5020 (lanes 2–6). Cells transfected with BUS-NLS lacking the DBD show no CAT activity over basal levels (lanes 7 and 8), but with BUS-DBD-NLS a dose-dependent increase in transcriptional activity is observed (lanes 9–18); at the highest plasmid concentrations (lanes 15–18), BUS-DBD-NLS constitutively activates transcription to levels comparable to those of hormone-activated, full-length B receptors. Thus, we conclude that this construct contains an autonomous third transactivation function, AF3.

In HeLa cells, transcription levels with BUS-DBD-NLS are equivalent to those seen with full-length B receptors, and are higher than those seen with A receptors. Additional studies show that BUS specifically requires an intact progesterone receptor DNA-binding domain in order to be transcriptionally active. DNA-binding domain mutants that cannot bind DNA, or whose DNA-binding specificity has been altered, cannot cooperate in BUS transcriptional activity. This suggests that autonomous AF3 activity resides in a discontinuous

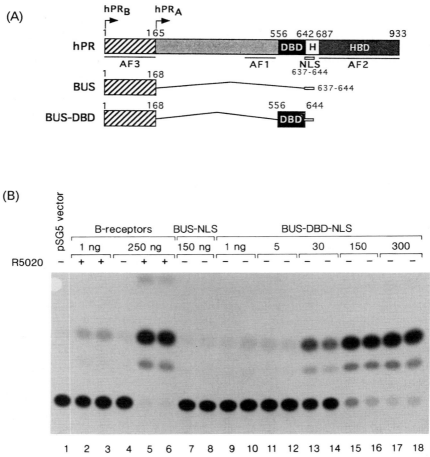

FIG. 3. The BUS-DBD-NLS construct contains a third human progesterone receptor (hPR) transcription activation function, AF3. (A) Structure of the fusion proteins tested. BUS, B-upstream segment; AF, activation domain; DBD, DNA-binding domain; NLS, nuclear localization signal; HBD, hormone-binding domain; H, hinge region. (B) HeLa cells were co-transfected with $2\,\mu g$ of $PRE_2\text{-}TATA_{tk}\text{-}CAT$ reporter construct and the empty expression vector pSG5 (lane 1), or the indicated concentrations of expression vectors encoding hPR B receptors (hPR_B) (lanes 1–6), BUS-NLS (lanes 7 and 8), or BUS-DBD-NLS (lanes 9–18). Twenty-four hours after transfection, the medium was replenished and cells were either untreated ($-$) or treated with $50\,nM$ of the agonist R5020 ($+$). Cell lysates were normalized to β-galactosidase activity and CAT (chloramphenicol acetyltransferase) assays were performed by thin-layer chromatography. (Reproduced with permission from Sartorius et al 1994b.)

domain formed from BUS and the progesterone receptor DNA-binding domain. We also find that the autonomous function of BUS-DBD-NLS is promoter and cell specific. BUS-DBD-NLS does not transactivate MMTV-CAT in HeLa cells, and only poorly transactivates PRE_2-$TATA_{tk}$-CAT in the progesterone receptor-negative T-47D-Y breast cancer cells. In the latter, however, transcription can be restored either by elevating cellular levels of cAMP, or by linking BUS to AF1 or AF2, each of which alone is also inactive in T-47D-Y cells. Thus, although in T-47D-Y cells each AF alone is weak, when AF3 is linked to either of the other two AFs (AF3 + AF1 or AF3 + AF2), strong transcriptional activation is regenerated, which is approximately equal to that obtained with B receptors. These data suggest that in the appropriate cell or promoter context, BUS can supply an important transactivation function in two different ways: either by autonomously activating transcription in the absence of the other two AFs, as it does in HeLa cells on the PRE_2-$TATA_{tk}$-CAT construct; or by synergizing with the other AFs on the progesterone receptor molecule, as it does in T-47D-Y cells on the PRE_2-$TATA_{tk}$-CAT construct (Sartorius et al 1994b). Is it the autonomous function that produces agonist-like effects from antagonist-occupied B receptors?

Summary

We are beginning to accept that one model alone cannot possibly describe the actions of steroid receptors. The conventional model, which depicts receptors as ligand-activated proteins that bind to specific DNA sequences at 'consensus' hormone response elements and activate transcription, is not incorrect. It is, however, oversimplified and we now appreciate that other models are also applicable. This should not be surprising given the complex regulatory demands on these receptors. These demands include requirements for both positive and negative transcriptional regulation, for tissue specificity of action and for regulation of composite and simple gene promoters. It should also not be surprising given the complex structural organization of these proteins. This includes multiple covalent modifications by phosphorylation, and multiple functional domains that control intramolecular contacts, intermolecular protein–protein interactions and DNA binding. Finally, given the fact that steroid antagonists are synthetic rather than natural hormones, it is perhaps not surprising that their binding produces structural alterations in the receptors that unveil additional novel interactive capabilities. Thus, although antiprogestins can indeed competitively inhibit agonists by forming non-productive receptor–DNA complexes, this is not their sole mechanism of action. Depending on the promoter and cell type regulated, antiprogestin effects may also be mediated by receptor interactions with co-activators whose function is in turn controlled by non-steroidal signals. Therefore, when two different signalling pathways are simultaneously activated, they can cooperate to produce unintended effects.

Additionally, it seems clear from several studies that antagonist-occupied receptors can act without binding to canonical progesterone response elements—or without binding to DNA at all—relying perhaps on tethering proteins. This may be a consequence of the unusual allosteric structure imparted on the receptors by synthetic ligands. Because of these and undoubtedly other mechanisms yet to be discovered, the most serious mistake that investigators in this field can make when studying antiprogestins is to assume that a specific mechanism is operating.

With respect to protein structure, we are only beginning to appreciate receptor complexity. For example, it appeared at first blush that the structural independence of functional domains permitted us to analyse receptor fragments by fusing them to heterologous proteins. However, we now know that important functional domains can overlap, that other functional domains may be discontinuous and that one domain can modulate the activity of another. This means that analysis of receptor fragments in chimeras is an incomplete test of domain function and that we need innovative experimental strategies to understand this intramolecular cross-talk. Finally, what could be more unexpected than finding that one receptor isoform can inhibit not just its mate, but even distantly related cousins! Stay tuned for more surprises from this fascinating protein family.

Acknowledgements

The studies described herein were generously supported by the National Institutes of Health through grants CA26869, CA55595 and DK48238, and by the National Foundation for Cancer Research.

References

Cato ACB, Miksicek R, Schütz G, Arnemann J, Beato M 1986 The hormone regulatory element of mouse mammary tumor virus mediates progesterone induction. EMBO J 5:2237–2240

Cato ACB, Henderson D, Ponta H 1987 The hormone response element of the mouse mammary tumor virus DNA mediates the progestin and androgen induction of transcription in the proviral long terminal repeat region. EMBO J 6:363–368

Diamond MI, Miner JN, Yoshinaga SK, Yamamoto KR 1990 Transcription factor interactions: selectors of positive or negative regulation from a single DNA element. Science 249:1266–1272

Gravanis A, Schaison G, George M et al 1985 Endometrial and pituitary responses to the steroidal antiprogestin RU486 in postmenopausal women. J Clin Endocrinol & Metab 60:156–163

Gruol DJ, Campbell NF, Bourgeois S 1986 Cyclic AMP-dependent protein kinase promotes glucocorticoid receptor function. J Biol Chem 261:4909–4914

Härd T, Kellenbach E, Boelens R et al 1990 Solution structure of the glucocorticoid receptor DNA-binding domain. Science 249:157–160

Horwitz KB 1992 The molecular biology of RU486. Is there a role for antiprogestins in the treatment of breast cancer? Endocr Rev 13:146–163

Horwitz KB 1993 Mechanisms of hormone resistance in breast cancer. Breast Cancer Res Treat 26:119–130

Jonat C, Rahmsdorf HJ, Park K-K et al 1990 Antitumor promotion and antiinflammation: down-modulation of AP-1 (fos/jun) activity by glucocorticoid hormone. Cell 62:1189–1204

Kutoh E, Stromstedt P-E, Poellinger L 1992 Functional interference between the ubiquitous and constitutive octamer transcription factor 1 (OTF-1) and the glucocorticoid receptor by direct protein–protein interaction involving the homeo subdomain of OTF-1. Mol Cell Biol 12:4960–4969

McDonnell DP, Goldman ME 1994 RU486 exerts antiestrogenic activities through a novel progesterone receptor A form-mediated mechanism. J Biol Chem 269:11945–11949

Meyer ME, Pomon A, Ji JW, Bocquel M-T, Chambon P, Gronemeyer H 1990 Agonistic and antagonist activities of RU486 on the functions of the human progesterone receptor. EMBO J 9:3923–3932

Miner JN, Yamamoto KR 1991 Regulatory cross-talk at composite response elements. Trends Biochem Sci 16:423–426

Mohamed KM, Tung L, Takimoto GS, Horwitz KB 1994 The leucine zippers of c-Fos and c-Jun for progesterone receptors dimerization: A-dominance in the A/B heterodimer. J Steroid Biochem Mol Biol 51:241–250

Oro AE, Hollenberg SM, Evans RM 1988 Transcriptional inhibition by a glucocorticoid receptor–β-galactosidase fusion protein. Cell 55:1109–1114

O'Shea EK, Rutkowski R, Stafford WF III, Kim PS 1989 Preferential heterodimer formation by isolated leucine zippers from Fos and Jun. Science 245:646–648

Sakai DD, Helm S, Carlstedt-Duke J, Gustafsson JÅ, Rottman FM, Yamamoto KR 1988 Hormone-mediated repression of transcription: a negative glucocorticoid response element from the bovine prolactin gene. Genes & Dev 2:1144–1154

Sartorius CA, Tung L, Takimoto GS, Horwitz KB 1993 Antagonist-occupied human progesterone receptors bound to DNA are functionally switched to transcriptional agonists by cAMP. J Biol Chem 5:9262–9266

Sartorius CA, Groshong SD, Miller LA et al 1994a New T47D breast cancer cell lines for the independent study of progesterone B- and A-receptors: only antiprogestin-occupied B-receptors are switched to transcriptional agonists by cAMP. Cancer Res 54:3868–3877

Sartorius CA, Melville MY, Hovland AR, Tung L, Takimoto GS, Horwitz KB 1994b A third transactivation function (AF3) of human progesterone receptors located in the unique, N-terminal segment of the B-isoform. Mol Endocrinol 8:1347–1360

Schüle R, Rangarajan P, Kliewer S et al 1990 Functional antagonism between oncoprotein c-Jun and the glucocorticoid receptor. Cell 62:1217–1226

Tung L, Mohamed KM, Hoeffler JP, Takimoto GS, Horwitz KB 1993 Antagonist-occupied human progesterone B-receptors activate transcription without binding to progesterone response elements, and are dominantly inhibited by A-receptors. Mol Endocrinol 7:1256–1265

Vegeto E, Shahbaz MM, Wen DX, Goldman ME, O'Malley BW, McDonnell DP 1993 Human progesterone receptor A form is a cell- and promoter-specific repressor of human progesterone receptor B function. Mol Endocrinol 7:1244–1255

Yang-Yen H-F, Chambard J-C, Sun Y-L et al 1990 Transcriptional interference between c-Jun and the glucocorticoid receptor: mutual inhibition of DNA binding due to direct protein–protein interaction. Cell 62:1205–1215

DISCUSSION

Muramatsu: I think this is beautiful work. It doesn't surprise me that the N-terminal region of the progesterone receptor has a transactivating capacity: several years ago, we found that the oestrogen receptor not only has a DNA-binding domain and a C-terminal activation domain, but also an 80 amino acid region at the N-terminus, in the middle of the B region, which is a transactivation domain (Imakado et al 1991). Generally, steroid receptors have several different activation regions which, in the cell, may form a 3D structure. Each region may thus adopt a special configuration and then accept a transactivator protein that acts on the general transcription factor. The next challenge will be to construct these complexes.

Horwitz: That is exactly the model we are proposing. In it, the different activation domains (AFs) of the receptor could act in synergism because the receptor molecule folds to form new contact sites for coactivators which then mediate transcription. These coactivators will undoubtedly turn out to be cell specific, so that one would observe AF synergism in some cells, but not in others.

Baulieu: I want to come to the accessory factor, which is so important if you postulate an effect not necessarily mediated by hormone response elements. Have you any idea of its molecular nature?

Horwitz: We don't yet know what the accessory factors are, other than the fact that they appear to be present in nuclei of breast cancer cells and HeLa cells. HMG-1 is believed to be involved in aiding steroid receptors to bind DNA, perhaps by influencing DNA bending. Other transcriptional mediators are being described (Jacq et al 1994, Halachmi et al 1994) and a number of factors have already been shown to interact with steroid receptors and influence the way the receptors bind to DNA and transactivate genes.

Baulieu: Would you exclude retinoic X receptor (RXR)?

Horwitz: I am sure that people have tried to look at RXR dimerization to the sex steroid receptors. I don't recall ever seeing a positive result, which may mean either that the experiments have been done and they've failed, or that they have not been done. My guess would be that the experiments have been done and have failed. We have not done them.

Baulieu: S. Liao recently showed me data on the U receptor. It's an orphan receptor-like molecule that he cloned and sequenced and which doesn't correspond to any other known receptor of the superfamily. It interacts and makes dimers with the RXR receptor and with the vitamin D receptor.

Horwitz: Is it ligand regulated?

Baulieu: He doesn't know.

Parker: In your model to account for how elevation of cAMP might modulate the response of receptors to antagonist, you suggested that cAMP response elements might be involved. But none of the reporters that you used contained binding sites for cAMP response elements, so I don't see that this model can account for the response. What other ideas have you had about how you might modulate the response of the receptor to cAMP?

Horwitz: It's possible that the cAMP-regulated protein is a phosphoprotein that is part of the general transcriptional apparatus. An alternative model would be that on a progesterone response element-containing promoter, the receptor is DNA bound, and the cAMP-regulated factor is tethered to the receptor in the absence of a cAMP response element.

Parker: It might be that the receptor targets a protein whose phosphorylation is affected by protein kinase A.

Horwitz: That's certainly possible. It is highly likely that we're talking about phosphoproteins and that cAMP is phosphorylating a protein that is interacting with the receptor. But what we don't think is happening—at least in our models—is that the cAMP is directly phosphorylating progesterone receptors. We have done a lot of experiments to try to rule this in or out: I can't say that they're absolutely conclusive, but we have considerable evidence to suggest that cAMP does not directly phosphorylate human progesterone receptors.

Parker: The model where you suggested that the factor that promotes DNA binding might be a protein that interacts with the DNA-binding domain is also unlikely, because there's no alteration in the mobility of the complex. So it seems likely that the nuclear extracts are involved in a post-translational modification or something of that sort.

Horwitz: We have done a series of DNA binding studies by gel mobility shift mobility assays using the BUS-DBD fragment. It does not bind to DNA strongly unless we add either a bivalent antibody, or a concentrated nuclear extract. As we add nuclear extracts, we see progressive DNA binding and then up-shifts of the BUS-DBD/DNA complex, suggesting that another factor is interacting with this complex and increases the overall molecular weight.

Parker: When you add the extract, do you know whether the protein is being phosphorylated?

Horwitz: No, I don't. The BUS-DBD complex is a highly phosphorylated fragment of the receptor. You will recall that we see triplet protein bands with wild-type B receptors that are due to phosphorylation. We see the same triplet bands with the BUS fragment. That fragment contains five Ser-Pro phosphorylation motifs, some or all of which are probably phosphorylated. We're now doing studies in which we have mutated all of these phosphorylation sites, to see whether we can knock out either the constitutive transcriptional activity of BUS-DBD or its ability to synergize with the other activation domains.

Beato: You have shown evidence that you may need something in the nuclear extracts to help binding of the constitutive receptor to DNA. What is the evidence that the wild-type progesterone receptor requires anything for DNA binding?

Horwitz: Very little. *In vitro* we get very good DNA binding with wild-type B receptors, but only when they are occupied by R5020 or RU486. With ZK98299 we do not get DNA binding under any of these conditions. That is, adding antibody or nuclear extracts does not restore DNA-binding capacity.

Beato: Those who are not experts in this are probably confused by these results. They see data indicating that one antagonist promotes binding of the receptor to DNA and another does not, while in another assay they have heard that neither antagonist leads to DNA binding of the receptors. It's very important to define the assay systems we are using. Imagine that you were using an extract from T-47D cells that contains the receptor probably in a complex with Hsp90. Under these conditions it's true that you need certain ligands to see DNA binding. However, if you purify the receptor from the same cells, it binds in the absence of any ligand. You can add RU486 or ZK98299 and it binds perfectly well to DNA. If you try *in vivo*—as I reported—you don't see binding with any antagonists. We are dealing with very different assays and we are looking at different things. With the purified receptor, we are looking at the behaviour of the isolated protein once it has been occupied by the ligand and separated from the other components of the non-activated receptor complex. This protein has a conformation that enables it to bind to DNA very efficiently *in vitro*. In the assays that Kate Horwitz described, the components involved in the non-activated receptor complex are all present. There the ligands have selective effects. Some of antagonistic ligands promote dissociation from other components of the non-activated complex. In this case, there is some debate as to whether ZK98299 may not (even at high doses) also promote DNA binding. But these *in vitro* results are completely artificial. We are offering the receptors exclusively DNA fragments with specific sequence, and they will even bind if they have very low affinity or unfavourable kinetics. *In vivo* the situation is different—the DNA is in chromatin, and the receptor has to find its way to the relevant sequences. There's probably no basic contradiction in the findings with the different assays, but we have to put them in context to understand what they are telling us.

Horwitz: Even *in vivo*, the models can result in contradictory data: there is some evidence to suggest that transcriptionally active chromatin is different from inactive chromatin; the former may not be entirely suppressed by nucleosomes. Furthermore, in the cell, transcriptionally active chromatin appears to be tethered to the nuclear matrix, where the DNA may be much more accessible to DNA-binding proteins than it is in more conventional, nucleosome-suppressed chromatin models. No model is perfect, and I think your caveat is an important one. It is important that we all understand the limitations of the models with which we are working.

Baulieu: In support of what Miguel Beato was saying, a number of laboratories, including ours, have found no evidence that a specific ligand—say

RU486—modifies the interaction of a receptor to which it binds with DNA. However, there is a tightening of the binding of Hsp90 with the receptor if RU486 is bound, as compared with an agonist. This has been demonstrated *in vitro* and can be reproduced with purified compounds (Groyer et al 1987)

Have you any idea of the progesterone receptor composition (in terms of A and B subunits) which may be important for agonist/antagonist activity in actual breast cancer? Is there any medical screening for this?

Horwitz: I suspect that many groups are now analysing A : B ratios in breast cancer and other target tissues. I have seen one abstract from Christine Clarke's laboratory in Australia. She screened a series of breast cancers and measured the A : B ratio by Western blotting. She showed that about 75% of the cancers had an excess of A over B receptors. However, there were no correlative data with treatment response. I think knowledge of this ratio is going to be very important in order to interpret the response of tissues to progestin or antiprogestin therapies.

Baulieu: Do all cells in a given tumour show the same receptor composition?

Horwitz: We have developed a progesterone receptor immunoflow cytometry assay that allows us to measure cell-to-cell heterogeneity. We see a huge range in the distribution of progesterone receptor levels among cells. Some cells can be progesterone receptor negative, whereas others have a million copies per cell, when all are growing in the same flask. It was because of this heterogeneity that we were able to isolate the progesterone receptor-negative T-47D-Y cells. I would suspect that the A : B ratio will also vary from cell to cell, but that is something that will be more difficult to demonstrate.

Baulieu: Does that make life simpler?

Horwitz: No, but it makes it more interesting!

References

Graham J, Harvey S, Balliene R et al 1994 Expression of progesterone receptor A and B proteins in human breast cancer. J Cell Biochem 18B:266(abstr)

Groyer A, Schweizer-Groyer G, Cadepond F, Mariller M, Baulieu EE 1987 Antiglucocorticosteroid effects suggest why steroid hormone is required for receptors to bind to DNA *in vivo* but not *in vitro*. Nature 328:624–626

Halachmi S, Mardeu E, Martin G, MacKay H, Abbondanza C, Brown M 1994 Estrogen-receptor associated proteins: possible mediators of hormone-induced transcription. Science 264:1455458

Imakado S, Koike S, Kondo S, Sakai M, Muramatsu M 1991 The N-terminal transactivation domain of rat estrogen receptor is localized in a hydrophobic domain of eighty amino acids. J Biochem 109:684–689

Jacq X, Brou C, Lutz Y, Davidson J, Chambon P, Tora L 1994 Human TAFII30 is present in a distinct TAFIID complex, and is required for transcriptional activation by estrogen receptor. Cell 79:107–117

Oestrogen- and anti-oestrogen-regulated genes in human breast cancer

Henri Rochefort

INSERM U148, Unité Hormones et Cancer, Université de Montpellier, Faculté de Médecine, 60 rue de Navacelles, 34090 Montpellier, France

Abstract. The study of several human breast cancer cell lines containing oestrogen receptors has allowed characterization of a number of oestrogen-induced proteins (e.g. progesterone receptor, cathepsin D, pS2, Hsp27, c-Myc). In primary tumours these markers have different prognostic significance for predicting whether the tumour will be hormone responsive (e.g. pS2, progesterone receptor) and whether it will metastasize (e.g. cathepsin D). The mechanism of regulation of gene expression by oestrogens and anti-oestrogens in breast cancer is complex and varies according to the nature of both the gene and the cell in which it is transcribed. Our laboratory has identified the sequences mediating oestrogen activity in the proximal region of cathepsin D, including a non-consensus oestrogen-responsive element located at -260 which acts in synergy with other regulatory elements. In addition to the classical effect of oestrogen receptor in stimulating transcription of genes controlled by the oestrogen-responsive element, we found that oestrogen receptor is able to modulate transcription of AP-1-responsive genes without interacting directly with DNA. Cross-talk between oestrogen receptor and members of the Fos/Jun family via protein–protein interactions may explain how anti-oestrogens inhibit the mitogenic effect of growth factors in the apparent absence of oestrogens and why tamoxifen is able to stimulate cathepsin D gene expression and induce apoptosis in certain oestrogen receptor-positive breast cancer cells. The nature and degree of this cross-talk appears to vary according to the gene, the cell type and the type of oestrogen receptor ligand involved. Studies of oestrogen-regulated genes are not only useful for classifying breast cancers according to their ability to metastasize and respond to therapies, but also should lead to new therapeutic approaches for hormone-dependent and hormone-resistant cancers.

1995 Non-reproductive actions of sex steroids. Wiley, Chichester (Ciba Foundation Symposium 191) p 254–268

Oestrogens have been proposed to stimulate the growth of breast cancers since the early works of Beatson (1886) and Lacassagne (1936). The mechanism behind this effect was then specified by the identification of the oestrogen receptor in mammary cancers (Jensen et al 1966) and by the study of hormonal regulation in oestrogen receptor-positive breast cancer cell lines (Lippman et al

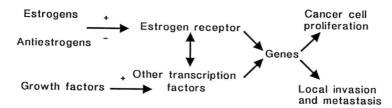

FIG. 1. The oestrogen receptor, activated or inhibited by its specific ligands, can control cancer cell proliferation and invasion by regulating oestrogen-responsive genes, generally controlled by oestrogen-responsive elements. In addition, the oestrogen receptor has the ability to interfere with the effect of growth factors via protein–protein interactions involving other transcription factors such as the Fos/Jun family.

1976). The critical steps leading to our present understanding of the role of oestrogens in breast cancer have been the identification of oestrogen-induced genes and proteins, the development of the corresponding specific antibodies and cDNA probes, and the application of these probes to tumour biopsy specimens in order to assess the prognostic and the predictive significance of oestrogen-induced proteins.

One current hypothesis is that oestrogens control the growth and invasiveness of oestrogen receptor-positive primary breast cancers by inducing oestrogen-regulated proteins that function as autocrine, paracrine or intracrine growth factors (Rochefort et al 1980, Chalbos et al 1982, Dickson et al 1986) or by inducing proteases associated with invasion and metastasis (reviewed in Rochefort et al 1990). In addition to the classical effects of oestrogens and anti-oestrogens via genes controlled by oestrogen-responsive elements (EREs), the oestrogen receptor ligands can modulate the action of transcription factors such as Fos/Jun via protein–protein interactions without necessarily interacting with EREs (Gaub et al 1990, Philips et al 1993) (Fig. 1).

In this short review, I will concentrate on some recent results from our laboratory concerning:

(1) The oestrogen-induced proteins and their different prognostic significance in breast cancer, with special emphasis on cathepsin D.
(2) The identification of the ERE in the cathepsin D proximal promoter region which acts in synergy with other transcriptional regulatory elements in mammary cells.
(3) The complex mode of action of tamoxifen, which is able to induce some oestrogen-regulated genes, such as cathepsin D, and to reduce breast cancer growth.
(4) The cell- and tissue-specific activities of the oestrogen receptor that involve interference with other transcription factors.

Different oestrogen-induced proteins have different prognostic significance

The first oestrogen-induced protein characterized and used in the clinic was the progesterone receptor (Horwitz & McGuire 1978). Our laboratory found that a 52 kDa oestrogen-induced protein secreted by MCF-7 human breast cancer cells (Westley & Rochefort 1980), initially proposed to be an autocrine growth factor involved in the control of cell proliferation, was in fact the precursor of cathepsin D, a ubiquitous lysosomal protease. Using differential screening of a cDNA library from oestrogen-treated MCF-7 cells, Pierre Chambon's laboratory described an mRNA corresponding to a secreted protein of 84 amino acids which they called pS2. This pS2 protein, usually expressed by gastric cells, is homologous to pancreatic spasmolytic protein (Rio et al 1988).

The 52 kDa protein is a good example of a protein that has proved to have a serendipitous clinical application. It was expected to be a circulating marker of hormone responsiveness, since it is regulated by oestrogen and is secreted. However, its concentration was not increased in plasma from breast cancer patients, but it was found to be a tissue marker which correlated with risk of metastasis (for review, see Rochefort 1992). Cathepsin D concentration did not correlate with oestrogen receptor status or pS2 status, even though the genes for both cathepsin D and pS2 are stimulated by oestrogen and growth factors in oestrogen receptor-positive breast cancer cells (Cavaillès et al 1989). By contrast, pS2 appears to correlate better with the progesterone receptor and is useful as a test for responsiveness to anti-oestrogen therapy. This apparent paradox may be explained by the fact that oestrogen receptor-negative breast cancers constitutively express cathepsin D but not pS2. Cathepsin D is the first example of a protease that has been transferred into clinical application as a prognostic marker. A convenient cytosolic immunoassay for cathepsin D as an independent tissue marker is available (standardized with quality control), which helps to predict prognosis in combination with other markers associated with cell differentiation and cell proliferation.

Mechanism of over-expression of cathepsin D and characterization of its oestrogen-responsive element

The unexpected clinical finding that the cytosolic concentration of cathepsin D (contrary to pS2) did not correlate with the status of the oestrogen or progesterone receptors, but rather with metastasis-free survival, raised a number of questions concerning the cause and the consequence of its over-expression in breast cancer tissue. The biological significance of high cathepsin D production in invasion, tissue remodelling and metastasis will not be discussed here (for a recent review, see Rochefort et al 1994).

Concerning the mechanism of cathepsin D over-expression in some breast tumours, we had to define:

(1) The cells responsible for this over-expression in the primary tumour. By immunohistochemistry and *in situ* hybridization, we showed that breast cancer cells, in addition to some macrophages, over-expressed this enzyme (Roger et al 1994).

(2) The molecular mechanism of regulation of cathepsin D expression in oestrogen receptor-positive breast cancers cells.

(3) The mechanism of its over-expression in oestrogen receptor-negative breast cancers.

Cathepsin D over-expression in breast cancer has been described at the mRNA and protein levels in cell lines as well as in patients. No amplification or major rearrangement of the cathepsin D gene has been detected (Augereau et al 1988), suggesting that altered RNA expression or dysregulation of a diploid gene is responsible for this over-expression. Our laboratory has initiated the characterization of the proximal promoter region of the cathepsin D gene in MCF-7 cells, and identified the central ERE involved in its regulation by oestrogens (Cavaillès et al 1993, Augerau et al 1994). The cathepsin D promoter has a compound structure with features of both a housekeeping gene (high G and C content, and several potential Sp1-binding sites) and a regulated gene (TATA box). Cavaillès et al (1993) showed that transcription was initiated at five transcription sites. In hormone-responsive breast cancer cells, oestradiol caused a six- to 10-fold increase in the level of RNAs initiated at the TSSI (first transcription start site), located about 28 bp downstream from the TATA box, and shifted the housekeeping transcription pattern to that of hormone-regulated transcription. Site-directed mutagenesis indicated that the TATA box is essential for initiation of cathepsin D gene transcription at the TSSI. In breast cancer biopsies, high levels of TATA-dependent transcription were correlated with over-expression of cathepsin D mRNA. This was also observed in some oestrogen receptor-negative tumours, suggesting that factors other than the oestrogen receptor can stimulate TATA-dependent transcription of this gene. Depending on the conditions, cathepsin D thus behaves as a housekeeping gene with multiple start sites or as a regulated gene that can be overexpressed via its TATA box. Oestrogens in breast cancer cells stimulate TATA-dependent cathepsin D transcription but not transcription initiated upstream of the TATA box. Independently, May et al (1993) came to similar conclusions concerning the dual nature of this promoter. We have now characterized the *cis*-acting sequences mediating the effect of oestrogen receptor. The proximal promoter region of the cathepsin D gene contains several putative oestrogen-responsive elements (Augereau et al 1994). Functional studies using a tk-CAT (thymidine kinase chloramphenicol acetyl-transferase) reporter gene plasmid fused with different sequences of the cathepsin D promoter region have identified a central ERE (named CD-E2) located at -261 (Augerau et al 1994, Fig. 2). In *in vitro* footprinting

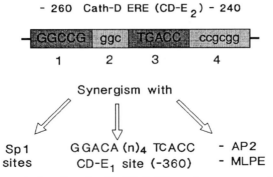

FIG. 2. The central cathepsin D (Cath-D) oestrogen-responsive element (ERE; CD-E$_2$) contains two mismatches. Its GC-rich tail is not required for oestrogen regulation. Synergism is necessary between the Cath-D ERE and at least one of the other *cis* elements indicated on the bottom line (Augereau et al 1994). MLPE, major late promoter element.

experiments, this ERE is protected by the mouse oestrogen receptor produced by baculovirus and also by MCF-7 nuclear extracts. Gel shift experiments have shown that the CD-E2 site can bind oestrogen receptor *in vitro*. However, the central cathepsin D ERE sequence (GGCCG(n)$_3$TGACC) contains two mismatches (*) compared with the vitellogenin A2 ERE consensus. As a result, the binding affinity of the oestrogen receptor for the cathepsin D ERE is decreased. Moreover, the CD-E2 ERE is necessary, but not sufficient, to mediate oestrogen responsiveness. Using different constructs, we have shown that CD-E2 needs to synergize with other sequences: either with another ERE-like sequence (CD-E1) located at -360, or with Sp1-binding sequences, and/or with proximal sequences containing MLPE (major late promoter element) and AP-2 sites located close to the TATA box. We currently do not know which synergism is the most important under biological conditions. Recently, results obtained by other laboratories using transfected recombinants have suggested that, as with the c-*myc* and creatinine kinase genes, synergism between ERE and Sp1 boxes is critical (Krishnan et al 1994). Although this synergism may be biologically important for a housekeeping gene such as cathepsin D, further experiments are required to ascertain which synergy between which sequences is operational *in vivo*.

Cathepsin D over-expression in oestrogen receptor-positive breast cancer cells may therefore be explained by a local increased sensitivity to oestrogen compared with normal resting mammary glands. This might be due to altered *trans*-acting factors (oestrogen receptor, other transcription factors, transcription intermediary proteins) or to altered *cis*-acting sequences. The mechanism by which the concentration of cathepsin D protein is increased in oestrogen receptor-negative breast cancer cells is still unknown.

Tamoxifen induces cathepsin D but inhibits growth factor and AP-1 activities in oestrogen receptor-positive breast cancer cells

The effect on cathepsin D of the non-steroidal anti-oestrogen tamoxifen, used in breast cancer treatment, is complex and appears to be paradoxical. We initially described that in MCF-7 cells tamoxifen inhibits the secretion of procathepsin D and displays no agonist activity for this protein (Westley & Rochefort 1980), while it inhibits MCF-7 cell growth. By contrast, in MCF-7 anti-oestrogen resistant variants (RTX6, R27), which grow in the presence of $1\,\mu M$ tamoxifen, this anti-oestrogen increased procathepsin D secretion (Westley et al 1984). This was recently confirmed in another variant cell line, LCC2 (Coopman et al 1994). The reason for the variable extent of procathepsin D secretion in these different cell lines is unknown.

When MCF-7 cells were cultured under conditions of strict oestrogen deprivation (stripped serum and phenol red-free medium), tamoxifen behaved as an oestrogen agonist in inducing cathepsin D mRNA production, whereas it inhibited pS2 production (Johnson et al 1989, Chalbos et al 1993). This cathepsin D-specific agonist activity was further increased by cAMP (Chalbos et al 1993). It is also consistent with clinical observations that three weeks after presurgical tamoxifen treatment, procathepsin D and cathepsin D mRNA levels were increased in breast cancer tumours (Maudelonde et al 1989, 1994). These results appear paradoxical—on the one hand, tamoxifen is efficient in the treatment of oestrogen receptor-positive breast cancers, and on the other hand it increases cathepsin D concentration in the tumour, an effect also induced by oestrogens and which generally indicates poor prognosis in untreated patients.

To explain this discrepancy, we propose that the over-expression of the cathepsin D gene may have opposite effects depending on whether growth factor activities are stimulated or inhibited. For instance, the oestrogen-induced stimulation of cathepsin D synthesis and growth factor activity would result in increased tumour growth, invasion and metastasis. In contrast, the tamoxifen-induced increase in cathepsin D, in conjunction with tamoxifen-dependent inhibition of growth factor activity, may result in apoptosis and increased autophagia. Two lines of evidence favour this hypothesis.

First, our laboratory has shown that in oestrogen receptor-positive breast cancer cell lines, the anti-oestrogens 4-hydroxytamoxifen and ICI 164384 inhibit the stimulatory effects of growth factors (insulin-like growth factor 1 and epidermal growth factor) and of high doses of insulin on cell proliferation (Vignon et al 1987) and on the induction of pS2 (Chalbos et al 1993) in the apparent absence of oestrogens.

Second, tamoxifen and its high-affinity metabolite 4-hydroxytamoxifen are not only cytostatic on oestrogen receptor-positive breast cancer cells, they are also cytotoxic and can induce apoptosis in some cells (Bardon et al 1987). In anti-oestrogen-treated MCF-7 cells, the number of dead cells, estimated by a

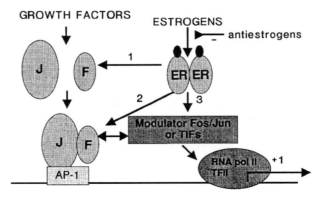

FIG. 3. Oestrogens stimulate and anti-oestrogens inhibit transcription of AP-1-responsive genes in oestrogen receptor-positive breast cancer cells. Via the nuclear oestrogen receptors (ERs), oestrogens can increase transcription of AP-1-responsive genes by increasing synthesis of c-Fos (F) and c-Jun (J) (path 1), increasing their efficacy in inducing AP-1-responsive genes either directly (path 2) or via interaction with modulators of Fos and Jun activity or with transcription intermediary factors (TIFs) (path 3). Anti-oestrogens have the opposite effect by inhibiting oestrogen action if residual oestrogens are present, or by stabilizing a different oestrogen receptor conformation in the total absence of oestrogens.

dye-exclusion technique, doubled within a few days and represented two-thirds of the total cell population at the end of treatment. Oestradiol prevented the hydroxytamoxifen-induced cell death. No anti-oestrogen-induced cell death could be detected in human breast cancer cells lacking oestrogen receptor. Our laboratory has recently confirmed, by the *in situ* nick-end labelling technique, that hydroxytamoxifen and a pure antagonist (ICI 164 384) are able to induce apoptosis in cultured MCF-7 cells. This technique allows quantification of apoptotic cells in tissue sections of human tumour biopsies to estimate whether apoptosis is increased *in vivo* by tamoxifen treatment (P. Roger, F. Vignon & H. Rochefort, unpublished results).

The mechanism of the oestrogen receptor-mediated inhibition of growth factor action by anti-oestrogens might involve several genes. In addition to induction of growth inhibitory factors, such as transforming growth factor β (Knabbe et al 1987) and protein tyrosine phosphatase (Freiss & Vignon 1994), it is important to consider that the anti-oestrogen–oestrogen receptor complex may also inhibit expression of AP-1-controlled reporter genes as suggested by studies in MCF-7 cells transfected with AP-1-tk-CAT recombinants (Philips et al 1993) (Fig. 3). It is unlikely that this inhibition results from the presence of residual oestrogens, because AP-1 activity is inhibited under conditions in which the anti-oestrogens have no anti-oestrogenic effect on basal transcription mediated by the ERE. Moreover, the two anti-oestrogens 4-hydroxytamoxifen and ICI 164 384 differ in their ability to inhibit transcription mediated by the two types of enhancers. Hydroxytamoxifen is more efficient at inhibiting ERE-mediated activity, whereas ICI 164 384 is more efficient at *trans*-repressing AP-1 activity. This also strongly suggests that the modulation of AP-1 activity by oestrogen receptor ligands is direct and does not require ERE-dependent regulation of an intermediary protein. There was a striking parallel between the effect of oestrogens and of anti-oestrogens on AP-1 activity and on the growth of MCF-7 cells. For instance, in MCF-7 cells the partial oestrogen agonist hydroxytamoxifen does not stimulate but rather inhibits AP-1 activity as it inhibits cell growth.

Cell- and tissue-specific oestrogen regulation in oestrogen target tissues

The oestrogen receptor efficacy varies according to the nature of the gene and of the target cell. In addition to gene specificity in the same cell (i.e. pS2 versus cathepsin D) the oestrogen receptor has different effects according to the cell type in which it has been activated by oestrogen or anti-oestrogen. For instance, in the human endometrium (Maudelonde et al 1990, Miralles et al 1994), the cathepsin D gene is unresponsive to oestrogen but is stimulated by progesterone, and anti-oestrogens stimulate endometrial cancer but inhibit breast cancer growth.

Tissue specificity may involve not only genes controlled by EREs, such as cathepsin D, but also AP-1-controlled genes, as illustrated by our recent studies on the cross-talk between transfected oestrogen receptor and AP-1 reporter genes (Philips et al 1993) (Fig. 3). We have shown in MCF-7 cells that oestradiol stimulates growth factor-induced AP-1 activity, measured by transient transfection of a CAT reporter plasmid controlled by four head-to-tail copies of the polyoma virus AP-1-PEA3 binding site upstream of the tk promoter (Philips et al 1993). The dose–response curves of oestradiol-dependent transcription of ERE-tk-CAT and (AP-1)X4-tk-CAT constructs could be superimposed, and co-transfection of oestrogen receptor amplified the stimulation, strongly suggesting that oestrogen receptor mediates the oestrogen effect. Similar results were obtained when an AP-1 response was induced by transfected c-*fos* and c-*jun* expression vectors, demonstrating that neither induction of growth factors nor induction of growth factor receptors were responsible for increased transcription of AP-1-controlled genes by oestradiol. We extended this finding to two other oestrogen receptor-positive human breast cancer cell lines (T-47D and ZR75) cultured in the presence of epidermal growth factor. However, under the same conditions, oestradiol decreased AP-1-mediated transcription in several oestrogen receptor-negative cell lines, MDA-MB231 (human breast cancer), NIH/3T3 (mouse fibroblast) and HeLa, which were transiently transfected with the oestrogen receptor expression plasmid (Chalbos et al 1994). The mechanism by which the oestrogen–oestrogen receptor complex has opposite effects on AP-1 activity in oestrogen receptor-negative and oestrogen receptor-positive cell lines is unknown. The DNA binding of oestrogen receptor does not seem to be needed for the oestradiol-induced increase in AP-1 activity, which suggests that a protein–protein interaction is responsible for interference between the two kinds of transcription factors (Chalbos et al 1994).

One practical application of this cell-specific cross-talk might be to change the aggressive oestrogen receptor-negative breast cancers into more differentiated and hormone-responsive breast cancers following stable transfection of the oestrogen receptor. First attempts in the oestrogen receptor-negative MDA-MB231 breast cancer cell line have, in fact, resulted in restoration of their sensitivity to oestrogen which, paradoxically, inhibited cell growth *in vitro* (Jiang & Jordan 1992) and metastasis *in vivo* (Garcia et al 1992), whereas the oestrogen receptor-positive breast cancer cells were stimulated by oestradiol under the same conditions. Thus, in oestrogen receptor-negative, as in oestrogen receptor-positive cancer cells, there is still a parallelism between the effect of oestrogens on AP-1 activities and their effects on tumour growth. This may guide us in identifying the factors responsible for this contrasting effect of oestrogens in the two types of cell, and to develop new therapeutic approaches.

Conclusions

The few points that we have considered illustrate the complexity of the mode of action of oestrogens and anti-oestrogens in mammalian cells. The oestrogen receptor, encoded by a single gene, apparently has the same amino acid sequence in all cells. However, the genes under its transcriptional control have different promoter regions and EREs, and synergism with other transcription factors might vary. This may explain the different prognostic significance of pS2 and cathepsin D in breast cancer.

Moreover, the oestrogen receptor can interact with other nuclear proteins and transcription factor(s) and modulate their activity. Therefore, one should also consider genes controlled by other regulatory elements, such as AP-1, which have the potential to be stimulated or inhibited by oestrogens and anti-oestrogens, respectively. This type of mechanism involves protein–protein interaction, it is based on transient transfection studies with reporter genes, and remains to be demonstrated *in vivo* on genes lacking any oestrogen-responsive elements. It is tempting to propose that the inhibitory effect of anti-oestrogens on AP-1 activity explains their anti-growth factor activity and possibly induction of apoptosis in some oestrogen receptor-positive breast cancers. The complexity is further increased by the cell and tissue specificity of oestrogen regulation that exists for genes controlled by the EREs, such as cathepsin D, as well as for AP-1 transcriptional interference by oestrogen receptor.

The case of oestrogen-induced genes in breast cancer illustrates how clinical applications, such as new prognostic markers and therapies, can be derived from basic studies. Conversely, these studies also show how paradoxical observations obtained in patients or in nude mice stimulate further mechanistic studies aimed at understanding these *in vivo* discrepancies.

Acknowledgements

This work was supported by the Institut National de la Santé et de la Recherche Médicale, the University of Montpellier 1, the Centre National de la Recherche Scientifique, the Ligue National Française de Lutte contre le Cancer and the Association pour la Recherche sur le Cancer. I thank M. Bouldoires for preparing the manuscript. I am indebted to my colleagues whose contributions (cited in the references) have been crucial.

References

Augereau P, Garcia M, Mattei MG et al 1988 Cloning and sequencing of the 52K cathepsin D cDNA of MCF7 breast cancer cells and mapping on chromosome 11. Mol Endocrinol 2:186–192

Augereau P, Miralles F, Cavaillès V, Gaudelet C, Parker M, Rochefort H 1994 Characterization of the proximal estrogen responsive element of human cathepsin D gene. Mol Endocrinol 8:693–703

Bardon S, Vignon F, Montcourrier P, Rochefort H 1987 Steroid-receptor mediated cytotoxicity of an antiestrogen and an antiprogestin in breast cancer cells. Cancer Res 47:1441–1448

Beatson GT 1886 On the treatment of inoperable cases of the carcinoma of the mamma. Suggestion for a new method of treatment with illustrative cases. Lancet II:162–165

Cavaillès V, Garcia M, Rochefort H 1989 Regulation of cathepsin D and pS2 gene expression by growth factors in MCF7 human breast cancer cells. Mol Endocrinol 3:552–558

Cavaillès V, Augereau P, Rochefort H 1993 Cathepsin D gene is controlled by a mixed promoter and estrogens stimulate only TATA dependent transcription in breast cancer cells. Proc Natl Acad Sci USA 90:203–207

Chalbos D, Vignon F, Keydar I, Rochefort H 1982 Estrogens stimulate cell proliferation and induce secretory proteins in a human breast cancer cell line (T47D). J Clin Endocrinol & Metab 55:276–283

Chalbos D, Philips A, Galtier F, Rochefort H 1993 Synthetic antiestrogens modulate induction of pS2 and cathepsin D messenger ribonucleic acid by growth factors and adenosine 3′,5′-monophosphate in MCF7 cells. Endocrinology 133:571–576

Chalbos D, Philips A, Galtier F, Rochefort H 1994 Differential cross talk between AP-1 and estrogen receptor in ER positive and ER negative breast cancer cell lines. Abstracts of Endocr Soc 76th Annu Meet, 15–18 June, Anaheim, CA. Abstr 1801

Coopman P, Garcia M, Brunner N, Derocq D, Clarke R, Rochefort H 1994 Anti-proliferative and anti-estrogenic effects of ICI 164,384 and ICI 182,780 in 4-OH-tamoxifen-resistant human breast cancer cells. Int J Cancer 56:295–300

Dickson RB, McManaway ME, Lippman ME 1986 Estrogen-induced factors of breast cells partially replace estrogen to promote tumor growth. Science 832:1540–1544

Freiss G, Vignon F 1994 Antiestrogens increase protein tyrosine phosphatase activity in human breast cancer cells. Mol Endocrinol 8:1389–1396

Garcia M, Derocq D, Freiss G, Rochefort H 1992 Activation of estrogen receptor transfected in a receptor negative breast cancer cell line decreases the metastatic and invasive potential of the cells. Proc Natl Acad Sci USA 89:11538–11542

Gaub MP, Bellard M, Scheuer I, Chambon P, Sassone-Corsi P 1990 Activation of the ovalbumin gene by the estrogen receptor involves the Fos–Jun complex. Cell 63:1267–1276

Horwitz KB, McGuire WL 1978 Estrogen control of progesterone receptor in human breast cancer. J Biol Chem 248:6351–6353

Jensen EV, Jacobson HI, Flesher JW et al 1966 Estrogen receptors in target tissues. In: Nakao T, Pincus G, Tait J (eds) Steroid dynamics. Academic Press, New York, p 133–157

Jiang SY, Jordan VC 1992 Growth regulation of estrogen receptor-negative breast cancer cells transfected with complementary DNAs for estrogen receptor. J Natl Cancer Inst 84:580–591

Johnson MD, Westley BR, May FEB 1989 Estrogenic activity of tamoxifen and its metabolites on gene regulation and cell proliferation in MCF-7 breast cancer cells. Br J Cancer 59:727–738

Knabbe C, Lippman ME, Wakefield LM et al 1987 Evidence that transforming growth factor-β is a hormonally regulated negative growth factor in human breast cancer cells. Cell 48:417–428

Krishnan V, Wang X, Safe S 1994 Estrogen receptor–Sp1 complexes mediate estrogen-induced cathepsin D gene expression in MCF-7 human breast cancer cells. J Biol Chem 269:15912–15917

Lacassagne A 1936 Hormonal pathogenesis of adenocarcinoma of the breast. Am J Cancer 27:217–228

Lippman ME, Bolan G, Huff K 1976 The effects of estrogens and anti-estrogens on hormone responsive human breast cancer in long-term tissue culture. Cancer Res 36:4595–4601

Maudelonde T, Domergue J, Henquel C et al 1989 Tamoxifen treatment increases the concentration of 52K-cathepsin D and its precursor in breast cancer tissue. Cancer 63:1265–1270

Maudelonde T, Martinez P, Brouillet JP, Laffargue F, Pages A, Rochefort H 1990 Cathepsin D in human endometrium: induction by progesterone and potential value as a tumor marker. J Clin Endocrinol & Metab 70:115–121

Maudelonde T, Escot C, Pujol P et al 1994 In vivo stimulation by tamoxifen of cathepsin D RNA level breast cancer. Eur J Cancer 30A:2049–2053

May FEB, Smith DJ, Westley BR 1993 The human cathepsin D-encoding gene is transcribed from an estrogen-regulated and a constitutive start point. Gene 134:277–282

Miralles F, Gaudelet C, Cavaillès V, Rochefort H, Augereau P 1994 Insensitivity of cathepsin D gene to estradiol in endometrial cells is determined by the sequence of its estrogen responsive element. Biochem Biophys Res Commun 203:711–718

Philips A, Chalbos D, Rochefort H 1993 Estradiol increases and antiestrogens antagonize the growth factor-induced AP-1 activity in MCF7 breast cancer cells without affecting c-*fos* and c-*jun* synthesis. J Biol Chem 268:14103–14108

Rio MC, Bellocq JP, Daniel JY et al 1988 Breast cancer-associated pS2 protein: synthesis and secretion by normal stomach mucosa. Science 241:705–708

Rochefort H, Coezy E, Joly E, Westley B, Vignon F 1980 Hormonal control of breast cancer in cell culture. In: Lacobelli S, King RJB, Lindner HR, Lippman ME (eds) Hormones and cancer: progress in cancer research and therapy. Raven Press, New York, vol 14:21–29

Rochefort H, Capony F, Garcia M 1990 Cathepsin D in breast cancer: from molecular and cellular biology to clinical applications. Cancer Cells 2:383–388

Rochefort H 1992 Cathepsin D in breast cancer: a tissue marker associated with metastasis. Eur J Cancer 28A:1780–1783

Rochefort H, Augereau P, Miralles F, Liaudet E, Montcourrier P, Garcia M 1994 Hormonal regulation and biological significance of cathepsin D in breast cancer metastasis. In: Motta M, Serio M (eds) Sex hormones and antihormones in endocrine dependent pathology: basic and clinical aspects. Elsevier Science, p 227–232

Roger P, Montcourrier P, Maudelonde T et al 1994 Cathepsin D immunostaining in paraffin-embedded breast cancer cells and macrophages. Correlation with cytosolic assay. Hum Pathol 25:863–871

Vignon F, Bouton MM, Rochefort H 1987 Antiestrogens inhibit the mitogenic effect of growth factors on breast cancer cells in total absence of estrogens. Biochem Biophys Res Commun 146:1502–1508

Westley B, Rochefort H 1980 A secreted glycoprotein induced by estrogen in human breast cancer cell lines. Cell 20:352–362

Westley B, May FEB, Brown AMC et al 1984 Effects of antiestrogens on the estrogen-regulated pS2 RNA, and the 52- and 160-kilodalton proteins in MCF7 cells and two tamoxifen-resistant sublines. J Biol Chem 259:10030–10035

DISCUSSION

Muramatsu: The most important issue is that of the interaction between the oestrogen receptor and AP-1. You suggested that oestrogen might interact with another protein which then activates AP-1. Do you find any direct interaction between AP-1 and oestrogen receptor?

Rochefort: We have no evidence for a direct interaction.

Muramatsu: What part of the oestrogen receptor molecule is involved in this interaction? You use the whole molecule. If it does not bind to the oestrogen response element, perhaps the DNA-binding domain is dispensable. You could test this by using different oestrogen receptor deletion mutants.

Rochefort: That question is easy to answer if you look for an effect in an oestrogen receptor-negative cell line. This kind of effect may not be the most important one to study, since in oestrogen receptor-negative cell lines transfected with the receptor we find a decrease in AP-1 activity and not an increase as in the MCF-7 cells. That's why we are trying to obtain MCF-7 or T-47D cell lines devoid of endogenous oestrogen receptor to do the experiments you are mentioning involving transfection with truncated receptors to find out which domain is involved in this positive cross-talk.

Horwitz: I just wanted to address one part of your comment, which is the assumption that if a receptor is active independently of hormone response element binding, then its DNA-binding domain is not needed. This may not be correct. Even though the DNA-binding domain might not be required for DNA binding, this domain has other functions. It is involved in dimerization, and it may also provide the surface for interactions with other proteins. There are data from the glucocorticoid receptor literature (e.g. Schüle et al 1990, Miner & Yamamoto 1991) and from our laboratory (Tung et al 1993) that an intact DNA-binding domain is required for functions that are independent of DNA binding.

Muramatsu: I still think it's worth testing.

Rochefort: It is not easy to test this in MCF-7 cells because these cells have endogenous oestrogen receptor, and you cannot truncate the endogenous receptor.

Beato: It's interesting that your cathepsin promoter has two weak oestrogen-response elements separated by the length of one nucleosome. You say that it probably has to synergize with Sp1-binding sites in order to induce cathepsin. I wanted to call your attention to the fact that Sp1 has now been shown to be a member of a family of proteins (consisting of Sp1, 2, 3 and 4) and that Sp3 is actually a repressor of Sp1. By studying the expression patterns of this Sp family in different cells, we might be able to explain the fact that in one particular cell there is a positive interaction and in another there isn't.

Parker: I find it amazing that the introduction of oestrogen receptors into many types of cells causes growth inhibition. In those situations, oestrogen receptors block AP-1 activity. In contrast, in breast cancer cells, which are stimulated to grow by oestrogens, oestrogen receptors synergize with AP-1, according to your results. So there's some change in those cells which respond to oestrogens by progressing through the cell cycle, so that the oestrogen receptor and AP-1 interact differently. The receptor in some way promotes cell proliferation, whereas in the majority of cells the receptor acts as a differentiation hormone and, in common with retinoids and glucocorticoids, it blocks cell growth. If you introduce the oestrogen receptor into those cells where oestrogen acts as a mitogen, you do not arrest cell growth. So there's something about MCF-7 cells that allows the receptor to synergize with AP-1 that's different from the majority of other oestrogen receptor-negative cells. This would obviously be worth looking for.

Rochefort: This is what we have been doing. Dany Chalbos, in our laboratory, has some preliminary evidence from gel shift assays that the transcription factors which bind AP-1 sequences are different in MCF-7 cells from those in MDA-MB231 cells.

Parker: So you think that perhaps the particular Fos/Jun family members that bind to the AP-1 site determine whether or not oestrogen acts as a mitogen.

Rochefort: Yes, different members of the Fos/Jun family are able to bind to AP-1 sequences and their dosage may vary according to the cell type.

Parker: But if you look in breast cancers, there should be a subset—maybe 30% of breast cancers—whose AP-1 family members differ from all oestrogen-insensitive tumours, if that is the case.

Rochefort: I agree, but we first have to try to understand what happens in cell lines before we go to patients. More generally, the complexity that was discussed earlier in this meeting concerning oestrogen and anti-oestrogen action in different tissues can also be explained by the complexity of the different proteins interacting with a single type of oestrogen receptor. I'm afraid I cannot give you a definite answer concerning which transcription factor is modulated by the oestrogen receptor. To add to this complexity, Masami Muramatsu has shown that Efp might play a role in mediating oestrogen activity (see p 43).

Bonewald: I have a question about the use of the presence of oestrogen receptor and expression of cathepsin D as prognostic factors. In theory you would want to treat oestrogen receptor-positive breast cancer with tamoxifen to inhibit proliferation. But you have shown that it acts as an agonist and stimulates cathepsin D production. Which is more important: proliferation of the cells or the increase in cathepsin D?

Rochefort: I have asked myself this question. There is apparently a contradiction here, because short-term treatment with tamoxifen increases cathepsin D levels in patients (Maudelonde et al 1989), whereas higher cathepsin D levels before treatment have a bad prognostic significance (Rochefort et al 1990). However, although tamoxifen increases cathepsin D, it also blocks growth factor action (Vignon et al 1987). There are many examples of processes underlying apoptosis where you have an induction of proteases and a blocking of growth factors. We have some preliminary observations, *in vivo* and *in vitro*, that tamoxifen could also induce apoptosis. The significance of the increase in cathepsin D in breast cancer might therefore depend on whether or not growth factors are active.

Beato: Have you looked at collagenase? Was collagenase also induced by growth factors in this system?

Rochefort: We are doing this now. Dany Chalbos is looking at the genes regulated by AP-1 which apparently lack oestrogen-responsive elements. The proteases involved in the proteolytic cascade triggered in metastasis are particularly important in this regard. Different collagenases, stromelysin and urokinase (a plasminogen activator which also has poor prognostic significance in breast cancer) are regulated by members of the AP-1 family triggered by growth factors and phorbol esters. It would be interesting to know whether the genes for these proteins are also regulated by the oestrogen receptor.

References

Maudelonde T, Domergue J, Henquel C et al 1989 Tamoxifen treatment increases the concentration of 52K-cathepsin D and its precursor in breast cancer tissue. Cancer 63:1265–1270

Miner JN, Yamamoto KR 1991 Regulatory cross-talk at composite response elements. Trends Biochem Sci 16:423–426

Rochefort H, Capony F, Garcia M 1990 Cathepsin D in breast cancer: from molecular and cellular biology to clinical applications. Cancer Cells 2:383–388

Schüle R, Rangarajan P, Kliewer S et al 1990 Functional antagonism between oncoprotein c-Jun and the glucocorticoid receptor. Cell 62:1217–1226

Tung L, Mohamed KM, Hoeffler JP, Takimoto GS, Horwitz KB 1993 Antagonist-occupied human progesterone B-receptors activate transcription without binding to progesterone response elements, and are dominantly inhibited by A-receptors. Mol Endocrinol 7:1256–1265

Vignon F, Bouton MM, Rochefort H 1987 Antiestrogens inhibit the mitogenic effect of growth factors on breast cancer cells in total absence of estrogens. Biochem Biophys Res Commun 146:1502–1508

Human prostate cancer: a direct role for oestrogens

Luigi A. M. Castagnetta*† and Giuseppe Carruba*

*Hormone Biochemistry Laboratories, University of Palermo and †Experimental Oncology & Molecular Endocrinology Units, Palermo Branch of the National Institute for Cancer Research of Genoa, Cancer Hospital Centre 'M. Ascoli', PO Box 636, Palermo, Italy

Abstract. We have studied the response to oestrogen and expression of oestrogen receptors in responsive LNCaP and androgen non-responsive PC3 human prostate cancer cell lines. Growth of LNCaP cells is significantly stimulated by physiological concentrations of oestradiol; this growth increase appears to be comparable to that induced by either testosterone or dihydrotestosterone. In contrast, oestradiol significantly inhibits the proliferation of PC3 cells. We also present novel evidence for functional oestrogen binding in LNCaP cells. This evidence was first obtained by means of radioligand binding assays and was further corroborated using: (a) immunocytochemical analysis of oestrogen and progesterone receptors; (b) reverse transcriptase polymerase chain reaction of oestrogen receptor mRNAs; and (c) immunofluorescence of the 27 kDa heat shock protein (Hsp27), which has been reported to be a marker of functional oestrogen receptors. There appeared to be significantly and consistently lower levels of oestrogen receptor expressed in PC3 cells than in LNCaP cells. The observation that oestradiol-induced growth of LNCaP cells is completely reversed by the pure anti-oestrogen ICI 182 780 clearly implies that the biological response of these cells to oestradiol is mediated mainly via its own receptor. On the other hand, use of a neutralizing antibody against transforming growth factor (TGF)-β1 results in a remarkable increase in the growth of PC3 cells; this effect is almost completely abolished after the addition of oestradiol. This suggests that the oestradiol-induced growth inhibition may be mediated by TGF-β1. These results suggest that the current model for hormone-dependence of human prostatic carcinoma should be revised. This is of special concern, because recent data indicate that prostate cancer has become the most prevalent cancer and the second principal cause of cancer death in western countries.

1995 Non-reproductive actions of sex steroids. Wiley, Chichester (Ciba Foundation Symposium 191) p 269–289

The incidence of prostate cancer varies greatly throughout the world; it is highest in African–Americans and lowest in the Asian populations of China and India (Zaridze et al 1984). Overall, statistical data indicate that during the last decade its incidence has been rising dramatically worldwide, to such a level that prostate cancer has become the most prevalent cancer and the second principal cause of cancer death (Carter & Coffey 1990). The most recent epidemiological studies give an alarming indication of the dimensions of the problem: in the USA, more than 200 000 new cases of prostatic cancer are to be expected in 1994, resulting in approximately 58 000 deaths (Garnick 1994).

Autopsy studies have revealed that the human prostate has a uniquely high occurrence of latent microcarcinomas, the latter being associated with the mortality rates of clinically apparent prostate cancer (Breslow et al 1977). It has been postulated that tumour initiation does not differ in populations having high or low mortality rates, but that geographical differences in both prevalence of latent prostatic cancer and mortality can be ascribed to the influence of diverse tumour-promoting factors, either environmental or dietary.

It is a widely held belief that human prostate cancer is essentially androgen dependent. This assumption has influenced strategies for endocrine therapy (including total androgen blockade) of prostate cancers. In contrast, experimental results from animal model systems have shown that long-term treatment with testosterone and/or oestradiol, but not dihydrotestosterone, results in the development of prostatic tumours (Leav et al 1978). This has led to the hypothesis of conjoint, androgen-supported oestrogen-enhanced induction of aberrant prostate growth (Leav et al 1989). It is worth noting that pharmacological doses of synthetic oestrogens, such as diethylstilbestrol (DES), have proved to be an effective therapy for advanced human prostatic cancer. The high response rates (70–80%) obtained in prostate cancer patients have been ascribed merely to the fact that, following DES administration, plasma levels of testosterone drop to those found in castrates (Turkes et al 1988). The latter, however, have been argued to be still compatible with the maintenance of malignant prostate growth (Labrie et al 1993). In addition, intravenous stilbestrol diphosphate, and oestrogens in general, have been reported to be beneficial even in patients with metastatic, hormone-refractory disease (Ferro 1991).

In vitro systems are frequently used as models for most hormone-dependent tumours, including those of the prostate. We have investigated both androgen-responsive LNCaP (Fig. 1A) and androgen-unresponsive PC3 (Fig. 1B) human prostate cancer cell lines to gain insight into mechanisms of oestrogen growth control in these systems.

Expression of oestrogen-binding proteins and mRNA transcripts

High-affinity sites of oestrogen binding

Several studies have explored the effects and the mechanisms of action of oestradiol and other oestrogen-related steroids in the growth regulation of

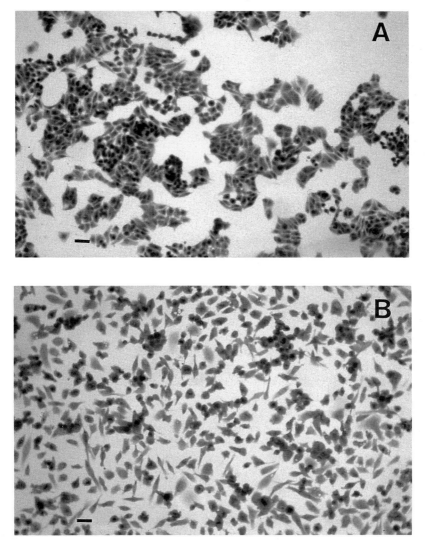

FIG. 1. Photomicrographs of (A) LNCaP and (B) PC3 cells in culture (bar = 500 μm; Papanicolau staining).

androgen-responsive LNCaP human prostate cancer cells. Investigators have mainly used radioligand binding assays to measure either the androgen or the oestrogen receptor content of this cell line.

Originally, Horoszewicz et al (1983) revealed the presence of oestrogen-binding sites in the cytosolic fraction of LNCaP cells, with dissociation

TABLE 1   Studies on androgen receptor status and relative binding affinity (RBA) for oestradiol in LNCaP human prostate cancer cells

Author	Cell fraction assayed	Radioligand concentration (nM)	Dissociation constant (nM)	Concentration (fmol/mg P)	RBA for oestradiol[a]
Berns et al 1986[b]	cytosol nuclear extract	10 R1881 25 DHT	NR	900 1679	ND
Schuurmans et al 1988	cytosol	0.5–10	0.4	920	4.3
Sonnenschein et al 1989	whole cell extracts	2–20	3.0	68	0.4
Veldscholte et al 1990b	cytosol	0.5–10	0.4	NR	2.4
Kirschenbaum et al 1993	whole cells	0.3–6	3.1	1366	6.0

[a]RBA was expressed (in percentage) as the ratio of amounts of unlabelled R1881 and competing oestradiol which are required to inhibit binding of tritiated R1881 by 50%. The RBA for R1881 was set as 100%.
[b]Single point assay.
P, protein; NR, not reported; ND, not determined.

constant (K_d) values in the nanomolar range (nearly 5 nM). Further studies, however, failed to detect oestrogen receptor in either nuclear extract or soluble fraction of LNCaP cells (Berns et al 1986, Sonnenschein et al 1989). Parallel work has indicated that LNCaP cells contain one point mutation in codon 868 of the androgen receptor gene; this mutation (A to G) results in the transition of a threonine into an alanine residue at the C-terminus of the steroid-binding domain of the receptor (Trapman et al 1990). It has been suggested that this abnormality confers to the androgen receptor system a broader steroid specificity, allowing binding of oestrogens, progestogens and several anti-androgens, and the resulting induction of gene expression (Veldscholte et al 1990a). Results from relevant studies are summarized in Table 1. As can be seen, the relative binding affinity (RBA) reported for oestradiol ranged from 0.4–6.0% of that of the synthetic androgen R1881. It must be emphasized that most of these studies used ligand concentrations (sometimes even as a single point assay), which do not permit the proper identification of high-affinity, low-capacity (type I) binding sites, but instead mostly involve type II sites, which have higher K_d values (in the nanomolar range). In keeping with the original definition of the biochemical and functional features relevant to distinct sites of oestrogen binding (Clark & Peck 1979), this may represent a critical shortcoming, especially in the light of the fact that only the type I oestrogen receptors have been shown to be of value in the prognosis of and for the

TABLE 2 Status of soluble and nuclear Type I and Type II oestrogen receptors in LNCaP cells

	Soluble fraction			Nuclear fraction		
	K_d (nM)	fmol DNA	Sites/cell	K_d (nM)	fmol DNA	Sites/cell
Type I	0.49 ±0.03	2124 ±368	79 755 ±4603	0.25 ±0.09	328.1 ±48.2	13 737 ±2006
Type II	7.24 ±0.30	11 785 ±3867	442 520 ±48 369	7.02 ±3.51	1942 ±513	81 308 ±21 350

Oestrogen receptor content and status of LNCaP cells was assessed in both soluble and nuclear compartments through radioligand binding assays (for details of methodology see Carruba et al 1993). Values represent means ± SD of $n=7$ assays.

response to endocrine treatment of breast and endometrial cancer patients (Castagnetta et al 1987a, 1992).

We have recently investigated the presence of oestrogen-binding sites in both LNCaP and PC3 cells by radioligand binding assay, using a ligand ([3H]oestradiol) concentration range of 0.1–5 nM (precisely, 0.1, 0.2, 0.3, 0.5, 0.75, 1, 2, 3 and 5 nM) and a 100-fold excess of unlabelled DES or R1881 for competition studies. As shown in Table 2, our data clearly indicate that both type I (high affinity, limited capacity) and type II (lower affinity, higher capacity) oestradiol-binding proteins are present in either the cytosolic or nuclear fraction of LNCaP cells and that their levels are comparable to those we have found in other oestrogen-responsive cancer tissues (Castagnetta et al 1987a,b) and cells (Lo Casto et al 1983, Castagnetta et al 1986). Conversely, PC3 cells did not show any detectable oestradiol-binding sites in the soluble fraction, whereas type I oestrogen receptors were found in the nuclear fraction. These had mean K_d and concentration values, respectively, of 0.16 ± 0.04 nM and 118.8 ± 22.3 fmol/mg DNA (mean ± SD, $n=5$). The results of competition studies in LNCaP cells are worth noting (see Fig. 2). Using a 100-fold excess of R1881, an appreciable increase of bound radioligand was observed in both cell fractions; conversely, DES competed effectively for both soluble and nuclear oestrogen binding, a consistent displacement of [3H] oestradiol being observed (ranges of 36.9–57.0% and 35.5–53.4%, respectively). Surprisingly, when R1881 was used as a competitor, Scatchard analysis revealed a substantial increase in type I oestrogen receptor content in both soluble (over 90%) and nuclear (nearly 30%) fractions of LNCaP cells. This evidence suggests that [3H]oestradiol may partly bind to type II androgen receptors; however, since the addition of excess R1881 displaces this type II androgen binding only, a considerable proportion of labelled oestradiol would be made available for further binding to type I oestrogen receptors. Type II binding sites may serve as

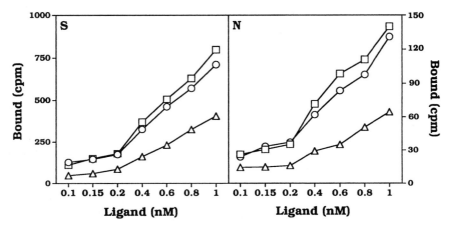

FIG. 2.   Ligand binding curves of oestradiol in LNCaP cells. The amount of bound oestradiol was measured through incubation of both soluble (S) and nuclear (N) cell fractions with increasing concentrations of [³H]oestradiol alone (○) and [³H]oestradiol plus a 100-fold excess of the synthetic oestrogen diethylstibesterol (△) or the synthetic androgen R1881 (□). cpm, counts per minute.

storage proteins, which have a reduced affinity and a broader specificity of binding for different steroids (including androgens, oestrogens and progestagens) and antisteroids. Since the conditions currently used for androgen receptor assay are compatible with a prevalent measurement of type II receptors, this may also account for the displacement of androgen binding observed in previous studies using R1881 as the radioligand and oestradiol as the competitor (Sonnenschein et al 1989, Veldscholte et al 1990a, Schuurmans et al 1991, Kirschenbaum et al 1993).

Another source of major concern, which is often mistakenly disregarded, originates from both the conditions used for routine maintenance of cells and the age of the culture. Concerning LNCaP cells, culture conditions vary greatly among several studies, including differences in the culture medium used (RPMI, DME, Ham's F12) and the percentage (5–15%) and treatment (heat-inactivated or not) of fetal calf serum. It is well documented that the parental LNCaP cell line may give rise to several sublines, also depending on steroid addition or depletion, in a relatively short time (van Steenbrugge et al 1991, Kirschenbaum et al 1993). The receptor studies herein discussed have used LNCaP cells which have a wide range of passage number (from the 28th to the 72nd passage *in vitro*), corresponding to between 100 and 400 population doublings after the original isolation. This, apart from the differences in the experimental conditions used, could make it difficult to obtain reliable or comparable results. In our experience, the use of cells at low, narrow-span passages may avoid this problem and limit the risks inherent in *in vitro* systems, whereby a heterogeneous cell line is invariably exposed to the selective pressures of an otherwise artificial environment.

Quantitative immunocytochemistry of oestrogen and progesterone receptors

Although it is widely recognized that oestradiol stimulates growth of the androgen receptor-positive LNCaP cells, studies on the oestrogen receptor content of this cell line have been surprisingly rare. In a previous report, no oestrogen or progesterone receptors were found in LNCaP cells by either ligand binding or immunocytochemical assays (Berns et al 1986). These authors used monoclonal antibodies directed against the oestogen receptor from MCF-7 human mammary cancer cells but, unfortunately, failed to include details of the methodological approach. More recently, Brolin et al (1992), using the same immunocytochemical assay, reported that both LNCaP and PC3 cells were found to be negative for oestrogen and progesterone receptors.

We have investigated the presence of both oestrogen and progesterone receptors in LNCaP, DU145 and PC3 cells by using modified versions of the commercially available Abbott immunocytochemical assay kits. We analysed the receptor staining using the CAS™200 image analysis system, which automatically estimates the percentage of positively stained nuclei and measures the intensity of staining; the latter is defined as the summed optical density for the receptor-positive nuclear area over the summed total optical density of all the nuclei, expressed as a percentage. Percentages $\leqslant 30\%$, from 30 to 60% and >60%, respectively, identified weak, moderate and strong intensities of staining. As illustrated in Plate II, LNCaP cells stained intensely for both receptors; the percentages of positive nuclei were 56.7–66.4% and 59.0–72.6%, respectively, for oestrogen and progesterone receptors. The intensity of staining was consistently strong (69.7–82.8% for oestrogen receptor and 79.8–87.9% for progesterone receptor), with a coefficient of variation of <15%. In contrast, PC3 cells displayed very low expression of oestrogen receptors, with about 25% of positive nuclei having a weak-to-moderate degree of staining, and all cells were progesterone receptor-negative. DU145 cells were negative for both oestrogen and progesterone receptors. Our results are at variance with those of Brolin et al (1992). However, this discrepancy could be ascribed simply to the different conditions used for exposure to primary antibodies (1 h at 37 °C as opposed to 24 h at 4 °C in our assay).

It is worth noting that progesterone receptors are commonly thought to associate with functional oestrogen receptors in human breast and endometrial cells; consequently, clinicians have been using their expression as a helpful indicator in the management of breast cancer patients. The presence of progesterone receptors in the LNCaP cell line therefore reinforces the

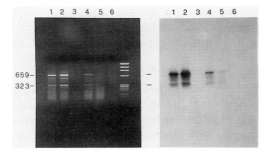

FIG. 3. Reverse transcriptase polymerase chain reaction (RT-PCR) of oestrogen receptors in LNCaP and PC3 prostate cancer cells. (*Left panel*) Ethidium bromide staining of agarose gel electrophoresis of PCR amplification products obtained from one milligram total RNA: lane 1, ZR75-1; lane 2, MCF-7; lane 3, MDA-MB231; lane 4, LNCaP; lane 5, PC3 cells; lane 6, RT-PCR of an RNA-free control sample; marker lane, φx-174-*Hae* III digest fragments (length given in bp). (*Right panel*) Southern blot analysis of the PCR-amplified DNA from the left panel. Samples were hybridized using a ^{32}P-labelled human oestrogen receptor cDNA as a probe.

conviction that these cells contain an apparently intact receptor machinery suited to mediate the biological effects of oestrogen.

Reverse transcriptase polymerase chain reaction of
oestrogen receptor mRNA transcripts

We investigated the expression of oestrogen receptor mRNAs in human prostate tumour cell lines using reverse transcriptase polymerase chain reaction (RT-PCR) amplification of total RNA extracted from LNCaP, DU145 and PC3 cells; oestrogen receptor-positive ZR75-1 and MCF-7 and oestrogen receptor-negative MDA-MB231 human mammary cancer cell lines were used for comparison. We performed PCR amplification using an oestrogen receptor cDNA (kindly supplied by Pierre Chambon, University of Strasbourg, France) as a template, with a sense primer corresponding to a sequence in exon 3 and an anti-sense primer corresponding to a sequence in exon 6 of the human oestrogen receptor gene. We separated the amplification products on a 1.4% agarose gel and confirmed their identity by Southern blot analysis, using the complete oestrogen receptor cDNA as a probe. As shown in Fig. 3, specific hybridization was seen in mammary ZR75-1 and MCF-7 cells, similar to that in prostate LNCaP and PC3 cells. In particular, two major bands (of 659 and 323 bp) and some minor bands in between were observed in ZR75-1, MCF-7, LNCaP and PC3 cells, whereas no detectable amplification product was found in the oestrogen non-responsive MDA-MB231 breast cancer cell line. The first band corresponds well to the expected length (nucleotides 937–1596 of the oestrogen receptor sequence) of a normal oestrogen receptor mRNA, as determined by comparison with the φx-*Hae* III marker fragments. The second band originates from a variant oestrogen receptor mRNA, lacking the entire exon 4; this variant, which is likely to represent a product of alternative splicing, has been isolated and characterized in our laboratories from human breast tumour cell lines (Pfeffer et al 1993). The bands of intermediate length are of unknown origin, although one amplification product shows a length compatible with a putative variant mRNA lacking exon 5, which would also be amplified by the primer set used. Using a nested primer set with an anti-sense primer in exon 4, we observed a single band in all cases; moreover, digestion of the reaction products by the restriction enzyme *Hin*d III, which makes one cut in the amplified region (oestrogen receptor exon 4), yielded identical fragments of the expected size (not shown).

For the first time, we have obtained a normal oestrogen receptor mRNA in LNCaP cells using the RT-PCR system. In addition, a rather abundant expression of a variant oestrogen receptor mRNA is also reported; this variant is jointly expressed with the normal messenger only in oestrogen receptor-positive, oestrogen-responsive MCF-7 and ZR75-1 mammary carcinoma cell lines, whereas both mRNAs are absent from oestrogen receptor-negative, non-responsive MDA-MB231 cells. It is worth mentioning that the amplification primers used (24-mers) show nine and 10 mismatches to the androgen receptor sequence and, if they were to give rise to any amplification product from that sequence, it would not be of the observed length.

Although RT-PCR is not a quantitative technique, the relative amounts of the oestrogen receptor mRNAs were reproducibly different in the cell lines studied (Fig. 3). The levels of expression of both oestrogen receptor mRNAs that were observed in the androgen-sensitive LNCaP cells were similar to those in the MCF-7 and ZR75-1 cells, whereas both mRNAs appeared to be expressed at far lower levels in the androgen non-responsive PC3 cells. The two mammary carcinoma cell lines showed equivalent expression levels, although this was slightly higher in MCF-7 than in ZR75-1 cells. Although the RNA isolated from the PC3 cell line gave rise to low levels of amplification products, no reaction could be observed in the oestrogen non-responsive human breast cancer cell line MDA-MB231. Therefore, the low expression of oestrogen receptor mRNA in PC3 cells is clearly significant and may also account for the apparent lack of soluble oestrogen receptor seen in this cell line in the radioligand binding assay and the absence of cytochemical staining for progesterone receptors.

The oestrogen receptor-associated 27 kDa heat shock protein

As with the larger heat shock proteins (Hsps), a group of small proteins, in the range of 24–30 kDa, have repeatedly been reported to be involved in cellular thermotolerance and response to a miscellany of both growth and differentiation factors. In particular, recent studies have revealed that the 28 kDa protein found in MCF-7 human mammary cancer cells corresponds to both the oestrogen-related 24 kDa protein and the mammalian Hsp27 (Faucher et al 1993). Interestingly, immunological evidence has indicated that the oestrogen-regulated Hsp27 and the oestrogen receptor-associated 29 kDa protein (p29) are the same molecule (Ciocca & Luque 1991). The p29 phosphoprotein was previously characterized as a cytosolic component of both breast and endometrial cancer tissues that is quantitatively and qualitatively related to the oestrogen receptor but not to other steroid receptors or other binding proteins (King et al 1987). It was proposed that p29 represents a non-hormone-binding component of the receptor mechanism which forms a complex with the oestrogen receptor under certain conditions, such as treatment with ammonium sulphate (Cano et al 1986), and that it may be used as a marker of oestrogen sensitivity in breast and endometrial epithelial cells (King et al 1987) and in cultured fetal Leydig cells (Tsai-Morris et al 1986).

Although several lines of evidence favour the idea that Hsp27 is involved in oestrogen action, it is still uncertain whether this protein actually participates in the processes of association and dissociation of oestrogen receptor heterocomplexes or, alternatively, whether it is involved in the receptor activation and DNA-binding phenomena.

On this basis, we have also inspected expression of Hsp27 in human prostate cancer cells. We immunostained the cells with a D5 monoclonal antibody (Coffer et al 1985) raised against Hsp27 which was visualized through a fluorescein isothiocyanate-conjugated secondary antiserum. Positive staining was observed in both LNCaP and PC3 cells. In LNCaP cells, the majority (over 85%) of cells showed a consistently intense staining, mostly located in the cytoplasm (Fig. 4A). Conversely, PC3 cells displayed a positive staining that was mainly nuclear, although some cytoplasmic staining was also seen (Fig. 4B); the percentage of positive cells was lower than that seen in LNCaP cells (approximately 50%), with minor discrepancies in staining intensity. Control wells from either cell line not receiving the primary antibody stained very poorly, indicating that the Hsp27 staining was specific.

Although a linear correlation between oestrogen receptor content and Hsp27 levels has been reported in human breast tumours, tissues with low levels of oestrogen receptor also exhibited a variable degree of Hsp27 staining (Cano et al 1986). Previous evidence from our laboratories indicated that Hsp27 is associated with nuclear oestrogen receptor in smooth muscle cells of human aorta (Campisi et al 1993). In addition, although the relationship between oestrogen receptor and Hsp27 has been established within individual target tissues, this cannot be true for certain other tissues, including prostate (King et al 1987). Recent studies revealed that presence of Hsp27 is associated with oestrogen receptor in human prostate tumour tissues and is related to the subsequent prognosis for the patients (Ohishi et al 1992).

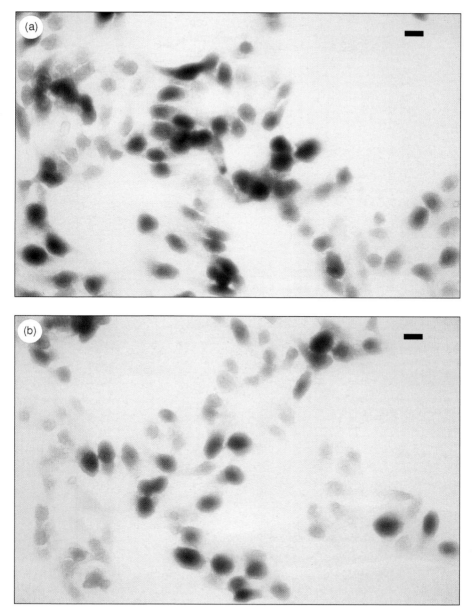

PLATE 2 Immunocytochemistry of oestrogen (a) and progesterone (b) receptors in LNCaP prostate cancer cells. Immunocytochemical assays of both receptors were carried out using modified versions of the oestrogen and progesterone receptor Abbott kits. Cells were incubated for 24 h at 4 °C with primary monoclonal anti-oestrogen receptor or monoclonal anti-progesterone receptor antibodies. The reaction was visualized by means of a peroxidase–anti-peroxidase (PAP) system using diaminobenzydine as the chromogen substrate (bar = 150 μm).

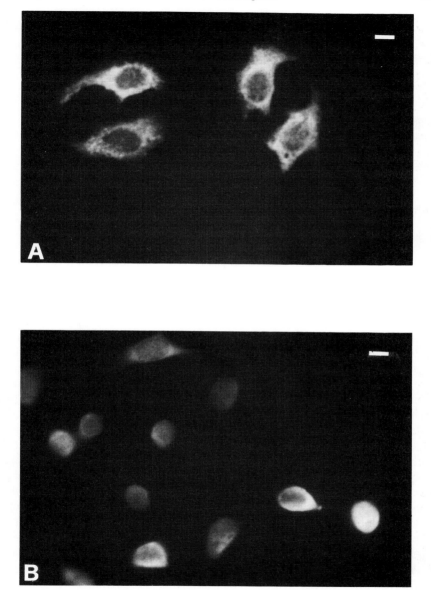

FIG. 4. Immunofluorescence of the 27 kDa heat shock protein (Hsp27) in (A) LNCaP and (B) PC3 prostate cancer cells. Expression of Hsp27 was visualized using a D5 primary monoclonal antibody and a secondary fluorescein isothiocyanate-conjugated antiserum (bar = 150 μm).

TABLE 3   **Reports on the oestradiol-induced growth stimulation of LNCaP cells**

Author	Year	Stimulation (-fold)	Concentration (nM)
Schuurmans et al	1988	3.5	10
Sonnenschein et al	1989	3.3	10
Veldscholte et al	1991	3.0	10
Kirschenbaum et al	1993	2.0	1–100
Castagnetta et al	1994	2.2	0.01–1

Growth response to oestradiol

Oestradiol-induced growth stimulation of LNCaP cells

Previous studies have consistently reported that oestradiol significantly stimulates proliferation of LNCaP cells (see Table 3) at levels comparable to or even greater than those produced by androgens (Schuurmans et al 1988, Sonnenschein et al 1989, Iguchi et al 1990, Castagnetta et al 1994). It has been proposed that the mitogenic effects of oestradiol on LNCaP cells may be mediated via the point-mutated form of the androgen receptor (Veldscholte et al 1990b). However, the very low RBA values of oestradiol for the androgen receptor (see Table 1) cannot explain its remarkable proliferative potency. Similarly, both progesterone and pregnenolone have been reported to be more potent inducers of growth of LNCaP cells than expected on the basis of their RBA values (Sonnenschein et al 1989). Therefore, the assumption that oestradiol acts via an abnormal androgen receptor appears to be unfounded. We have measured [^{3}H]thymidine incorporation in LNCaP cells cultured for a week in steroid-deprived medium and then exposed for six days to different concentrations of oestradiol (0.01–100 nM), in the presence or absence of the pure anti-oestrogen ICI 182 780 (Wakeling et al 1991). As reported in Table 4, growth of LNCaP cells was significantly increased by all doses of oestradiol. A peak stimulation of over 120% above control values was seen at the 0.01 nM dose ($P < 10^{-6}$). This is not surprising, at least in the present experimental conditions. The simultaneous addition of 100 nM ICI 182 780 completely abolished the oestradiol-induced growth stimulation at any oestradiol concentration, whereas ICI 182 780 itself did not significantly affect cell proliferation. This evidence strongly supports the view that oestradiol acts via its own receptor. This finding is also in accord with a previous observation that the pure anti-oestrogen EM 139 not only reversed the androgen-induced increase in LNCaP proliferation, but also significantly inhibited basal cell growth (de Launoit et al 1991).

TABLE 4   Effects of oestradiol and ICI 182 780 on tritiated thymidine uptake by LNCaP cells

Treatment	[³H]thymidine uptake (dpm)
None	38 538 ± 1799
0.01 nM oestradiol	87 255 ± 9459
1 nM oestradiol	64 766 ± 5297
100 nM oestradiol	54 789 ± 2928
ICI 182 780[a]	37 267 ± 1933
ICI 182 780[a] + 0.01 nM oestradiol	36 932 ± 1760
ICI 182 780[a] + 1 nM oestradiol	36 832 ± 1706
ICI 182 780[a] + 100 nM oestradiol	37 586 ± 1202

[a]100 nM.

Values represent mean disintegrations per minute (dpm) ± SD; $n = 4$. Cells, grown in steroid-deprived medium, were exposed to increasing concentrations of oestradiol, with or without ICI 182 780, or to ICI 182 780 alone. Cell proliferation was measured by means of a six-hour pulse of [³H]thymidine and an estimate of the acid-precipitable counts.

Oestradiol-induced growth inhibition of PC3 cells

Further interesting evidence comes from the androgen-unresponsive PC3 cells. As illustrated in Fig. 5, oestradiol exerts a clear inhibitory activity on growth of these cells. As can be seen, after six days' exposure, PC3 cells displayed a dose-related inhibition of growth, which was maximal at 10^{-7} M oestradiol (55.2% with respect to control; $P < 10^{-6}$) but still significant at physiological concentrations. At higher doses of oestradiol (1 nM or more) this effect became evident after 72 h incubation (not shown). This negative growth regulation is cognate to that observed in PC3 cells after the addition of 1 ng/ml transforming growth factor (TGF)-β1 (54.4% of control) under exactly the same experimental conditions. Consequently, we have explored the possibility that the oestradiol-induced growth inhibition of PC3 cells is mediated by TGF-β. Interestingly, the addition of a neutralizing anti-TGF-β1 antibody in stringent experimental conditions provoked a remarkable increase (close to 300% of control) in cell proliferation (Fig. 5), suggesting the presence of high levels of endogenous TGF-β1 in PC3 cells. This effect was opposed by the addition of oestradiol (10^{-11} to 10^{-7} M), and was almost completely reversed at the 100 nM dose (Fig. 5). However, data from Northern blot analysis indicate that different oestradiol doses (10^{-11} to 10^{-7} M) do not affect TGF-β1 mRNA expression in PC3 cells after various (six, 24 or 72 h) incubation times (Carruba et al 1994a). This is not unexpected, since Knabbe et al (1987) found that anti-oestrogens such as tamoxifen have no influence on TGF-β1 mRNA in the oestrogen-dependent MCF-7 human breast cancer cells, although tamoxifen

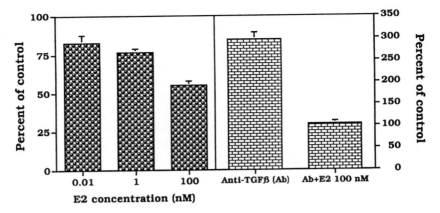

FIG. 5.   Growth effects of oestradiol and anti-transforming growth factor (TGF)-β1 antibody in PC3 cells. Cells were exposed for six days to increasing concentrations (0.01 to 100 nM) of 17β-oestradiol (E$_2$) or anti-TGF-β1 (Ab) antibody (9.2 μg/ml) $\pm$ 100 nM E$_2$. Values represent means $\pm$ SEM (bars) of $n=3$ experiments. Degrees of significance (P) with respect to controls (two-tailed Student's t-test, 95% confidence limits): 0.01 nM E$_2$, $P<0.0002$; 1 nM, $P<0.00002$; 100 nM, $P<10^{-6}$; Ab, $P<0.00001$; Ab + 100 nM E$_2$, $P=0.435$ (no significant difference).

treatment increased TGF-β1 production (up to 27-fold) in this cell line. More recent studies have revealed that levels of TGF-β1 mRNA in either oestrogen receptor-positive breast cancer cell lines (MCF-7 or ZR75-1), or oestrogen receptor-negative cancer cells (such as mammary MDA-MB231 [Arrick et al 1990] and endometrial HEC-50 [Anzai et al 1992]) remain unchanged following oestradiol administration. It is noteworthy that the very same oestrogen receptor-negative human cancer cells have been reported to be exquisitely sensitive to TGF-β1 (Arteaga et al 1988). Equally, growth of PC3 cells has been reported to be remarkably affected by TGF-β1, contrary to what occurs in LNCaP cells (Wilding et al 1989, Carruba et al 1994b).

Perspectives

Previous evidence from either *in vivo* or *in vitro* studies clearly suggests that oestrogens may play a role in the regulation of prostate cancer cell growth. In addition, the presence of high-affinity binding proteins for oestradiol has been revealed in both hyperplastic and cancerous human prostate tissues (Murphy et al 1980, Donnelly et al 1983, Ekman et al 1983). The results presented here show that oestradiol negatively regulates the growth of PC3 cells, but in LNCaP cells it behaves as a mitogen at physiological (10^{-11} to 10^{-9} M) concentrations; in addition, exposure of LNCaP cells to higher doses of oestradiol may produce, in stringent experimental conditions, a decrease in cell proliferation (L. Castagnetta et al, unpublished work). This biphasic response of LNCaP cells to oestradiol parallels that repeatedly observed for androgens by different research groups, including our own (Sonnenschein et al 1989, Carruba et al 1994b).

The oestradiol-induced growth stimulation of LNCaP cells appears to be mediated via specific, high-affinity binding proteins that are refractory to any interfering androgen, such as R1881. In addition, the consistent expression of both the progesterone receptor and Hsp27 supports the presence of an intact action mechanism of oestrogen. This view is also corroborated by evidence that the addition of the pure anti-oestrogen ICI 182 780 entirely abolishes the mitogenic effect of oestradiol on LNCaP cells. The latter evidence, coupled with the inconsistency between RBA values and relative proliferative potency of oestradiol, leads us to conclude that the effects of oestradiol on the growth of LNCaP cells are totally independent of the point-mutated form of the androgen receptor. Incidentally, type I and II oestrogen receptor content is remarkably (over fivefold) higher than that of the androgen receptor we have recently measured in LNCaP cells (Carruba et al 1994b).

The overall functions of oestrogen in human prostate cancer, at least in these *in vitro* systems, appear to be more complex and disparate than expected. On the one hand, mechanisms responsible for the oestradiol-induced growth inhibition of PC3 cells may help explain the beneficial effects of pharmacological doses of oestrogen in advanced human prostate cancer, even in the androgen-independent state. On the other hand, the evidence that oestradiol stimulates growth of LNCaP cells may account for results previously obtained in animal model systems, favouring the resulting concept of an androgen-supported oestrogen-enhanced stimulation of prostate epithelium as an essential requirement for overcoming factors that normally limit cell proliferation (Leav et al 1989). The question of whether the different oestrogen receptor expression levels observed in LNCaP and PC3 cells may somehow be related to their distinct biological response to oestradiol is worthy of further investigation.

Overall, the varied activities triggered by oestradiol strongly suggest, at least *in vitro*, a direct role for oestrogen function in the regulation of the growth of prostate cancer cells. Consequently, the current criterion of hormone-dependence of human prostatic cancer should be reassessed.

Acknowledgements

The authors are indebted to P. Chambon (University of Strasbourg, France) and to R. J. B. King (University of Surrey, Guildford, UK) for their generous provision of the oestrogen receptor cDNA and the D5 monoclonal antibody, respectively. The authors also wish to thank M. Cutolo (Department of Rheumatology, Genova, Italy) for technical help with the immunofluorescence experiments, and U. Pfeffer (Molecular Biology Unit, IST Genova, Italy) for assistance with the RT-PCR system. These studies have been partly funded by the Italian Association for Cancer Research (AIRC) and by the National Research Council (CNR, Special Project Ageing, c.no. 93.435.PF40).

References

Anzai Y, Gong Y, Holinka CF et al 1992 Effects of transforming growth factors and regulation of their mRNA levels in two human endometrial adenocarcinoma cell lines. J Steroid Biochem Mol Biol 42:449–455

Arrick BA, Korc M, Derynck R 1990 Differential regulation of expression of three transforming growth factor β species in human breast cancer cell lines by estradiol. Cancer Res 50:299–303

Arteaga CL, Tandon AK, von Hoff DD, Osborne CK 1988 Transforming growth factor β: potential autocrine growth inhibitor of estrogen receptor-negative human breast cancer cells. Cancer Res 48:3898–3904

Berns EMJJ, de Boer W, Mulder E 1986 Androgen-dependent growth regulation of and release of specific protein(s) by the androgen receptor containing human prostate tumour cell line LNCaP. Prostate 9:247–259

Breslow N, Chan CW, Dhom G et al 1977 Latent carcinoma of prostate at autopsy in seven areas. Int J Cancer 20:680–688

Brolin J, Skoog L, Ekman P 1992 Immunohistochemistry and biochemistry in detection of androgen, progesterone, and estrogen receptors in benign and malignant human prostatic tissue. Prostate 20:281–295

Campisi D, Cutolo M, Carruba G et al 1993 Evidence for soluble and nuclear site I binding of estrogens in human aorta. Atherosclerosis 103:267–277

Cano A, Coffer AI, Adatia R, Millis RR, Rubens RD, King RJB 1986 Histochemical studies with an estrogen receptor-related protein in human breast tumours. Cancer Res 46:6475–6480

Carruba G, Pfeffer U, Fecarotta E et al 1994a Estradiol inhibits growth of hormone non-responsive PC3 human prostate cancer cells. Cancer Res 54:1190–1193

Carruba G, Leake RE, Rinaldi F et al 1994b Steroid-growth factor interaction in human prostate cancer. 1. Short-term effects of transforming growth factors on growth of human prostate cancer cells. Steroids 59:412–420

Carter HB, Coffey DS 1990 The prostate: an increasing medical problem. Prostate 16:39–48

Castagnetta L, Granata OM, Lo Casto M, Miserendino V, Calò M, Carruba G 1986 Estrone conversion rates by human endometrial cancer cell lines. J Steroid Biochem 25:803–809

Castagnetta L, Lo Casto M, Granata OM, Calabrò M, Ciaccio M, Leake RE 1987a Soluble and nuclear oestrogen receptor status of advanced endometrial cancer in relation to subsequent clinical prognosis. Br J Cancer 55:543–546

Castagnetta L, Traina A, Di Carlo A, Latteri AM, Carruba G, Leake RE 1987b Heterogeneity of soluble and nuclear oestrogen receptor status of involved nodes in relation to primary breast cancer. Eur J Cancer Clin Oncol 23:31–35

Castagnetta L, Traina A, Carruba G, Fecarotta E, Palazzotto G, Leake RE 1992 The prognosis of breast cancer patients in relation to the oestrogen receptor status of both primary disease and involved nodes. Br J Cancer 65:167–170

Castagnetta L, Miceli MD, Sorci MG et al 1994 Growth of LNCaP cells is stimulated by estradiol via its own receptor. Endocrinology 136, in press

Ciocca DR, Luque EH 1991 Immunological evidence for the identity between the hsp27 estrogen-regulated heat shock protein and the p29 estrogen-receptor associated protein in breast and endometrial cancer. Breast Cancer Res Treat 20:33–42

Clark JH, Peck EJ Jr 1979 Characteristics of cytoplasmic and nuclear receptor forms. In: Gross F, Labhart A, Mann T, Zander J (eds) Monographs on endocrinology, vol 14: Female sex steroids receptors and function. Springer-Verlag, Berlin, p 46–69

Coffer AI, Lewis KM, Brockas AJ, King RJB 1985 Monoclonal antibodies against a component related to soluble estrogen receptors. Cancer Res 45:3686–3693

de Launoit Y, Veilleux R, Dufour M, Simard J, Labrie F 1991 Characteristics of the biphasic action of androgens and of the potent antiproliferative effects of the new pure antiestrogen EM-139 on cell cycle kinetic parameters in LNCaP human prostatic cancer cells. Cancer Res 51:5165–5170

Donnelly BJ, Lakey WH, McBlain WA 1983 Estrogen receptor in human benign prostatic hyperplasia. J Urol 130:183–187

Ekman P, Barrack ER, Greene GL, Jensen EV, Walsh PC 1983 Estrogen receptors in human prostate: evidence for multiple binding sites. J Clin Endocrinol & Metab 57:166–176

Faucher C, Capdevielle J, Canal I et al 1993 The 28-kDa protein whose phosphorylation is induced by protein kinase C activators in MCF-7 cells belongs to the family of low molecular mass heat shock proteins and is the estrogen-regulated 24-kDa protein. J Biol Chem 268:15168–15173

Ferro MA 1991 Use of intravenous stilbestrol diphosphate in patients with prostatic carcinoma refractory to conventional hormonal manipulation. Urol Clin North Am 18:139–143

Garnick MB 1994 The dilemmas of prostate cancer. Sci Am 270:52–59

Horoszewicz JS, Leoung SS, Kawinski E et al 1983 LNCaP model of human prostatic carcinoma. Cancer Res 43:1809–1818

Iguchi T, Fukazawa Y, Tani N et al 1990 Effect of some hormonally active steroids upon the growth of LNCaP human prostate tumour cells *in vitro*. Cancer J 3:184–191

King RBJ, Finlay JR, Coffer AE, Millis RR, Rubens RD 1987 Characterization and biological relevance of a 29-kDa oestrogen-related protein. J Steroid Biochem 27:471–475

Kirschenbaum A, Ren M, Levine AC 1993 Enhanced androgen sensitivity in serum-free medium of a subline of the LNCaP human prostate cancer cell line. Steroids 58:439–444

Knabbe C, Lippman ME, Wakefield LM et al 1987 Evidence that transforming growth factor-β is a hormonally regulated negative growth factor in human breast cancer cells. Cell 48:417–428

Labrie F, Belanger A, Simard J, Labrie C, Dupont A 1993 Combination therapy for prostate cancer. Endocrine and biological basis of its choice as new standard first-line therapy. Cancer (suppl 3) 71:1059–1067

Leav I, Merk FB, Ofner P et al 1978 Bipotentiality of response to sex hormones by the prostate of castrated or hypophysectomized dogs: direct effects of estrogen. Am J Pathol 93:69–92

Leav I, Merk FB, Kwan PW-L, Ho S-M 1989 Androgen-supported estrogen-enhanced epithelial proliferation in the prostates of intact Noble rats. Prostate 15:23–40

Lo Casto M, Granata OM, Castagnetta L 1983 Characterization of oestrogen receptors in some long-term tissue cultures. Ital J Biochem 32:211–213

Murphy JB, Emmott RC, Hicks LL, Walsh PC 1980 Estrogen receptors in the human prostate, seminal vesicle, epididymis, testis and genital skin: a marker for estrogen-responsive tissues? J Clin Endocrinol & Metab 50:938–947

Ohishi M, Miyake K, Koshikawa T, Asai J, Murase T 1992 Determination of ER-D5 (estrogen receptor related antigen) in prostatic cancer and its significance. Hinyokika Kiyo 38:789–796

Pfeffer U, Fecarotta E, Castagnetta L, Vidali G 1993 Estrogen receptor variant messenger RNA lacking exon 4 in estrogen-responsive human breast cancer cell lines. Cancer Res 53:741–743

Schuurmans ALG, Bolt J, Voorhorst MM, Blankenstein RA, Mulder E 1988 Regulation of growth and epidermal growth factor receptor levels of LNCaP prostate tumor cells by different steroids. Int J Cancer 42:917–922

Schuurmans ALG, Bolt J, Veldscholte J, Mulder E 1991 Regulation of growth of LNCaP human prostate tumor cells by growth factors and steroid hormones. J Steroid Biochem Mol Biol 40:193–197

Sonnenschein C, Olea N, Pasanen ME, Soto AM 1989 Negative controls of cell proliferation: human prostate cancer cells and androgens. Cancer Res 49:3474–3481

Trapman J, Ris-Stalpers C, van der Korput JA et al 1990 The androgen receptor: functional structure and expression in transplanted human prostate tumors and prostate tumor cell lines. J Steroid Biochem Mol Biol 37:837–842

Tsai-Morris C-H, Knox G, Luna S, Dufau ML 1986 Acquisition of estradiol-mediated regulatory mechanism of steroidogenesis in cultured fetal Leydig cells. J Biol Chem 261:3471–3474

Turkes AO, Peeling WB, Wilson DW, Griffiths K 1988 Evaluation of different endocrine approaches in the treatment of prostatic carcinoma. In: Klosterhalfen H (ed) New developments in biosciences, vol 4: Endocrine management of prostatic carcinoma. Walter de Gruyter, Berlin, p 75–86

van Steenbrugge GJ, van Uffelin CJG, Bolt J, Schroder FH 1991 The human prostatic cancer cell line LNCaP and its derived sublines: an *in vitro* model for the study of androgen sensitivity. J Steroid Biochem Mol Biol 40:207–214

Veldscholte J, Voorhorst-Ogink MM, Bolt-de Vries J, van Rooij HCJ, Trapman J, Mulder 1990a Unusual specificity of the androgen receptor in the human prostate tumor cell line LNCaP: high affinity for progestagenic and estrogenic steroids. Biochim Biophys Acta 1052:187–194

Veldscholte J, Ris-Stalpers C, Kuiper GG et al 1990b A mutation in the ligand binding domain of the androgen receptor of human LNCaP cells affects steroid binding characteristics and response to anti-androgens. Biochem Biophys Res Commun 173:534–540

Wakeling AE, Dukes M, Bowler J 1991 A potent specific pure antiestrogen with clinical potential. Cancer Res 51:3867–3873

Wilding G, Zugmeier G, Knabbe C, Flanders K, Gelmann E 1989 Differential effects of transforming growth factor β on human prostate cancer cells *in vitro*. Mol Cell Endocrinol 62:79–87

Zaridze DG, Boyle P, Smans M 1984 International trends in prostatic cancer. Int J Cancer 33:223–230

DISCUSSION

Baulieu: I was surprised at the relatively low numbers of prostatic cancers in the USA. I thought that almost all men over the age of 80 developed prostatic cancer. You quoted a figure of about 200 000.

Castagnetta: That was 200 000 new cases per year of clinically manifest human prostatic carcinoma, i.e. twice the number expected in 1990 (Garnick 1994).

Robel: Have you found significant amounts of oestrogen receptor in the tumours of your patients?

Castagnetta: Yes, we have clear evidence for both type I (high affinity, low capacity) and type II (lower affinity, greater capacity) sites of oestrogen binding in benign and malignant human prostate, although receptor content and distribution vary considerably in different tissues.

Baulieu: You haven't considered that, as well as epithelial cells, the prostate is made up of mesenchymal cells, and that these cells may or may not have the oestrogen receptor. You haven't considered the development of the tumour in the prostate in terms of other cells responding to all sorts of signals, including steroids.

Castagnetta: That is an interesting point, but it is mostly outside the aims of this study. We have evidence that oestrogen actively stimulates growth of LNCaP cells via an oestrogen receptor-mediated mechanism (Castagnetta et al 1994); additionally, oestrogen induces a dose-dependent inhibition of the proliferative activity of the androgen non-responsive PC3 cells (Carruba et al 1994). This would imply that oestrogen may be effective in the treatment of advanced, hormone refractory disease. On the other hand, the oestrogen-induced stimulation of growth may also be counteracted by progesterone via a progesterone receptor-mediated mechanism. This is another indication as to how prostate cancer cells can be regulated.

Baulieu: Some people have used progestogens to treat prostatic cancer. Can you explain the scientific basis for this sort of therapy?

Castagnetta: Not from this study. However, there is evidence that progesterone down-modulates oestrogen receptors in human breast cancer cells.

Oelkers: It could suppress luteinizing hormone instead of gonadal functions.

Baulieu: But this is progesterone administered locally, not orally.

Jensen: You have used two cell lines, PC3 and LNCaP. Which do you consider to be most representative of human prostatic cancers? Do either of these cell lines synthesize prostate-specific antigen (PSA)?

Castagnetta: Maybe both, if we consider that human prostate cancer is commonly composed of both androgen-dependent and androgen-insensitive cell clones. We have found that expression of PSA is very low in LNCaP cells unless they are stimulated with androgens or dihydroxyvitamin D_3 (Skowronski et al 1993). In contrast, PC3 cells do not stain for PSA under any conditions.

Jensen: Then, in answer to my question, the LNCaP cells seem to be more like the human cancers, especially the androgen-dependent ones.

Castagnetta: I agree that the LNCaP cells are much closer to those of a steroid-responsive prostatic carcinoma. However, we have evidence that primary cultures of prostate tumour cells with high levels of PSA may rapidly become PSA negative when they are cultured.

Rochefort: If you think that oestrogen stimulates the growth of prostate cancer, would you assume that anti-oestrogen therapy would be as effective as or more effective than anti-androgen therapy in prostate cancer?

Castagnetta: This study suggests that the combination of anti-oestrogens and anti-androgens could provide additional benefit in the management of at least a subset of prostate cancer patients. Evidently, one should place very wide confidence intervals around the extrapolation of data from the *in vitro* to the *in vivo* condition, and vice versa.

Rochefort: We have to explain the *in vivo* situation first.

Castagnetta: Of course. I'm hoping we will get some answers in the next two years from clinical trials using anti-oestrogen therapy, alone or in combination, as first-line treatment for prostate cancer patients. The idea is to embark on a multicentre study in the framework of the Italian Task Force on Human Prostate Cancer (PONCAP).

Baulieu: And along these lines, have you measured circulating oestrogens in prostatic cancer?

Castagnetta: No. We thought it more appropriate to look at tissue content rather than plasma levels of relevant steroids.

Manolagas: Why do prostate cancer cells metastasize so readily to bone?

Castagnetta: Bone is a target tissue for oestrogens. One could speculate that treatment of advanced prostate cancer metastatic to bone with anti-oestrogens would be beneficial; anti-oestrogens such as tamoxifen, have been reported to stimulate TGF-β production in human breast tumour cells (Knabbe et al 1987).

Manolagas: The only suggestion I had is that bone is loaded with TGF-β. If prostate cancer is so sensitive to and dependent on TGF-β, this might be one possibility.

Castagnetta: It ought to be emphasized that oestrogens regulate TGF-β synthesis in bone cells, such as osteoblasts.

Bonewald: The interesting thing about prostate cancer is that not only does it make TGF-β, but it apparently also makes a latent form that lacks the TGF-β-binding protein (Eklöv et al 1993). This may play a role in what you're seeing—you may have an activation process going on and that's why you're seeing a TGF-β response.

More importantly, bone morphogenetic proteins are produced by prostate cancers (Harris et al 1994). These proteins are also expressed in normal prostate tissues. It may be that in the cancerous state there is over-expression of certain forms of the bone morphogenetic proteins, which predisposes this tissue to bone.

Jensen: It has long been known that prostatic cancers usually proliferate rather slowly, especially in the older patient, but once metastases are established in bone, they grow very rapidly. Recently, it has been shown that transferrin, produced in the bone marrow, acts as a stimulatory factor for the growth of prostate cancer cells (Rossi & Zetter 1992).

References

Carruba G, Pfeffer U, Fecarotta E et al 1994 Estradiol inhibits growth of hormone non-responsive PC3 human prostate cancer cells. Cancer Res 54:1190–1193

Castagnetta L, Miceli MD, Sorci MG et al 1994 Growth of LNCaP cells is stimulated by estradiol via its own receptor. Endocrinology 136, in press

Eklöv S, Funa K, Nordgren H et al 1993 Lack of the latent transforming growth factor β binding protein in malignant, but not benign prostatic tissue. Cancer Res 53: zm3193–3197

Garnick MB 1994 The dilemmas of prostate cancer. Sci Am 270:52–59

Harris SE, Harris MA, Mahy P, Wozney J, Feng JQ, Mundy GR 1994 Expression of bone messenger RNAs by normal rat and human prostate and prostate cancer cells. Prostate 24:204–211

Knabbe C, Lippman ME, Wakefield LM et al 1987 Evidence that transforming growth factor-β is a hormonally regulated negative growth factor in human breast cancer cells. Cell 48:417–428

Rossi MC, Zetter BR 1992 Selective stimulation of prostatic carcinoma cell proliferation by transferrin. Proc Natl Acad Sci USA 89:6197–6201

Skowronski RJ, Peehl DM, Feldman D 1993 Vitamin D and prostate cancer: 1,25-dihydroxyvitamin D_3 receptors and actions in human prostate cancer cell lines. Endocrinology 132:1952–1960

Final discussion

Beato: I wanted to comment on the last three talks and, in particular, on the mechanisms by which cells become hormone resistant.

My impression has been that there are two groups of findings. In many cases resistance is related to the receptors—for instance, progesterone receptor variants that may change the behaviour of the cells. But in other cases resistance may arise without receptors being affected, for instance because different family members of the AP-1 complex are found in different cells, which then respond in a different way. This a very interesting possibility that could be discussed.

Horwitz: All hormone-dependent cancers, and that includes breast and prostate cancers, become resistant to endocrine treatments after a period of positive response. Stage IV breast cancers are destined to become resistant to hormone treatments, which commonly means that they become resistant to tamoxifen. Why does this occur? In the past, the simple explanation was that tumours lose hormone receptors: hence, they lose their ability to respond to the anti-oestrogen. However, this explanation is untenable, since 50% of hormone-resistant breast cancers retain oestrogen receptors. An alternative explanation, which is relevant to steroid antagonists like tamoxifen and the antiprogestins that I have been discussing, is that tumour cells change. As a tumour progresses, new protein factors are expressed and new genes are activated. If these factors are transcriptional coactivators for steroid receptors, or if the genes are receptor regulated, then new responses to the antagonist can be expected. Thus, the anti-oestrogen or antiprogestin may be perceived as an agonist, not an antagonist in this advanced tumour. One can imagine that there would be selection pressure in favour of this progression—cells that respond to the antihormone as an inhibitor would be suppressed, whereas cells that respond to the antihormone as an agonist would proliferate. The latter cells would be classified as 'hormone resistant', when in fact they remain quite sensitive to the hormone. Thus, the hormone has not changed and the receptor has not changed: it is the cancer cells that have changed. Parenthetically, this is referred to in the clinical literature as a 'tamoxifen withdrawal response'. Patients whose tumours are initially inhibited by tamoxifen, but then resume proliferation while they are still receiving tamoxifen, often experience a remission when tamoxifen is withdrawn. Thus, tamoxifen has switched from being an antagonist to being an agonist. This switch can occur without invoking changes in the receptors or the ligands. On the other hand, in some tumours, mutations in the receptors could account for functional switching, and this too has been described (Horwitz 1994).

Baulieu: How are we going to transfer this sort information to the clinic? It is a very difficult task. I remember that at the beginning of the RU486 story it seemed that it would be very simple to use this antiprogestin for pregnancy interruption. Soon after, we published that it was also an agonist (Gravanis et al 1985). Interestingly, the people at Roussel were furious that we published such a 'complexity', fearing it would destroy the image of the compound. We had some conflict with those who wanted to simplify the story and eventually mislead the public. After all, our final goal is to help people, not to praise the so-called wonder properties of a drug.

Manolagas: A huge problem in clinical medicine is the resistance of multiple myeloma to glucocorticoids. Practically every multiple myeloma will respond very well to glucocorticoid steroids for the first six to eight months, and then most of them lose responsiveness. Several leukaemias do the same thing. This concept and a lot of the discussions that have taken place over the past three days have led to a realization that we cannot work in vacuum. We cannot just take a gene, a receptor and a ligand and try to explain the whole story; we should be aware of interactions within the cell, as well as outside the cell. There is an integration of signals coming from numerous transcription factors within the cell and, as our data now suggest, also from outside the cell. The immune, endocrine and neural systems 'talk' all the time. Rather than being abstractionist, we have to be more constructionist and consider the whole organism. We have to try to go as fast as we can from the *in vitro* to the *in vivo*, from the animal to the human.

Baulieu: This has been demonstrated for several peptides, such as somatostatin (which is produced in the brain and pancreas) and now also for steroids produced in the brain. It seems that many molecules are (1) produced at different sites from the same genome, and not only in some cells originally discovered as synthesizing cells, and (2) more and more we have local systems of distribution. These are extremely difficult to work with because we don't know how to explore them in an integrated organism. I think we are in a sort of crisis; there are too many properties already known which we cannot apply for sure to what we are interested in *in vivo*, and particularly in pathological states.

Beato: I want to add a word of caution. In the glucocorticoid field, we have been biased by the original findings on selection of lymphoid cells resistant to killing by glucocorticoids (Gehring 1986). Ninety-nine per cent of all the resistant cells had receptor mutants: either they were receptor deficient, or they had receptors that didn't move to the nucleus properly, or some other defect. It was virtually impossible at that time to find other partners involved in the cytostatic reaction. The impression was that the main player in the action of glucocorticoids on the lymphoid system was the receptor. And yet we recently looked at chronic lymphatic leukaemia patients who have become resistant to glucocorticoid therapy. We carried out polymerase chain reaction sequencing of the receptor, and there was not a single case of mutation of the receptor (Soufi et al 1994).

Manolagas: Are you surprised that systemic hormones, which come from a distance, require local control to modify their actions? If you consider that a hormone coming from a gland can affect many different tissues, you will see that its action needs to be modified locally to accommodate local needs. So the cross-talk between systemic and local signals is something we should have been expecting.

Baulieu: One aspect we often neglect is that of the access of signalling molecules to tissues. It is not just a matter of diffusion, but also of permeability of vessels, binding of protein, or lipid accumulation. Another aspect is the effects of continuous versus pulsatile hormone treatment.

McEwen: There is an example of pulsatile versus continuous oestrogen treatment and the activation of sexual behaviour: it has been claimed that intermittent application of oestradiol to male rats activates sexual behaviour whereas continuous treatment doesn't (Södersten et al 1989, Moreines et al 1986).

Baulieu: That's very difficult to understand *in vitro*, because we have to take into account not only the receptor and its regulation, but also the half-lives of all the biological consequences of its activation.

Oelkers: Another exciting new mechanism involving local factors that regulate access of steroids to their receptors is the role of 11-hydroxysteroid dehydrogenase (11-HSD). The human aldosterone receptor has been recently cloned and it was found that it cannot distinguish between hydrocortisone and aldosterone. However, the receptor is governed by aldosterone, although concentrations of free cortisol in plasma are 100 times higher. Then it was found that the collecting duct cells of the kidney that harbour the receptor are very rich in the enzyme 11-HSD that converts cortisol to cortisone, which cannot bind to the receptor. Aldosterone is not a substrate of this enzyme. Thus, substrate or steroid specificity is conferred to the cell by 11-HSD. This mechanism is also important for the type I corticosteroid receptor in the brain.

Bonewald: With reference to Miguel Beato's comment about being misled by the individual cell systems that we're looking at, often when we want to address a problem we find a window and that's where we look. Perhaps we're not looking through the right window. Maybe we need to be looking at the normal processes: we need to be looking at the primary cell, and how oestrogen functions in the normal immune response or on the normal mammary cell first, and then extrapolate from the normal to the perturbed system.

Baulieu: It isn't very easy to do quantitative work with normal cells *in vitro*, because primary cultures often give different results when we repeat the experiments.

Bonewald: A prime example is the prostate cell lines. We have looked at three prostate cell lines and we get totally different profiles when we look at factors that induce bone formation, for example. For this reason we are trying to go back to the normal prostate tissue, but it's hard to find normal prostate to work with.

Rochefort: Concerning *in vitro* versus *in vivo* results, recent reports have suggested that the shape of cells is important. The results we can get from growing cells on plastic may not be totally relevant to what happens *in vivo*. Cells grown on extracellular matrix-like matrigel, for instance, behave differently from those grown on plastic culture plates.

Baulieu: The anti-oestrogen story is very clear—anti-oestrogen activity with no agonistic activity is easily obtained in many cells *in vitro*, whereas the corresponding cells *in vivo* may undergo some agonistic response.

Rochefort: The different regulation of cathepsin D by the anti-oestrogen tamoxifen *in vivo* versus *in vitro* may be explained at the post-translational level rather than at the transcriptional level. More generally, breast cancer cell lines like to grow on plastic, since they have been selected for this ability and cell polarity may not be important. Normal mammary cells grown on plastic lose their polarity and cannot organize themselves into a glandular structure as in a three-dimensional matrix or as *in vivo*. There is an interesting molecule called calreticulin, which has a consensus amino acid sequence which recognizes both the intracellular domain of the integrin α-subunit at the plasma membrane level, and the region between the two zinc fingers of nuclear steroid receptors (Dehar et al 1994, Burns et al 1994). This molecule may provide an explanation of how cells can respond differently at the gene level depending on whether they are growing on plastic or growing in a three-dimensional extracellular matrix.

Horwitz: With regard to model systems, I think you have all heard the analogy of the three blind men, each trying to describe the same elephant. One feels the tusk and concludes that the elephant is smooth and has a sharp point. Another feels the trunk and concludes that the elephant is wrinkled. The third feels the tail and concludes that it is skinny and hairy. None of them, of course, sees the big picture. In a similar way, we're all looking at parts of the elephant, and some day we hope to be able to integrate all of our information and come up with an accurate picture of the whole elephant. None of us, in our microsphere of activity, can come up with the entire story—we have to listen to one another if we are to assemble the whole from the parts.

Baulieu: This meeting has been rather good in this respect, because people have presented their results in terms which are comprehensible to scientists from diverse backgrounds. We have been concerned with both basic and medically applicable concepts. Certainly, there is still a long way to go, but I think we should avoid being too sceptical about the results already obtained.

One issue that we haven't tackled is how we should respond to proposals to give anti-hormones prophylactically to healthy people 'at risk' of breast cancer. From from what we know, if we take literally what we have discussed here, we shouldn't do anything. On the other hand, it can be argued that we don't have the right to do nothing. Rather pompously, people like to call these issues 'ethics'. I don't know who is going to discuss this and make a decision, but clearly it takes courage. People often don't realize that scientists are not computers, but we have to make difficult 'human' decisions. Nobody doubts that the background for making these decisions is the sort of results that have been reported here. It is not enough, but it is necessary.

References

Burns K, Duggan B, Atkinson EA et al 1994 Modulation of gene expression by calreticulin binding to the glucocorticoid receptor. Nature 367:476–480

Dehar S, Rennie PS, Shago M et al 1994 Inhibition of nuclear hormone receptor activity by calreticulin. Nature 367:480–483

Gehring U 1986 Genetics of glucocorticoid receptors. Mol Cell Endocrinol 48:89–96

Gravanis A, Schaison G, George M et al 1985 Endometrial and pituitary responses to the steroidal antiprogestin RU486 in postmenopausal women. J Clin Endocrinol & Metab 60:156–163

Horwitz KB 1994 How do breast cancers become hormone resistant? J Steroid Biochem Mol Biol 49:295–302

Moreines J, McEwen B, Pfaff D 1986 Sex differences in response to discrete estradiol injections. Horm Behav 20:445–451

Södersten P, Eneroth P, Hanson S 1989 Induction of sexual receptivity by pulse administration of estradiol 17β. Physiol & Behav 14:159–164

Soufi M, Kaiser U, Schneider A, Beato M, Westphal HM 1994 The DNA and steroid binding domains of the glucocorticoid receptor are not altered in treated CLL patients. Exp Clin Endocrinol, in press

Index of contributors

Subject index